Springer Tracts in Civil Engineering

Series Editors

Sheng-Hong Chen, School of Water Resources and Hydropower Engineering, Wuhan University, Wuhan, China

Marco di Prisco, Politecnico di Milano, Milano, Italy

Ioannis Vayas, Institute of Steel Structures, National Technical University of Athens, Athens, Greece

Springer Tracts in Civil Engineering (STCE) publishes the latest developments in Civil Engineering - quickly, informally and in top quality. The series scope includes monographs, professional books, graduate textbooks and edited volumes, as well as outstanding PhD theses. Its goal is to cover all the main branches of civil engineering, both theoretical and applied, including:

- Construction and Structural Mechanics
- Building Materials
- Concrete, Steel and Timber Structures
- Geotechnical Engineering
- Earthquake Engineering
- Coastal Engineering; Ocean and Offshore Engineering
- Hydraulics, Hydrology and Water Resources Engineering
- Environmental Engineering and Sustainability
- Structural Health and Monitoring
- Surveying and Geographical Information Systems
- Heating, Ventilation and Air Conditioning (HVAC)
- Transportation and Traffic
- Risk Analysis
- Safety and Security

Indexed by Scopus

To submit a proposal or request further information, please contact:
Pierpaolo Riva at Pierpaolo.Riva@springer.com (Europe and Americas) Wayne Hu at wayne.hu@springer.com (China)

Carlos Chastre · José Neves · Diogo Ribeiro ·
Maria Graça Neves · Paulina Faria
Editors

Advances on Testing and Experimentation in Civil Engineering

Geotechnics, Transportation, Hydraulics
and Natural Resources

 Springer

Editors
Carlos Chastre ⓘ
CERIS, NOVA School of Science
and Technology
NOVA University Lisbon
Lisbon, Portugal

Diogo Ribeiro ⓘ
CONSTRUCT-LESE, School
of Engineering, Polytechnic of Porto
Porto, Portugal

Paulina Faria ⓘ
CERIS, NOVA School of Science
and Technology
NOVA University Lisbon
Lisbon, Portugal

José Neves ⓘ
CERIS, Instituto Superior Técnico
Universidade de Lisboa
Lisboa, Portugal

Maria Graça Neves ⓘ
NOVA School of Science and Technology
NOVA University Lisbon
Lisbon, Portugal

ISSN 2366-259X   ISSN 2366-2603 (electronic)
Springer Tracts in Civil Engineering
ISBN 978-3-031-05877-6   ISBN 978-3-031-05875-2 (eBook)
https://doi.org/10.1007/978-3-031-05875-2

This Springer imprint is published by the registered company Springer Nature Switzerland AG
The registered company address is: Gewerbestrasse 11, 6330 Cham, Switzerland

Foreword by António Gomes Correia

Testing and experiments, in general, are an important part of the new digital era, building artificial intelligence ecosystems for modern societies to be safer and more sustainable. The application to geotechnics and transportation can take different forms and occur at different times and scales, solving issues and improving solutions in a given application sector, including validation and demonstration.

This collection of contributions provides an update on the recent developments regarding some of these issues and presents case histories involving the application of testing and experiment approaches.

In the geotechnics theme, it covers topics such as the role of engineering geology mapping and gis-based tools in geotechnical practice, in situ geotechnical investigations, laboratory and field testing of rock masses for civil engineering infrastructures, testing and monitoring of earth structures and offshore wind foundation monitoring and inspection.

In the second theme of transportation, the contributions highlight topics such as tests and surveillance on pavement surface characteristics, full-scale accelerated pavement testing and instrumentation, monitoring of pavement structural characteristics, intelligent traffic monitoring systems and testing and monitoring in railway tracks.

I believe that the present contributions are a valuable resource for researchers as well as practising engineers in geotechnics, transportation and other branches of civil engineering.

I would like to convey my deepest gratitude to the editors for their efforts, to the authors for their contributions and to the reviewers for their careful reviews, all of which have contributed to the quality of this book series in Springer Tracts in Civil Engineering.

António Gomes Correia
Emeritus Professor in Civil Engineering
University of Minho
Braga, Portugal

Foreword by Javier L. Lara

All through history, progress in engineering has made possible the development of mankind, allowing over the years a significant improvement in people's quality of life. In this sense, civil engineering has played a fundamental role in the ability of human beings to adapt to their environment thanks to the construction of infrastructures that have facilitated the transport of materials and people, communication, the capture of natural resources or their own residence and coexistence. This has been possible thanks to advances in knowledge, construction techniques and the application of new technologies based, firstly, on physical experimentation and, in recent decades, on the use of computational techniques, which in a complementary approach and based on solid theoretical foundations, has allowed the progress of technology and the construction of large infrastructures for society.

Civil engineering currently has a series of challenges that it will face during this century, which are linked to a series of social and economic demands that it must meet within a framework of sustainable development. Aspects such as the growth and development of medium-sized cities that will become large cities at the end of the century, generating new demands for management models, the inclusion of artificial intelligence in the management of complex systems associated with the construction, operation and maintenance of civil infrastructures, innovation in construction processes and the use of new materials and engineering solutions within a framework of sustainable development in agreement with nature are some of the challenges that civil engineering will have to face, all of them within an environmental framework of climate change.

All these challenges, and others that will appear throughout the century, will therefore require the development of new science and new knowledge. It is in this area where experimental techniques play a very relevant role. Computational techniques have developed considerably in the last decades and are now an everyday element of the engineer's work. Additionally, in recent years, techniques based on artificial intelligence have emerged with great efficiency showing their high applicability in the field of civil engineering. However, experimental techniques are still a very important approach to generate knowledge and as a complementary element for the understanding of complex systems in engineering, either on their own or as a complement to the above.

This book presents a good compilation of the latest advances and experimental techniques in four major fields in which civil engineering works: geotechnics, transportation, hydraulics and natural resources. Throughout 15 chapters, novel aspects in these four disciplines are presented, through the development and application of experimental techniques in different fields, which aim to show how experimentation can be a very valid tool to obtain the adequate knowledge that civil engineering needs for its development. The book is an ideal choice for curious professionals and students, both inside and outside civil engineering, and provides a cross-disciplinary view of the use of the latest experimental techniques in civil engineering and how they can be applied to generate new science and increase the understanding of physical processes.

Javier L. Lara
Professor of Civil Engineering
Instituto de Hidráulica Ambiental
Universidad de Cantabria
Santander, Spain

Preface

Testing and experimentation are essential to support the design of civil engineering infrastructures and buildings and to understand the phenomena involved. Over the last years, testing and experimentation in civil engineering have assumed increasing importance in a wide range of applications, providing reliable information for the decision-making process during the projects' life cycle. The experimental activities are performed in the laboratory or in situ. They typically allow an enhanced characterization of the behaviour of systems and their components, supporting the stakeholders to find more reliable and sustainable solutions. Currently, the literature lacks a contribution that provides general and foundational knowledge across these topics, and this book aims to fill this gap.

The book presents the recent advances on testing and experimentation in civil engineering, especially in the branches of materials, structures, buildings, geotechnics, transportation, hydraulics and natural resources. It is divided into two volumes, one dedicated to materials, structures and buildings and the other one dealing with geotechnics, transportation, hydraulics and natural resources. Both volumes include advances in physical modelling, monitoring techniques, data acquisition and analysis and provide an invaluable contribution for the installation of new civil engineering experimental facilities.

This volume covers the areas of geotechnics, transportation, hydraulics and natural resources and starts by pointing out the most recent advances in testing and experimentation in the main domains of geotechnics: soil mechanics and geotechnical engineering, rock mechanics and rock engineering and engineering geology. The initial part is dedicated to new developments in surveying acquisition for applied mapping and in situ geotechnical investigations. Laboratory and in situ tests to estimate the relevant parameters required to model the behaviour of rock masses and earth structures are presented next, updating the most important technological advances. The last part describes monitoring and inspection techniques designed for offshore wind foundations.

The second part of the book highlights the relevance of testing and monitoring in transportation. Full-scale accelerated pavement testing and instrumentation become even more important nowadays when, for sustainability purposes, non-traditional materials are used in road and airfield pavements. Innovation in testing and monitoring pavements and railway tracks is also developed in this part of the book. Intelligent traffic systems are the new traffic management paradigm, and an overview of new solutions is addressed.

Finally, trends in the field and laboratory measurements and corresponding data analysis are presented according to the different hydraulic domains addressed in this publication, namely maritime hydraulics, surface water and river hydraulics and urban water. These chapters of the book provide a holistic and comprehensive overview of hydraulic testing and experimentation and are addressed to professionals willing further improvement in their scientific and technical understanding and skills in this specific area of civil engineering.

Lisbon, Portugal   Carlos Chastre
Lisbon, Portugal   José Neves
Porto, Portugal   Diogo Ribeiro
Lisbon, Portugal   Maria Graça Neves
Lisbon, Portugal   Paulina Faria

Acknowledgments

The editors would like to express their sincere gratitude to professionals and organizations that directly or indirectly contribute for the realization of this book: to the *Springer Tracts in Civil Engineering* Editors, Prof. Shenghong Chen, Prof. Marco di Prisco and Prof. Ioannis Vayas, for accepting this book proposal in a challenging and innovative thematic in the field of experimentation in civil engineering; to the Springer Editor, Dr. Pierpaolo Riva, for all the advice and prompt support in addressing all the questions during the book preparation; to the experts, Prof. António Gomes Correia and Prof. Javier Lopez Lara, for addressing the foreword and critical review of the book contents; to the educational institutions NOVA School of Science and Technology (FCT NOVA), Instituto Superior Técnico (IST), School of Engineering of Polytechnic of Porto (ISEP) and National Laboratory for Civil Engineering (LNEC), as well as the Civil Engineering Research and Innovation for Sustainability (CERIS) and Institute of R&D in Structures and Construction (CONSTRUCT), which are financed by Foundation for Science and Technology (Portugal) through UIDB/04625/2020 and UIDB/04708/2020, respectively, for providing the necessary conditions for the development of the book edition; to all the authors of the chapters, for the extraordinary commitment with the objectives of the publication and their enthusiasm to be part of this project. Finally, the co-editors would like to thank all the readers of this book for their interest and passion in the most recent advances in testing and experimentation in civil engineering.

Contents

Contributors

Al-Qadi Imad L. Illinois Center for Transportation, Department of Civil and Environmental Engineering, University of Illinois at Urbana, Champaign, USA

Amaral Sílvia National Laboratory for Civil Engineering (LNEC), Lisbon, Portugal

Andriolo Umberto Department of Electrical and Computer Engineering, INESC Coimbra, Coimbra, Portugal

Barros Ablenya Faculty of Applied Engineering, EMIB Research Group, Antwerpen, Belgium

Brito Rita Salgado Laboratório Nacional de Engenharia Civil, Lisboa, Portugal

Cardoso Rafaela CERIS, Instituto Superior Técnico, DECivil, University of Lisbon, Lisbon, Portugal

Caruso Marco Materials Testing Laboratory, Politecnico Di Milano, Milan, Italy

Cerezo Véronique AME-EASE, Univ Gustave Eiffel, IFSTTAR, Univ Lyon, Lyon, France

Chaminé Helder I. Laboratory of Cartography and Applied Geology (Labcarga), Department of Geotechnical Engineering, School of Engineering (ISEP), Polytechnic of Porto, Porto, Portugal;
Centre GeoBioTEcIUA, Aveiro, Portugal

Chiapponi Luca Dipartimento di Ingegneria e Architettura (DIA), Università di Parma, Parma, Italy

Clavero María Andalusian Institute for Earth System Research, University of Granada, Granada, Spain

do Céu Almeida Maria Laboratório Nacional de Engenharia Civil, Lisboa, Portugal

Fernandes Isabel IDL—Instituto Dom Luiz, Department of Geology, Faculty of Sciences, University of Lisbon, Lisbon, Portugal

Ferreira Rui M. L. Instituto Superior Técnico (IST), Universidade de Lisboa (NHRI), Lisbon, Portugal

Fontul Simona LNEC, National Laboratory for Civil Engineering, Lisbon, Portugal

Fortunato Eduardo LNEC, National Laboratory for Civil Engineering, Lisbon, Portugal

Freire Ana Cristina LNEC, Nacional Laboratory for Civil Engineering, Lisboa, Portugal

Freitas Elisabete University of Minho, ISISE, Guimarães, Portugal

Gomes Sandra Vieira LNEC, National Laboratory for Civil Engineering, Lisbon, Portugal

Goubert Luc Belgian Road Research Centre, Brussels, Belgium

Jommi Cristina Materials Testing Laboratory, Politecnico Di Milano, Milan, Italy; Department Geoscience and Engineering, Delft University of Technology, Delft, The Netherlands

Lamas Luís National Laboratory for Civil Engineering (LNEC), Lisbon, Portugal

Longo Sandro Dipartimento di Ingegneria e Architettura (DIA), Università di Parma, Parma, Italy

Losada Miguel A. Andalusian Institute for Earth System Research, University of Granada, Granada, Spain

Lourenço Sérgio Department of Civil Engineering, The University of Hong Kong, Hong Kong, China

Mendes Diogo Geosciences Department and Centre of Environmental and Marine Studies (CESAM), University of Aveiro, Aveiro, Portugal;
HAEDES, Santarem, Portugal

Mendes João Faculty of Engineering and Environment, Department of Mechanical and Construction Engineering, Northumbria University, Newcastle, UK

Muralha José National Laboratory for Civil Engineering (LNEC), Lisbon, Portugal

Neves José CERIS, Department of Civil Engineering, Architecture and Georesources, Instituto Superior Técnico, Universidade de Lisboa, Lisboa, Portugal

Neves Maria Graça NOVA School of Science and Technology, NOVA University Lisbon, Lisbon, Portugal

Paixão André LNEC, National Laboratory for Civil Engineering, Lisbon, Portugal

Qamhia Issam Illinois Center for Transportation, Department of Civil and Environmental Engineering, University of Illinois at Urbana, Champaign, USA

Ramon-Tarragona Anna CIMNE Barcelona, and Division of Geotechnical Engineering and Geosciences, Department of Civil and Environmental Engineering, Universitat Politècnica de Catalunya, Barcelona, Spain

Ribeiro Alvaro Silva Laboratório Nacional de Engenharia Civil, Lisboa, Portugal

Rodrigues Rui Raposo National Laboratory for Civil Engineering (LNEC), Lisbon, Portugal

Santos Jaime A. CERIS, Department of Civil Engineering, Architecture and Georesources, Instituto Superior Técnico, Universidade de Lisboa, Lisboa, Portugal

Siedler Simon Ramboll, Hamburg, Germany

Sjögren Leif Swedish National Road and Transport Research Institute, Linköping, Sweden

Sousa Jorge University of Minho, ISISE, Guimarães, Portugal

Todo-Bom Luís Berenguer Ramboll, Hamburg, Germany

Tutumluer Erol Illinois Center for Transportation, Department of Civil and Environmental Engineering, University of Illinois at Urbana, Champaign, USA

Viseu Teresa National Laboratory for Civil Engineering (LNEC), Lisbon, Portugal

Vuye Cedric Faculty of Applied Engineering, EMIB Research Group, Antwerpen, Belgium

Zhao Gensheng Nanjing Hydraulic Research Institute (NHRI), Nanjing, China

Geotechnics

The Role of Engineering Geology Mapping and GIS-Based Tools in Geotechnical Practice

Helder I. Chaminé and **Isabel Fernandes**

Abstract Maps are of fundamental topical importance in the geology and engineering practice, mainly in field data surveys, synthesis and communication related to several domains, such as applied geomorphology, engineering geology, hydrogeology, soil and rock geotechnics, slope geotechnics and site investigation. The preparation of geological maps and plans specifically for engineering purposes is still a challenging task. The application of Geographic Information Systems (GIS) and geovisualisation techniques for geosciences and engineering are gaining increasing relevance. New developments in surveying acquisition for applied mapping—sketch or general maps, engineering geological maps and geotechnical maps, at diverse scales—take on critical importance in further ground investigations and modelling stages. It is also essential to highlight the value and cost-effectiveness of accurate mapping for geotechnical practice. The present chapter summarises the state of the art regarding engineering geology mapping techniques, methods, and models. Additionally, it intends to focus on an insightful geotechnical mapping reasoning concept established on advanced methods such as geomatic techniques, geovisualisation techniques, unmanned aerial vehicles for detailed surveys, and ground and numerical modelling.

Keywords Applied mapping · Engineering geology · Geotechnics · GIS · Geovisualisation techniques

H. I. Chaminé (✉)
Laboratory of Cartography and Applied Geology (Labcarga), Department of Geotechnical Engineering, School of Engineering (ISEP), Polytechnic of Porto, Porto, Portugal
e-mail: hic@isep.ipp.pt

Centre GeoBioTEc|UA, Aveiro, Portugal

I. Fernandes
IDL—Instituto Dom Luiz, Department of Geology, Faculty of Sciences, University of Lisbon, Lisbon, Portugal
e-mail: mifernandes@fc.ul.pt

1   Introduction

The thoughts from Rocha [1] are still topical: "The engineer cannot undertake the characterisation of the rock masses only from test results. In fact, first of all, the definition, in position and number, of the sites to be tested must be guided by knowledge of the geology of the formations and, secondly, due to the referred complexity, the vast legacy of qualitative knowledge contained in geological information plays a decisive role in the characterisation of the rock mass, which implies the extrapolation of results from the sites tested for other areas". This impressive quotation is the basis for the engineering geological mapping role in field site investigations for ground engineering purposes. Nowadays, any skilled professional (e.g., geologist, engineering geologist, geological engineer, geotechnical engineer, civil engineer, mining engineer, or military geologist/engineer) involved in the geotechnics practice should keep this in mind to reduce intrinsic uncertainties and geological variabilities. Thus, geological mapping for engineering purposes is vital in geotechnical practice in a double way: (i) it plays a crucial role in desk studies, on-site investigations, design, and work stages; (ii) it supports the decision-making activities, and it is a valuable communication tool with practitioners, contractors, researchers, geotechnical-related professionals, and society.

Delgado et al. [2] pointed out that many geologists were usually engaged in geological mapping for engineering at the surface and underground. However, these authors also highlighted the remarkable example carried out by the field geologist Paul Choffat in 1899 for the Lisbon downtown railroad tunnel (with 2600 m long) through Miocene soils and rock formations. Choffat [3] produced a comprehensive geological report for engineering purposes, particularly the excavation surfaces' geological mapping, and recorded the numerous water inflows in the tunnel. That study was of great help in the design of the rehabilitation works required for the stability of the tunnel 80 years later [2] (Fig. 1).

Rocks and soils are natural materials that form the Earth's crust. The geological materials are formed in a continuous geodynamic cycle—comprising numerous internal and external processes—through geological time. In engineering practice, geomaterials can be divided into hard rocks (unweathered, compact, and durable), soft rocks (weak and easily deformable) and soils (unconsolidated and friable materials overlying bedrock) [4–7]. Recently, the so-called anthropic rocks have been considered a collective term for those rocks and geomaterials modified or moved by humans [8]. Geological materials may be observed, collected, and analysed in many settings: (i) at the surface and subsurface (outcrops, cliffs, slopes, quarries), (ii) underground (tunnels, mines, cavities, boreholes) (Fig. 2).

The potential for geology to support engineering activities, particularly civil and mining engineering, occurs at every scale, from regional geological structures to microscale [5, 6]. However, geological data input for engineering purposes is only adequate if supported by the appropriate geotechnical parameters [11]. Rock engineering deals with large volumes of rock masses that will contain variable amounts of fluid in interlocking discontinuities, such as joints, fractures, faults, sedimentary

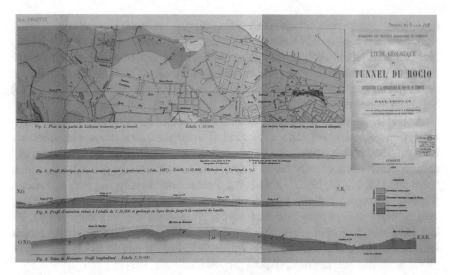

Fig. 1 An example of the impressive geological mapping and cross-sections for engineering purposes (Rossio tunnel, Lisbon downtown, Portugal) mapped by Choffat [3]

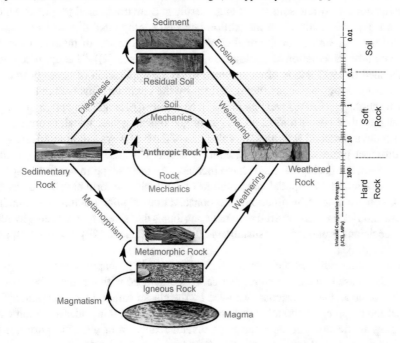

Fig. 2 The geological cycle in the perspective of the engineering geosciences framework (updated from [9] and revised from [10]): an outlook for rock mechanics and soil mechanics

or tectonometamorphic structures (bedding planes, schistosity, shear zones, folds, etc.), veins and geological contacts. These natural structures represent the rock mass discontinuous fabric [6, 7, 12–15]. In geotechnics, discontinuity refers to any break in the rock continuum having little or no tensile strength [16]. Intact rock refers to the unfractured blocks (ranging from a few millimetres to several metres in size) between discontinuities in a rock mass [11, 15, 17].

A clear structural geology framework is essential for investigating any rock engineering project [16, 18]. The geological structures of rock masses significantly influence their hydrogeomechanical properties, such as permeability, resistance, and deformability. Consequently, the rock mass structure controls the ground engineering behaviour [13, 15, 17, 19]. Therefore, it is fundamental to mapping rock exposures, whether outcrops, road and railway cuttings, underground excavations, or quarry operations, to collect structural data supported by field mapping surveys [10, 13, 20]. Conclusively, a civil or mining engineering work success is related to the proper knowledge of the site investigations framework, mainly related to the geological, morphotectonic, and hydrogeological ground conditions [21–26].

In the statutes of the International Association for Engineering Geology and the Environment (IAEG), engineering geology is defined as the science devoted to the investigation, study and solution of engineering and environmental problems which may arise as the result of the interaction between geology and the works or activities of man, as well as of the prediction of and development of measures for the prevention or remediation of geological hazards (details in [2]). That approach was established in 1992 and is often assumed to be the application of geology to civil engineering design and construction driven by the continued involvement of engineering geologists in site investigation [27]. Nowadays, geological mapping for engineering practice covers almost all ranges of geological, geotechnical, georesources and geoenvironmental activities. It is also one of the best ways to communicate with the different professionals in geotechnics practice. Culshaw [28] pointed out that the engineering geological maps have increased in their scope to cover all aspects of the collecting and spatial presentation of geological information for development, construction, rehabilitation, and conservation of infrastructures, embracing hazard and risk assessment studies. That contributes decisively to balanced geotechnical decisions grounded in a sustainable design with nature [29], society [30] and geohazards [31].

The chapter outlines the role of engineering geology mapping techniques and methods focused on engineering practice. It is organised with a general framework on the theme and key concepts, followed by the classification, type, components of maps, and mapping methodologies and techniques. Lastly, it includes a section with mapping applications in engineering at several scales, from planning purposes to detailed geotechnical studies, including field testing investigations.

2 Concept

IAEG–CEGM [32] defined an engineering geological map as *"(...) a type of geological map which provided a generalized representation of all those components of a geological environment of significance in land-use planning, and in design, construction and maintenance as applied to civil and mining engineering"*. That engineering geological mapping, even in the form of sketches mapping, is a longstanding and successful track record in applied geology and geotechnical practice, as pointed out by several authors [4, 25, 27, 33–38].

Lately, González de Vallejo and Ferrer [39] state that engineering geological maps present comprehensive geological and geotechnical information for land use and planning, constructing, and maintenance of engineering infrastructures. They also provide reliable data on the characteristics and properties of the ground to enable its behaviour to be evaluated and predict geological and geotechnical issues. Therefore, mapping—general or sketch geological maps, engineering geology maps and geotechnical maps, at diverse scales—undertakes fundamental value in further geotechnical investigations, design, and modelling stages.

The engineering geological maps or plans are a powerful tool for understanding the nature of the ground conditions at a study site and in the design and project development stages. Besides, they are resourceful databases of ground information on lithology, structure, morphology, soil and rock mechanics, hydrology and groundwater, and in-situ investigation conditions. A variety of best practices in the preparation of applied geological maps for engineering purposes have been highlighted in topical publications in the last half-century [4, 25, 27, 34, 35, 38–49]. Griffiths [50] and Culshaw [28] outline the historical milestones of the engineering geological mapping concept development evolution.

The review of principles, methods, and techniques for geological mapping for general engineering purposes or geotechnical surveys can be complemented by pivotal reading [6, 13, 32, 36, 51–58], among others. Furthermore, the appraisal of the study site for geotechnical practice shall follow the recommendations of IAEG— International Association for Engineering Geology and the Environment (www.iaeg. info); GSE|GSL—Engineering Geology Group of the Geological Society of London (www.geolsoc.org.uk); ISRM—International Society for Rock Mechanics and Rock Engineering (www.isrm.net); ISSMGE—International Society for Soil Mechanics and Geotechnical Engineering (www.issmge.org); CFCFF—Committee on Fracture Characterization and Fluid Flow (www.nap.edu), and European Eurocodes, particularly geotechnical design standards (https://eurocodes.jrc.ec.europa.eu).

3   Classification, Type and Components of Maps

The engineering geological maps should include a description of geological materials, quantitative data based on the physical and mechanical properties, and interpretative information for their geotechnical application. Also, they must reflect: (i) the geological nature of the ground in terms of lithology, structure, and geomorphology; (ii) quantitative data on the geomechanical properties of the materials, their heterogeneity and anisotropy; (iii) the dynamic nature and behaviour of the ground; (iv) the hydrogeological conditions of the geological medium.

The review of principles, methods, and techniques of geological mapping for general engineering purposes or geotechnical surveys can be complemented by reading IAEG–CEGM [32] and Dearman [34]. Griffiths [35, 36] concluded that scale is a critical aspect of all maps. Engineering geological maps are prepared on different scales suitable to their rationale, i.e., basic geological and geotechnical data for general use or specific applications. The maps can be classified according to their purpose, content, and scale [32, 39], as shown in Table 1. That is the basis for the concept of map or plan terms. Table 2 outlines the classification of engineering geological maps based on purpose, content and scale mainly derived from the IAEG–CEGM [32].

The purpose of the maps or plans depends on scale, such as [10, 25, 28, 30, 35, 39, 48]:

Table 1 Classification of engineering geological maps according to their scale (adapted from [39])

Type and Scale	Content	Method	Applications
Regional Desk study and fieldwork >1:10,000	Geological data, lithological groups, tectonic structures, geomorphology; general information and interpretations of geotechnical interest	Remote Sensing, aerial images, published topographic and geological maps; field observations	Preliminary studies and planning; regional information, and types of material and geomorphological processes
Local Desk study phase 1:10,000 to 1:500	Description and classification of soils and rocks, structures, morphology, hydrogeology, geodynamic processes, location of possible construction materials	Remote Sensing, aerial photography, ground trothing field surveys; field data and analysis	Planning and viability of works and detailed site reconnaissance. Basic design
Local/In Situ Ground investigation phase 1:5,000 to 1:500/1:1000 to 1:50	Material properties, geotechnical conditions, and other aspects essential for a specific construction project	All previous data plus data from boreholes and trial pits, geophysical methods, in situ and laboratory tests	Detailed information on sites and geological-geotechnical problems. Detailed design and analyses

Table 2 Classification of engineering geological maps based on purpose and content (adapted from [39])

Criterion	Type
Purpose	*Special:* Providing information either on one specific aspect of engineering geology or for a particular purpose
	Multi-purpose: Providing information covering many aspects of engineering geology for a variety of planning and engineering purposes
Content	*Analytical:* Giving details of or evaluating individual components of the geological environment. Their content is, as a rule, expressed in the title, for example, map of weathering grades; jointing map; seismic hazard map
	Comprehensive: Two types—maps of geological engineering conditions depicting all the principal components of the engineering geological environment or maps of geological engineering zoning, evaluating and classifying individual territorial units based on uniformity of their geological engineering conditions. The two types may be combined on small-scale maps
	Auxiliary: Presenting factual data and are, for example, documentation maps, structural contour maps, isopachyte maps
	Complementary: Geological, tectonic, geomorphological, pedological, geophysical, and hydrogeological maps They are maps of basic data that are sometimes included with a set of engineering geological maps

(i)   *Detailed field survey studies, plans and cross-sections (large scale):* typically 1:500 to 1:1000 (recording the details of trenches, pits, slopes, and other excavations). Engineering geological plans are created for specific purposes on a large scale, either during site investigation or during the construction stage [35, 44]. They are not intended to replace suitable detailed site-specific desk studies or ground investigations [39];

(ii) *Local/general maps or plans (large to medium-scale):* 1:500 to 1:10,000 (small reservoirs, dam sites, borrow areas, tunnels, large buildings, airfields, port facilities, bridges);

(iii) *Regional maps and planning purposes (medium to small-scale):* >1:10,000 (land-use and planning).

(iv) Regarding the production of engineering geological maps, the basic information shall include [39]: (i) topography, hydrography, toponymy, and terrain data, including geodiversity, geoheritage and archaeological aspects; (ii) lithological heterogeneity, tectonic framework and description of the geological units focused on the specific engineering purpose; (iii) soil horizons, superficial deposits and weathered rocks depths; (iv) discontinuities and structural data; (v) geotechnical classification of soils and rocks; (vi) geomechanical properties of soils and rocks; (vi) hydrogeological and geomorphological conditions for engineering purposes; (vii) dynamic processes; (viii) gather all data from previous geotechnical surveys and in situ investigations; (ix) identification of geological hazards and geoenvironmental issues.

4 Mapping Methods, Techniques and Models

Over the decades, map-making methods, techniques, design, and conceptualisation have developed dramatically [59–62]. Brown [59] stated that the earliest maps were based on personal experience and familiarity with a local situation. Hence, maps have been a medium to represent spatial data in visual form [63] and are a powerful graphic technique that provides a comprehensive key for unlocking the brain's potential [64]. Kraak [65] highlighted an essential issue: cartography, first of all, means "maps". However, encoding and decoding maps as a practice is a challenging task. New developments exploring geovisualisation analysis combine approaches from diverse backgrounds, including scientific visualisation, image analysis, information visualisation, exploratory data analysis and GIScience [61–63]. According to [65], geovisualisation combines the power capabilities of the computer (automated analysis techniques and geo-computation) and the human being (interactive visualisations for practical understanding, reasoning and decision making). Additionally, geovisual analytics concentrates on visual interfaces of analytical and computational methods that support reasoning with and about geo-information. Geographic Information System (GIS) techniques created new insights into geosciences mapping for geotechnics.

Engineering geological and geotechnical mapping is of great significance in site investigations, engineering works or contributing to urban planning issues, providing essential information about the geological features and geological materials and the geotechnical properties of soils and intact rock and the soil and rock masses. Norbury [58] offers a valuable and readable introduction to soil and rock description in engineering practice. Engineering geological and geotechnical mapping needs to advance toward an insightful cartographic reasoning concept established, among others, in geomatic techniques, fieldwork, site investigations conceptualisation and numerical modelling [10]. Nevertheless, about the valuable information of field surveys and geotechnical mapping for engineering purposes, Price [6] specified several key practical points: (i) it must never be forgotten to produce engineering maps that are of immediate use to the engineer or other technicians; (ii) maps must be user friendly, quickly understood and easily read; (iii) mapping for in situ engineering practice requires large scale maps (typically, in detailed surveys: 1:500–1:1000 to 1:5000). Furthermore, the method of walking-over, observing and mapping the whole site is a most valuable discipline, i.e., the core is to map form, describe materials and record evidence of the geodynamic processes [22].

Griffiths [36] and González de Vallejo and Ferrer [39] present an excellent summary of the basic elements of geotechnical maps to represent the suggested recommendations and standards of geological, geomorphological, hydrogeological, and geotechnical graphic symbols. Griffiths [35] also highlighted that the symbols used must be fully explained. Additionally, that might take the form of an expanded legend or a short report, either accompanying the map or printed on the bottom of the map. Finally, as a general recommendation, the map should stand alone and be understood by any potential user without referring to a separate report.

Dearman [34] underlined the value of the concept of engineering geological zoning in geotechnical mapping practice. This recognises areas on the map with roughly homogenous engineering geological behaviour and geotechnical conditions. Such zones would generally be derived from the accurate data collected on the base map and consequently should not form part of the original map but can be produced as an overlay [35].

The geologic and geotechnical data and the ground information of interest can also be supported by interpretative cross-sections. These complement the engineering maps and present valuable information such as variations of the soils and rock material properties with depth, the limits of weathered zones and depth of bedrock, boreholes and other subsoil investigation techniques [39]. The map legend used on an engineering geological map should be developed to suit the purpose of the study [35]. Additionally, Chapter 22 (In situ geotechnical investigations) and Chapter 23 (Laboratory and field testing of rock masses for civil engineering infrastructures) of this book make an insightful state of the art about site investigation, testing, and monitoring geotechnics practice interrelated with engineering geological mapping.

Culshaw [28] highlights a relevant question regarding the widespread use of digital mapping, particularly GIS-based mapping techniques and geomatic techniques. Consequently, the scale has become comparatively flexible in these digital maps, i.e., it can be very easily increased or decreased in scale. Nevertheless, this can create problems if users expand a map to a larger scale without fully understanding the potential implications.

Combining data from digital photogrammetry, unmanned aircraft vehicles (UAV), high-resolution global positioning systems (GPS), laser scanner technologies, geovisualisation techniques, geo-database systems, 3D and 4D modelling, and GIS-based mapping are even more recognised as powerful tools in supporting the map-making [30, 66, 67], Fig. 3. Thus, GIS applications are becoming part of almost all technical-scientific studies and are being used to collect, store, process, and analyse georeferenced data. Nowadays, several commercial GIS software or open-source software tools are operational, fast and accessible to be used in applied geosciences and geotechnics [10, 30, 68]. GIS techniques, supported by high-resolution GPS, permit the use of large amounts of data to be overlaid and analysed in a dynamical geo-database. GIS-based mapping also supports thematic maps, spatial data analysis, and data geovisualisation, such as assessing discontinuous rock mass systems or geotechnical mapping for urban planning or construction works [10, 30, 69]. The use of computer software and GIS technologies for digital mapping allows [39]: (i) processing and analysis of data; (ii) mapping of individual and combined factors or elements; (iii) creation and preparation of dynamic databases; (iv) ongoing updating of map data and information; (v) creation of 3D models and simulation of on-site action.

Table 3 shows the data collection methods used for creating engineering geological maps, and Table 4 lists the key information to be recorded on an engineering geological map.

New methodologies such as Building Information Modelling (BIM) are applied in geotechnics practice. BIM is an insightful multidimensional model-based process and

Fig. 3 Examples of geotechnologies used to support field mapping and in situ testing sites (images a) to (**c**) from LABCARGA∥ISEP archive; (**d**) kindly shared by José Filinto Trigo, NEC∥ISEP): **a, b** high-resolution global positioning system (GPS) in rock slope geotechnics studies and total station in surveying studies to support digital photogrammetry techniques (**b**); **c** unmanned aircraft vehicles (UAV) to support detailed mapping and terrain modelling with 3D rock slope model resulting from UAV aerial surveys (**d**)

Table 3 Data collection methods for geotechnical mapping (adapted from [39]

Data collection methods for engineering geology and geotechnical mapping

Method	Data
Remote Sensing, photogeology and terrain analysis	Topography and morphology Soils and rocks units Geological structures Hydrography and catchments Dynamic processes
Surveying and collecting field data	Geological and geomorphological analysis Geological and geotechnical data and assessment supported by scanline rock surveys
Geophysical methods	Electrical resistivity Seismic methods
Boreholes, trial pits and sampling	Provide representative samples Direct inspection of materials Physical properties and characteristics of the ground Hydrogeological conditions
In situ tests	Stress and strain properties In situ stress Permeability, porosity, water pressure Data borehole tests
Laboratory tests	Physical and mechanical properties of materials

information tool to plan, design, construct and manage infrastructure more efficiently. That approach is a powerful tool in design, time, and cost-effectiveness [70]. Finally, there are developments in integrating GIS and BIM technologies; maybe that will be a significant step in the digital mapping for engineering geology and geotechnical engineering.

The models in practice have been in use for a while because of their usefulness and importance [4, 37, 38, 47, 48, 71]. Baynes et al. [72] pointed out that the models in engineering geology comprise both conceptual ideas and observational data. The observational data is related to inherent heterogeneity, which can be reduced by acquiring more observations. The conceptual ideas are associated with epistemic uncertainty, which can only be reduced if more knowledge is incorporated into the model. Indeed, a model is a powerful tool to help solve engineering problems and support the decision-making and managing of geotechnical hazards and risks. Norbury [71] also stated that the generation and use of ground models should be fundamental for any geotechnical project. Such models are essential for engineering quality control and clearly identify project-specific, critical ground risks and parameters. In part, that is achieved following ground investigations' technical codes and standards, as well as incorporating the geotechnical baseline reports in ground models [73, 74]. The publications of [25, 37, 38, 48, 71, 75] clearly explained the importance, effectiveness and value of the models in engineering geology and

Table 4 Information to be recorded on an engineering geological map (updated from [35, 36] and revised from [30]

Engineering geological maps: basic data in a GIS-based environment and using engineering codes of practice or description of soils, rocks and groundwater

• *Topographical data*

Topography and geographical features, including terrain and elevation data; Hydrography

Toponyms, geodiversity, geoheritage and cultural aspects

Remote sensing analysis and geovisualisation techniques

• *Geological data*

Mappable units (based on descriptive engineering geological terms)

Geological boundaries (with accuracy indicated)

Description of soils and rocks (using engineering codes of practice)

Description of exposures (cross-referenced to field notebooks)

Description of the state of weathering and alteration (notes depth and degree of weathering)

Description of rock discontinuities based on scanline surveys

Subsurface conditions (provision of subsurface information if possible, e.g. rockhead isopachytes)

• *Hydrogeological data*

Availability of information (a reference to existing maps, well logs, abstraction data, etc.)

General groundwater conditions (flow lines; piezometry; water quality; artesian conditions; potability, etc.)

Hydrogeological properties of rocks and soils (aquifers, aquicludes and aquitards; permeability; perched water tables, etc.)

Springs, seepages and inflows (flows to be quantified wherever possible)

Hydrogeomechanical properties and behaviour

• *Geomorphological data*

General geomorphological features (ground morphology, landforms, processes)

Ground movement features (landslides, subsidence, solifluction lobes; cambering)

• *Geohazards and Geoenvironment*

Mass movement (extent and nature of landslides, type and frequency of landsliding, possible estimates of runout hazard)

Rock slope stability and assessment

Flooding (areas at risk, flood magnitude and frequency, coastal or river flooding)

Coastal zones (cliff form, rate of coastal retreat, coastal processes, types of coastal protection)

Seismicity (seismic hazard assessment)

Vulcanicity (volcanic hazard assessment)

geotechnical practice. Recently, Turner et al. [76] argued the insightful importance of the applied multidimensional geological modelling in the mapping practice (3-D to 5-D or maybe soon 6-D, i.e., 6-D geological models; three spatial coordinates + time + scale + uncertainty).

Understanding the complexity of Earth systems is possible using models mainly based on topographic, geologic, geomorphological and hydrogeological data [77]. Griffiths [37] emphasized that the model should be dynamic and part of the ground investigations and involve design, construction and monitoring phases. Griffiths [25] addresses that the natural soils and rocks have an inherent heterogeneity (i.e., aleatory uncertainty) that must be understood and allowed to develop the ground model and establish the material parameters to be adopted in design. Thus, a typical site characterisation should be outlined based on Earth systems analysis which forms the core for building models to create scenarios using different approaches, such as [10, 25, 30, 37, 39, 71, 75] and references therein):

(i) *Geological model*: general models that include the bedrock and shallow geology observations and data integrating the basic topography, geomorphology, hydrogeology, seismotectonic and applied geology;

(ii) *Engineering geological model*: geological models with engineering parameters from ground investigations and other available data;

(iii) *Ground model*: geotechnical models with predicted performance based on design parameters;

(iv) *Geomechanical model*: geotechnical models based on mathematical modelling focused on materials behaviour.

Norbury [71] made the significant observation that the information obtained from geological maps and field mapping at all project stages should be developed and include the ground model. Additionally, the cost-effectiveness of field mapping in site investigation must be recognised [25].

Figure 4 illustrates the role of engineering geological mapping in building models in geotechnics studies, and Table 5 shows the main techniques of a site investigation.

5 Engineering Mapping Applications: Some Examples

Engineering geology and geotechnical maps have assumed over time an essential role in geosciences and geotechnics practice, mainly in applied geology, geological engineering, geotechnical engineering, and civil engineering. Mapping—general or sketch maps, geological maps, engineering geology maps and geotechnical maps, at diverse scales—assumes crucial importance in further stages of geotechnical in situ investigations, modelling and design [30]. It is also essential to highlight the value and cost-effectiveness of geological mapping for engineering purposes compared with other activities [25, 35, 44, 50]. Thus, mapping plays a central role as a standard technique in engineering geosciences for land-use and planning, in situ geotechnical

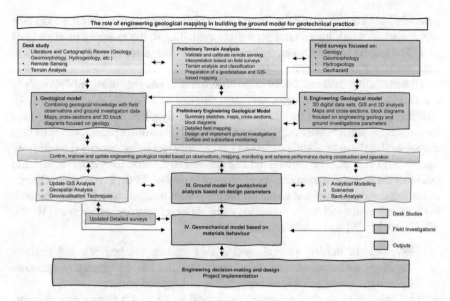

Fig. 4 The role of engineering geological mapping in building models in geotechnical studies (adapted and updated from [26])

investigations, ground modelling, hydrogeology, georesources, geohazards, and military works and operations, among others [10, 26, 30, 31, 35, 36, 45, 46, 50, 69, 77]. González de Vallejo and Ferrer [39] emphasize the importance and application of engineering geological maps and plans mainly oriented to land and urban planning, civil engineering, and geohazards. However, they could also be extended to mining and rock engineering studies and design related to the quarry, open-pit mining, underground excavations, and mines [10].

The advantage of preparing engineering geoscience maps and plans, particularly for urban engineering purposes, is still a challenging task, particularly for end-users and planners. Chaminé et al. [78] underline the significance of urban geoscience evolving to a holistic paradigm of smart urban geoscience, particularly linking geology, hydrology, groundwater, rock and soil geotechnics, natural resources, environment, geohazards, heritage and geoarchaeology issues. That approach includes integrating numerous data about all features of urban areas, such as transport, environment, economy, housing, culture, science, population, architecture, heritage, etc. Therefore, the small to medium scale maps produced for land-use and urban planning shall be comprehensive, multipurpose maps that provide reliable information on different geologic and geotechnical aspects for various urban engineering purposes for regional and local planning [39]. Besides, the layout of a multipurpose engineering geological map depends on its primary purpose: the requirements of the end-users and the need to communicate information to all agents involved, mainly practitioners, researchers, stakeholders, decision-makers and the public [30].

Table 5 The field mapping vs techniques to be employed in a site investigation: from desk studies until building the ground model (adapted from [25])

Technique	Comment	
Desk studies (including remote sensing, terrain analysis, and GIS-based mapping)	The starting process is the reconnaissance of the field constraints and collecting information at the earliest stage of the work. A critical requirement from the desk study stage is the creation of mechanisms for handling data. Currently, there is widespread acceptance of the value of GIS platforms oriented to compiling and analysing spatial data from a site investigation	
Walkover survey or ground reconnaissance	Collecting field data at an early stage requires a systematic approach, which is best provided by a clearly defined mapping programme A more recent development has been using ground-based remote sensing systems such as LiDAR and the developments in interferometry, which offer many new data analysis opportunities	
Ground investigations	It includes all forms of exploration holes, material sampling, field and laboratory testing, in situ testing and instrumentation, and geophysics. In addition, the standard techniques of drilling are tried and tested	
Building the ground model	Developing a conceptual ground model is the best method of compiling and visualizing 3D and 4D ground data to understand ground conditions. The model allows the potential behaviour of the ground under the changing stress conditions associated with construction and landscape development to be understood. Therefore, all ground investigation work should aim to produce ground models as a critical risk assessment component	

Figure 5 shows a conceptual flow-chart of the engineering geological map for urban geoscience and municipal planning. In addition, Fig. 6 exemplifies geotechnical mapping for urban planning purposes in Porto urban area (NW Portugal). Mapping contributes decisively to balanced urban planning decisions grounded in a sustainable map design with nature, urban heritage and potential geohazards. Engineering geological maps are also produced to support specific civil and mining engineering works, such as site investigations for geotechnical preliminary or feasibility studies, project, design and construction works (e.g. roads, railways, tunnels,

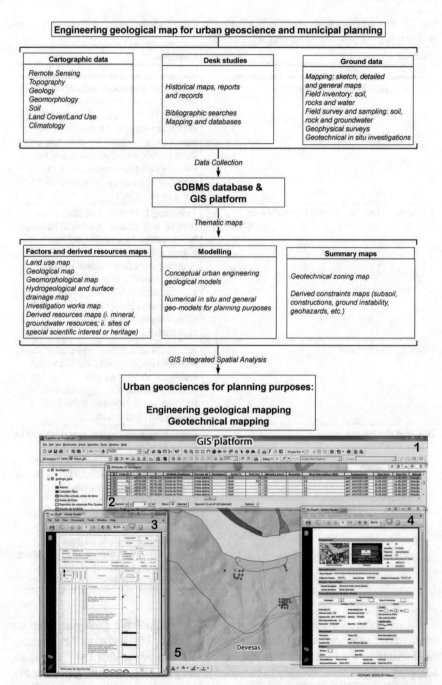

Fig. 5 Example of a conceptual flow-chart of the engineering geological map methodology for urban geoscience and municipal planning [30]: 1. GIS application tool creates hyperlinks between features (line, point or polygon) and other files; 2. Geo-database related to ground investigations information; 3. Geotechnical borehole datasheet; 4. Geotechnical inventory datasheet (field and desk data); 5. Detailed in situ investigation works mapping with geological background information

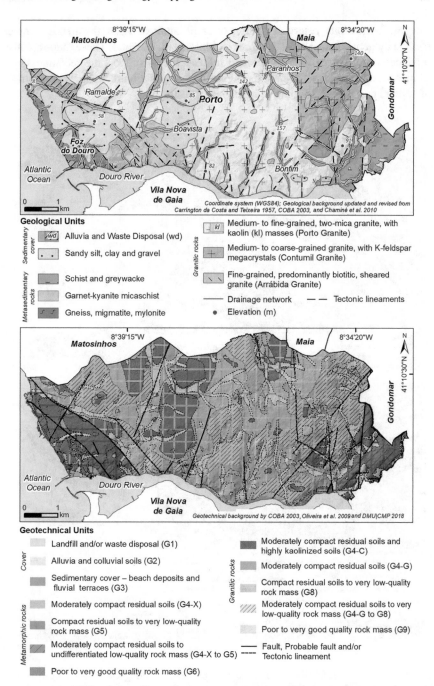

Geological Units

Sedimentary cover

wd Alluvia and Waste Disposal (wd)

Sandy silt, clay and gravel

Metasedimentary rocks

Schist and greywacke

Garnet-kyanite micaschist

Gneiss, migmatite, mylonite

Granitic rocks

kl Medium- to fine-grained, two-mica granite, with kaolin (kl) masses (Porto Granite)

Medium- to coarse-grained granite, with K-feldspar megacrystals (Contumil Granite)

Fine-grained, predominantly biotitic, sheared granite (Arrábida Granite)

—— Drainage network   — — Tectonic lineaments

● Elevation (m)

Geotechnical Units

Cover

Landfill and/or waste disposal (G1)

Alluvia and colluvial soils (G2)

Sedimentary cover – beach deposits and fluvial terraces (G3)

Metamorphic rocks

Moderately compact residual soils (G4-X)

Compact residual soils to very low-quality rock mass (G5)

Moderately compact residual soils to undifferentiated low-quality rock mass (G4-X to G5)

Poor to very good quality rock mass (G6)

Granitic rocks

Moderately compact residual soils and highly kaolinized soils (G4-C)

Moderately compact residual soils (G4-G)

Compact residual soils to very low-quality rock mass (G8)

Moderately compact residual soils to very low-quality rock mass (G4-G to G8)

Poor to very good quality rock mass (G9)

—— Fault, Probable fault and/or
---- Tectonic lineament

Fig. 6 Example of an engineering geological map for urban planning (Porto urban area, NW Portugal): Geological Map (adapted from [79] and Geotechnical Map (details in [80, 81], and adapted from [82]. The engineering geological map of the city of Porto consists of 9 thematic maps (seven-factor maps and two syntheses) and a dynamic geotechnical database that permits spatial analysis [81]

caverns, foundations, reservoirs) and geohazards studies. Often, the map is complemented by geotechnical cross-sections and informative legend. These maps are also
produced on large scales with detailed information to support the reconnaissance
studies, design and development of engineering works. Figure 7 shows an example
of applied geological mapping of an on-site investigation.

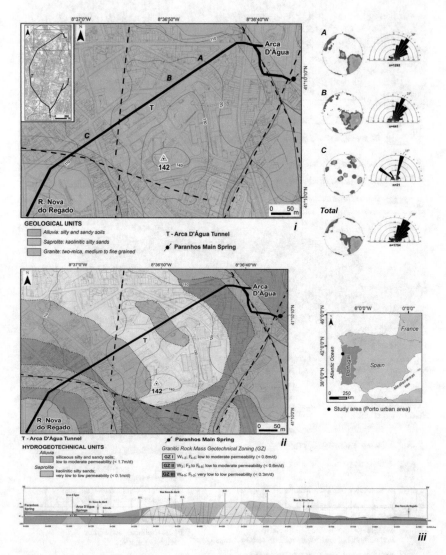

Fig. 7 Hydrogeotechnical GIS-based mapping of the Paranhos spring site—Arca d'Água Tunnel,
Porto urban area (details in [79]: **i**) geological map (A, B, C, Total: contour and rose diagrams
related to discontinuity analysis of fissured rock mass surveys); **ii** hydrogeotechnical units map; **iii**
cross-section with hydrogeotechnical zoning (adapted from [83])

Lastly, engineering geological mapping is a reliable tool to support geotechnical survey studies on heterogenous fissured rock media. Scanline surveys are a valuable technique based on collecting the basic geological, geotechnical and geomechanical parameters data. Linear or circular sampling or sampling within windows along a scanline are accurate approaches to the systematic record of discontinuities [13, 20, 69, 84]. Coupling engineering geological mapping and scanline sampling methods are consistent approaches in which a line is drawn over an outcropped rock surface. All the discontinuities crossing the line are mapped, measured and described in geotechnical and geomechanical terms ([12, 13, 20, 52, 54, 85]). Scanline survey methods for rock mass characterisation and monitoring will provide reliable information concerning structural geology, petrophysical and geotechnical features of rock masses, either in boreholes or exposed rock surfaces [20, 69]. Figure 8 exemplifies the usefulness of the scanline surveys in geotechnical practice.

6 Concluding Remarks and Outlook

> To be successful with both the hard and soft is based on the pattern of the ground. (Lao Tzu, ca. 5th century B.C.)

The engineering geological maps are helpful for the applied geologist or engineering geologist and geological, geotechnical, civil, mining, or environmental engineering practice. Engineering geology and geotechnical mapping have widespread applications in military operations, energy, mining and rock engineering, geotechnical engineering, environment, and planning. Engineering geological maps are very valuable for presenting properties, variations, and patterns of ground. They are derived from geological maps and aim to show the composition and structure of the subsoil influencing its behaviour. In short, geotechnical-related activities are considerably improved by terrain mapping methods, including remote sensing, photogrammetry, geographic information systems, building information modelling and geovisualisation techniques. The conceptualisation of ground systems must be developed on geological and ground-based models with design parameters and mathematical modelling based on geomechanical parameters to outline predicting scenarios. All the models must be robust, calibrated and supported on a permanent back-analysis scale based on a logical understanding of the natural ground behaviour. The models must incorporate earth-based systems, including the intrinsic geological ground variability and uncertainty and geological risk management in a multi-hazard environment approach.

Field surveying has been the backbone of geological studies. Field maps are crucial in geotechnical practice, particularly in data acquisition, synthesis, analysis and communication. The remarks of Wallace [86] are even topical: *"There is no substitute for the geological map and section—absolutely none. There never was, and there never will be. The basic geology still must come first—and if it is wrong, everything that follows will probably be wrong."* (p. 34). This impressive thought

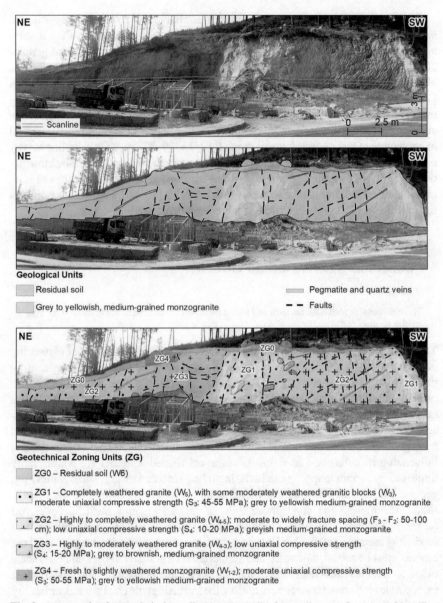

Geological Units

| | Residual soil | | | Pegmatite and quartz veins |
| | Grey to yellowish, medium-grained monzogranite | | | – – Faults |

Geotechnical Zoning Units (ZG)

ZG0 – Residual soil (W6)

ZG1 – Completely weathered granite (W_5), with some moderately weathered granitic blocks (W_3), moderate uniaxial compressive strength (S_3: 45-55 MPa); grey to yellowish medium-grained monzogranite

ZG2 – Highly to completely weathered granite (W_{4-5}); moderate to widely fracture spacing (F_3 - F_2: 50-100 cm); low uniaxial compressive strength (S_4: 10-20 MPa); greyish medium-grained monzogranite

ZG3 – Highly to moderately weathered granite (W_{4-3}); low uniaxial compressive strength (S_4: 15-20 MPa); grey to brownish, medium-grained monzogranite

ZG4 – Fresh to slightly weathered monzogranite (W_{1-2}); moderate uniaxial compressive strength (S_3: 50-55 MPa); grey to yellowish medium-grained monzogranite

Fig. 8 An example of geotechnical scanline survey studies in a rock mass slope (near Mourilhe, Cinfães, North Portugal)

is perfectly complemented by the words of Şengör [87]: *"Properly made geologic maps are the most quantitative data in geoscience: while we may debate the nature of a contact, the contact and dip-strike measurements, if properly located, should be there 100–200 years hence and are therefore both quantitative and reproducible, something that cannot be said of experiments in some of the other sciences."* (p. 44).

The geological mapping reasoning must be insightful and often inspiring in understanding earth-based systems. Therefore, ground conditions should be reliable in comprehensive geology for any geoengineering study or project. Perhaps one challenging issue is creating the total geological concept for ground site conditions required in the investigation stages following the early desk studies to present a comprehensive picture of the ground conditions, including a multidimensional modelling approach [30, 38, 76, 88]. Thus, establishing multidisciplinary approaches during all stages and different geo-professionals backgrounds is essential to safeguard the mutual exchange of experience. There is currently a temptation for digital mapping methods and techniques in ground investigations, but it shall encourage a balance of fieldwork experience and digital methodologies to achieve sound reasoning mapping and multidimensional modelling of the ground conditions. Indeed, the inspirational words of VanDine et al. [89] are still present: *"An engineering geologist knows a dam site better"*. Also, in the impressive words of Culshaw et al. [90], "A geological map is not a piece of paper.".

Acknowledgements H.I. Chaminé was partially supported by several projects: LABCARGA|ISEP re-equipment program (IPP-ISEP| PAD'2007/08), and Centre GeoBioTec|UA (UID/GEO/04035/2020). This work was funded by the Portuguese Fundação para a Ciência e a Tecnologia (FCT) I.P./MCTES through national funds (PIDDAC) – UIDB/50019/2020. Special thanks to L. Freitas (LABCARGA|ISEP) for supporting the final editing of the figures and tables. Our appreciation to the colleague editors, particularly J. Neves, for the challenging invitation and full support during the several editorial stages. Our appreciation to reviewers for their valuable comments that helped to improve the manuscript focus. Finally, we dedicated this chapter to some outstanding colleagues who have always been an inspiration in applied geology mapping practice for engineering and professional geology purposes: J.M. Cotelo Neiva, R. Oliveira, A. Gomes Coelho, A. Costa Pereira, J. Martins Carvalho, N. Plasencia, J. Chacón, L.I. González de Vallejo and J.S. Griffiths.

References

1. Rocha, M.: Mecânica Das Rochas. LNEC, Laboratório Nacional de Engenharia Civil, Lisboa (1971)
2. Delgado, C., Dupray, S., Marinos, P., Oliveira, R. (eds.): The International Association for Engineering Geology and the Environment: 50 years (1964–2014)—A Reflection on the Past, Present and Future of Engineering Geology and the Association. Science Press, Beijing (2014)
3. Choffat, P.: Étude géologique du tunnel du Rocio: contribution à la connaissance du sous-sol de Lisbonne. In : Commission des Travaux Géologiques du Portugal, Imprimerie del'Académie Royale des Sciences, Lisbonne (1889)
4. Fookes, P.: Geology for engineers: the geological model, prediction, and performance. Quart. J. Eng. Geol. **30**, 293–424 (1997). http://dx.doi.org/10.1144/GSL.QJEG.1997.030.P4.02

5. De Freitas, M.H.: Geology: its principles, practice and potential for geotechnics. Quart. J. Eng. Geol. Hydrogeol. **42**, 397–441 (2009). http://dx.doi.org/10.1144/1470-9236/09-014
6. Price, D.G.: Engineering Geology: Principles and Practice. Springer, Berlin (2009)
7. Hencher, S.: Practical Engineering Geology. Spon Press, Taylor & Francis Group, Abingdon (2012)
8. Underwood, J.: Anthropic rocks as a fourth basic class. Envir. Eng. Geosci. **7**(1), 104–110 (2001). https://doi.org/10.2113/gseegeosci.7.1.104
9. Dobereiner, L., De Freitas, M.H.: Geotechnical properties of weak sandstones. Géotechnique **36**(1), 79–94 (1986). https://doi.org/10.1680/geot.1986.36.1.79
10. Chaminé, H.I., Afonso, M.J., Teixeira, J., Ramos, L., Fonseca, L., Pinheiro, R., Galiza, A.C.: Using engineering geosciences mapping and GIS-based tools for georesources management: lessons learned from rock quarrying. Eur. Geol. J. **36**, 27–33 (2013)
11. Zhang, L.: Engineering Properties of Rocks, 2nd edn. Elsevier, Butterworth-Heinemann, Oxford (2017)
12. Terzaghi, R.D.: Sources of errors in joint surveys. Géotechnique **15**, 287–304 (1965). https://doi.org/10.1680/geot.1965.15.3.287
13. Priest, S.D.: Discontinuity Analysis for Rock Engineering. Chapman and Hall, London (1993)
14. Barton, N., Quadros, E.: Anisotropy is everywhere, to see, to measure and to model. Rock Mech. Rock Eng. **48**, 1323–1339 (2015). https://doi.org/10.1007/s00603-014-0632-7
15. Hudson, H.: Engineering in fractured rock masses. ISRM News J. **15**, 53–58 (2015)
16. Hudson, J.A., Cosgrove, J.W.: Integrated structural geology and engineering rock mechanics approach to site characterisation. Int. J. Rock Mech. Min. Sci. Geomech. Abstr. **34**(3/4):136.1–136.15 (1997). https://doi.org/10.1016/S1365-1609(97)00018-X
17. Barton, N.: From empiricism, through theory, to problem solving in rock engineering: a shortened version of the 6th Müller lecture. ISRM News J. **14**, 60–66 (2012)
18. Cosgrove, J.W., Hudson, J.A.: Structural Geology and Rock Engineering. Imperial College Press, London (2016)
19. Hoek, E.: Putting numbers to geology: an engineer's viewpoint. Quart. J. Eng. Geol. Hydrogeol. **32**(1), 1–19 (1999). https://doi.org/10.1144/GSL.QJEG.1999.032.P1.01
20. Chaminé, H.I., Afonso, M.J., Ramos, L., Pinheiro, R.: Scanline sampling techniques for rock engineering surveys: insights from intrinsic geologic variability and uncertainty. In: Giordan, D., Thuro, K., Carranza-Torres, C., Wu, F., Marinos, P., Delgado, C. (eds.) Engineering Geology for Society and Territory—Applied Geology for Major Engineering Projects, IAEG **6**, 357–361. Springer, Cham (2015). https://doi.org/10.1007/978-3-319-09060-3_61
21. Oliveira, R.: Engineering geological investigations of rock masses for civil engineering projects and mining operations. Memórias LNEC, Lisbon **693**, 1–28 (1987)
22. Hutchinson, J.N.: Reading the ground: Morphology and geology in site appraisal. Quart. J. Eng. Geol. Hydrogeol. **34**, 7–50 (2001). https://doi.org/10.1144/qjegh.34.1.7
23. Gustafson, G.: Hydrogeology for Rock Engineers. BeFo and ISRM Edition, Stockholm (2012)
24. Rocha, M.: Mecânica das Rochas. Edição no âmbito das comemorações do centenário do nascimento do Engenheiro Manuel Rocha (1913–2013). Laboratório Nacional de Engenharia Civil, Lisboa (2013)
25. Griffiths, J.S.: Feet on the ground: engineering geology past, present and future. Quart. J. Eng. Geol. Hydrogeol. **47**(2), 116–143 (2014). http://dx.doi.org/10.1144/qjegh2013-087
26. Hearn, G.J.: Geomorphology in engineering geological mapping and modelling. Bull. Eng. Geol. Envir. **78**, 723–742 (2019). https://doi.org/10.1007/s10064-017-1166-5
27. Culshaw, M.G.: From concept towards reality: Developing the attributed 3D model of the shallow subsurface. Quart. J. Eng. Geol. Hydrogeol. **38**, 231–284 (2005). http://dx.doi.org/10.1144/1470-9236/04-072
28. Culshaw, M.G.: Engineering geological maps. In: Bobrowsky, P.T., Marker, B. (eds.) Encyclopedia of engineering geology. Encyclopedia of earth sciences series, pp. 265–277. Springer, Cham (2018). https://doi.org/10.1007/978-3-319-73568-9_106
29. McHarg, I.L.: Design with nature. 25th anniversary edition, Wiley Series in Sustainable Design. Wiley, New York (1992)

30. Chaminé, H.I., Teixeira, J., Freitas, L., Pires, A., Silva, R.S., Pinho, T., Monteiro, R., Costa, A.L., Abreu, T., Trigo, J.F., Afonso, M.J., Carvalho, J.M.: From engineering geosciences mapping towards sustainable urban planning. Eur. Geol. J. **41**, 16–25 (2016)
31. González de Vallejo, L.I.: Design with geohazards: an integrated approach from engineering geological methods. Soils Rocks Int. J. Geotech. Geoenvir. Eng. **35**(1):1–28 (2012)
32. IAEG–CEGM: Engineering geological maps: a guide to their preparation. Earth Sciences Series No. 15, Commission on Engineering Geological Maps of the International Association of Engineering Geology, IAEG-CEGM, The UNESCO Press, Paris (1976)
33. Glossop, R.: The rise of geotechnology and its influence on engineering practice. Géotechnique **18**, 105–150 (1968)
34. Dearman, W.R.: Engineering geological mapping. Butterworth-Heinemann, Oxford (1991)
35. Griffiths, J.S.: Engineering geological mapping. In: Griffiths, J.S., Circus, D. (eds.) Land Surface Evaluation for Engineering Practice, Geological Society, London, Engineering Geology Special Publications **18**(1):39–42 (2001)
36. Griffiths, J.S. (coord.): Mapping in engineering geology. In: Key Issues in Earth Sciences Series, vol. 1. The Geological Society of London, London (2002a)
37. Griffiths, J.S.: Incorporating geomorphology in engineering geological ground models. In: Eggers, M.J., Griffiths, J.S., Parry, S., Culshaw, M.G. (eds.) Developments in Engineering Geology, Engineering Geology Special Publications 27, pp. 159–168. Geological Society, London (2016)
38. Fookes, P., Pettifer, G., Waltham, T.: Geomodels in Engineering Geology: An Introduction. Whittles Publishing, Dunbeath (2015)
39. González de Vallejo, L.I., Ferrer, M.: Geological Engineering. Taylor-Francis group, Boca Raton, CRC Press (2011)
40. Dearman, W.R.: Present state of engineering geological mapping in the United Kingdom. Bull. Int. Assoc. Eng. Geol. **4**(1), 15–20 (1971). https://doi.org/10.1007/BF02635377
41. Radbruch, D.H.: Status of engineering geologic and environmental geologic mapping in the United States. Bull. Int. Assoc. Eng. Geol. **4**(1), 4–14 (1971). https://doi.org/10.1007/BF0263 5376
42. Varnes, D.J.: The logic of geological maps, with reference to their interpretation and use for engineering purposes. U.S. Geological Survey Professional Paper 837, USGS, Washington (1974)
43. Montgomery, H.B.: What kinds of geologic maps for what purposes?. In: Ferguson, H.F. (ed) Geologic Mapping for Environmental Purposes, Engineering Geology Case Histories, vol. 10, ppp. 1–8. Geological Society of America, Boulder, Colorado (1974)
44. Dearman, W.R., Fookes, P.G.: Engineering geological mapping for civil engineering practice in the United Kingdom. Quart. J. Eng. Geol. Hydrogeol. **7**(3), 223–256 (1974). http://dx.doi.org/10.1144/GSL.QJEG.1974.007.03.01
45. Kiersch, G.A.: Engineering geosciences and military operations. Eng. Geol. **49**(2), 123–176 (1998). https://doi.org/10.1016/S0013-7952(97)00080-X
46. Zuquette, L., Gandolfi, N.: Cartografia geotécnica. Oficina de Textos, São Paulo, Brazil (2004)
47. Griffiths, J.S., Mather, A.E., Stokes, M.: Mapping landslides at different scales. Quart. J. Eng. Geol. Hydrogeol. **48**(1), 29–40 (2015). https://doi.org/10.1144/qjegh2014-038
48. De Freitas, M.H.: Future developments for ground models. Quart. J. Eng. Geol. Hydrogeol. **54**(2), 2020–2034 (2021). https://doi.org/10.1144/qjegh2020-034
49. Gilder, C.E.L., Geach, M., Vardanega, P.J., Holcombe, E.A., Nowak, P.: Capturing the views of geoscientists on data sharing: A focus on the geotechnical community. Quart. J. Eng. Geol. Hydrogeol. **54**(2), (2021). https://doi.org/10.1144/qjegh2019-138
50. Griffiths, J.S.: Mapping and engineering geology: an introduction. In: Griffiths, J.S. (coord.) Mapping in Engineering Geology. Key Issues in Earth Sciences Series, vol. 1, pp. 1–5. The Geological Society of London, London (2002b)
51. Matula, M., Radbruch, D.H., Dearman, W.R.: Report of the first meeting of the IAEG working group on engineering geological mapping, September 10 and 11, 1970. Paris. Bull. Int. Assoc. Eng. Geol. **3**, 3–6 (1971). https://doi.org/10.1007/BF02635389

52. Goodman, R.E.: Methods of Geological Engineering in Discontinuous Rocks. West Publishing Company, New York (1976)
53. ISRM—International society for rock mechanics: Suggested methods for the quantitative description of discontinuities in rock masses. Int. J. Rock Mech. Min. Sci. Geomech. Abstr. **15**(6):319–368 (1978). https://doi.org/10.1016/0148-9062(78)91472-9
54. ISRM—International society for rock mechanics: Basic geotechnical description of rock masses. Int. J. Rock Mech. Min. Sci. Geomech. Abstr. **18**:85–110 (1981)
55. GSE—Geological society engineering group working party report: The description and classification of weathered rocks for engineering purposes. Quart. J. Eng. Geol. Hydrogeol. **28**(3):207–242 (1995). https://doi.org/10.1144/GSL.QJEGH.1995.028.P3.02
56. CFCFF—Committee on Fracture Characterization and Fluid Flow: Rock Fractures and Fluid Flow: Contemporary Understanding and Applications. National Research Council, National Academy Press, Washington DC (1996). https://doi.org/10.17226/2309
57. NAP—National Academies of Sciences, Engineering, and Medicine: Characterization, modeling, monitoring, and remediation of fractured rock. The National Academies Press, Washington (2020). https://doi.org/10.17226/21742
58. Norbury, D.: Soil and Rock Description in Engineering Practice. 3rd Rev. Edn., Whittles Publishing, Caithness (2020)
59. Brown, L.A.: The Story of Maps. Bonanza Books, New York (1949)
60. Andrews, J.H.: What was a map?: The lexicographers reply. Cart. Int. J. Geog. Inf. Geovisual. **33**(4), 1–12 (1996). https://doi.org/10.3138/NJ8V-8514-871T-221K
61. Dykes, J., MacEachren, A.M., Kraak, M.-J.: Exploring Geovisualization. Elsevier Science, Amsterdam (2005)
62. Schiewe, J.: Geovisualization and geovisual analytics: The interdisciplinary perspective on cartography. J. Cartogr. Geogr. Inf. **63**, 122–126 (2013). https://doi.org/10.1007/BF03546122
63. Hogräfer, M., Heitzler, M., Schulz, H.: The state of the art in map-like visualization. Comp. Graph. For. **39**(3), 647–674 (2020). https://doi.org/10.1111/cgf.14031
64. Rao, M.R., Chowdary, P.S.: Geotechnical engineering through mind maps. In: Ilamparuthi, K., Robinson, R. (eds.) Geotechnical Design and Practice. Developments in Geotechnical Engineering, pp. 249–261. Springer, Singapore (2019)
65. Kraak, M.-J.: A cartographer, shaped by context and challenged by classics. Cart. J. **50**(2), 112–116 (2013). https://doi.org/10.1179/0008704113Z.00000000075
66. Russell, T.M., Stacey, T.R.: Using laser scanner face mapping to improve geotechnical data confidence at Sishen mine. J. Sout. Afric. Inst. Min. Metall. **119**(1), 11–20 (2019). http://dx.doi.org/10.17159/2411-9717/2019/v119n1a2
67. Giordan, D., Adams, M., Aicardi, I., Alicandro, M., Allasia, P., Baldo, M., De Berardinis, P., Dominici, D., Godone, D., Hobbs, P., Lechner, V., Niedzielski, T., Piras, M., Rotilio, M., Salvini, R., Segor, V., Sotier, B., Troilo, F.: The use of unmanned aerial vehicles (UAVs) for engineering geology applications. Bull. Eng. Geol. Environ. **79**(7), 3437–3481 (2020). https://doi.org/10.1007/s10064-020-01766-2
68. Chacón, J., Irigaray, C., Fernández, T., El Hamdouni, R.: Engineering geology maps: landslides and geographical information systems. Bull. Eng. Geol. Envir. **65**(4), 341–411 (2006). https://doi.org/10.1007/s10064-006-0064-z
69. Chaminé, H.I., Afonso, M.J., Trigo, J.F., Freitas, L., Ramos, L., Carvalho, J.M.: Site appraisal in fractured rock media: Coupling engineering geological mapping and geotechnical modelling. Eur. Geol. J. **51**, 31–38 (2021). https://doi.org/10.5281/zenodo.4948771
70. Gondar, J., Pinto, A., Fartaria, C.: The use of BIM technology in geotechnical engineering. In: Sigursteinsson, A., Erlingsson, S., Bessason, B. (eds.) Proceedings of the XVII ECSMGE-2019, Geotechnical Engineering Foundation of the Future. Reykjavík, Iceland, pp. 1–8 (2019)
71. Norbury, D.: Ground models: a brief overview. Quart. J. Eng. Geol. Hydrogeol. **54**(2), (2021). https://doi.org/10.1144/qjegh2020-018
72. Baynes, F.J., Parry, S., Novotný, J.: Engineering geological models, projects and geotechnical risk. Quart. J. Eng. Geol. Hydrogeol. **54**(2), (2021). https://doi.org/10.1144/qjegh2020-080

73. Norbury, D.: Standards and quality in ground investigation; squaring the circle. Quart. J. Eng. Geol. Hydrogeol. **50**(3), 212–230 (2017). https://doi.org/10.1144/qjegh2017-011
74. Davis, J.A.: Geotechnical baseline reports: ground models you can just make up? Quart. J. Eng. Geol. Hydrogeol. **54**(2), (2021). https://doi.org/10.1144/qjegh2020-019
75. Parry, S., Baynes, F.J., Culshaw, M.G., Eggers, M., Keaton, J.F., Lentfer, K., Novotny, J., Paul, D.: Engineering geological models: An introduction: IAEG commission 25. Bull. Int. Assoc. Eng. Geol. **73**, 689–706 (2014). https://doi.org/10.1007/s10064-014-0576-x
76. Turner, A.K., Kessler, H., van der Meulen, M.J.: Applied Multidimensional Geological Modeling: Informing Sustainable Human Interactions with the Shallow Subsurface. Wiley-Blackwell, Chichester, UK (2021)
77. Griffiths, J.S., Stokes, M.: Engineering geomorphological input to ground models: an approach based on earth systems. Quart. J. Eng. Geol. Hydrogeol. **41**, 73–91 (2008). https://doi.org/10.1144/1470-9236/07-010
78. Chaminé, H.I., Afonso, M.J., Freitas, L.: From historical hydrogeological inventory through GIS mapping to problem solving in urban groundwater systems. Eur. Geol. J. **38**, 33–39 (2014)
79. Chaminé, H.I., Afonso, M.J., Robalo, P.M., Rodrigues, P., Cortez, C., Monteiro Santos, F.A., Plancha, J.P., Fonseca, P.E., Gomes, A., Devy-Vareta, N.F., Marques, J.M., Lopes, M.E., Fontes, G., Pires, A., Rocha, F.: Urban speleology applied to groundwater and geo-engineering studies: underground topographic surveying of the ancient Arca D'Água galleries catchworks (Porto, NW Portugal). Int. J. Speleol. **39**(1), 1–14 (2010). http://dx.doi.org/10.5038/1827-806X.39.1.1
80. COBA Consultores de Engenharia e Ambiente, SA: Notícia explicativa da Carta Geotécnica do Porto. 2ª edição, COBA/FCUP/Câmara Municipal do Porto, Porto (2003)
81. Oliveira, R., Gomes C., Guimarães S.: Engineering geological map of Oporto: a municipal tool for planning and awareness of urban geoscience. In: Culshaw, M.G., Reeves, H.J., Jefferson, I., Spink, T.W. (eds.) Engineering Geology for Tomorrow's Cities: Proceedings of the 10th IAEG Congress, Geological Society, vol. 22, pp. 1–7. Engineering Geology Special Publications, London (2009)
82. DMU|CMP—Direção Municipal Urbanismo|Câmara Municipal do Porto: Suporte biofísico e ambiente caraterização biofísica: relatório de caraterização e diagnóstico. Revisão do Plano Diretor Municipal, Câmara Municipal do Porto, Porto (2018)
83. Afonso, M.J., Chaminé, H.I., Moreira, P.F., Marques, J.M.: The role of hydrogeotechnical mapping on the sustainable management of urban groundwater. In: Williams, A.L., Pinches, G.M., Chin, C.Y., McMorran, T.J., Massey, C.I. (eds.) Proceedings of the 11th Congress of the International Association for Engineering Geology, IAEG'2010, Geologically Active, pp. 595–1602. CRC Press, Taylor & Francis Group, Auckland (2010)
84. Palmström, A.: Collection and use of geological data in rock engineering. ISRM News J. 21–25 (1997)
85. Priest, S.D.: Determination of discontinuity size distributions from scanline data. Rock Mech. Rock Eng. **37**(5), 347–368 (2004). https://doi.org/10.1007/s00603-004-0035-2
86. Wallace, S.R.: The Henderson ore body-elements of discovery: Reflections. Min. Eng. **27**(6), 34–36 (1975)
87. Sengör, A.M.C.: How scientometry is killing science. GSA Today **24**(12), 44–45 (2014). http://dx.doi.org/10.1130/GSATG226GW.1
88. Oliveira, R.: Cartografia geológica de túneis. Memórias LNEC, Lisboa **489**, 1–12 (1977)
89. VanDine, D.F., Nasmith, H.W., Ripley, C.F.: The emergence of engineering geology in British Columbia 'An engineering geologist knows a dam site better.' Gov. Brit. Columbia, Ministry Mines Pet. Res. Geol. Surv. Open File **1992–19**, 22–27 (2006)
90. Culshaw, M., Jackson, I., Peach, D., van der Meulen, M.J., Berg, R., Thorleifson, H.: Geological survey data and the move from 2-D to 4-D. In: Turner, A.K., Kessler, H., van der Meulen, M.J. (eds.) Applied Multidimensional Geological Modeling: Informing Sustainable Human Interactions with the Shallow Subsurface, pp. 13–34. Wiley-Blackwell, Chichester (2021)

In Situ Geotechnical Investigations

Isabel Fernandes⬤ and Helder I. Chaminé⬤

Abstract The geological environment is in continuous evolution due to natural processes that can be accelerated by anthropogenic activities. The uncertainty associated with estimating the properties of the ground has a relevant impact on the design and mainly on construction safety. Poor site investigations and savings on the budget for the ground studies have proved to be false economies, causing increased costs and time overruns on the ground design and construction projects. The extent and depth of the investigation required for any project depend on several factors, namely, the type, dimensions, and location of the construction (e.g., at the surface or underground, urban or rural), geologic and geomechanical conditions, geological hazards and the site's features or characteristics. A comprehensive program of site investigation includes engineering geological mapping at an adequate scale, the use of indirect (geophysical techniques) and direct methods (boreholes and excavation techniques), as well as in situ tests in soils and rock masses. Although most of these methods have been created at the beginning of the twentieth century, they have evolved dramatically in the last decades to comply with the challenges of larger and complex constructions and the development of information technologies. The present chapter reviews the state of the art regarding site investigation methods, their applicability and objectives. It highlights the main achievements in the last decades and their response to increasingly demanding engineering works.

Keywords Direct and indirect methods · Engineering geology · Geotechnical parameters · In situ tests · Methodology

---

I. Fernandes (✉)
IDL—Instituto Dom Luiz, Department of Geology, Faculty of Sciences, University of Lisbon, Lisbon, Portugal
e-mail: mifernandes@fc.ul.pt

H. I. Chaminé
Laboratory of Cartography and Applied Geology, Department of Geotechnical Engineering, School of Engineering (ISEP), Polytechnic of Porto, Porto, Portugal
e-mail: hic@isep.ipp.pt

Centre GeoBioTEc|UA, Aveiro, Portugal

1 Introduction

1.1 Methodology

The insufficient knowledge of ground conditions is amongst the most significant causes of foundation failure [1], with unpredictable costly and time delays that are summarised by ICE [2] as "you pay for a site investigation whether you have one or not". Therefore, the design of any structure demands that an adequate site investigation is performed, including many research activities, although usually constrained by financial and time issues. A vast number of publications, guidelines and standards [3–5] can be followed to develop a site characterisation program.

Hatheway [6] defines site characterisation as "three-dimensional engineering geologic description of the surface and subsurface of the location for intended construction of engineered works, for habitation, commerce, resource development, mitigation of natural hazards or conduct of groundwater protection, waste management or environmental remediation". To establish an adequate site investigation plan, it is mandatory to have a deep knowledge of the geological principles and the methods applicable to each site. Moreover, the engineering geologist and geotechnical engineer must be aware of the loads imposed by each type of structure on the ground. In this concern, soils and rocks have different reactions to the loads applied, depending on their hydrological and mechanical properties. The depth of the geotechnical investigation depends on the complexity of the site, the type and scale of the project, and the design stage [4, 7]. It is worth defining and distinguish rock masses, rocks and soils. According to BS 5930 [3], rock refers to all solid material of natural geological origin that cannot be broken down by hand, while soil refers to any naturally-formed earth material or fill that can be broken down by hand and includes rock that has weathered in situ to the condition of an engineering soil. A rock mass comprises a system of rock blocks and fragments separated by discontinuities forming a material in which all elements behave in mutual dependence as a unit [8]. Rocks properties, such as mineral composition and texture, anisotropy, the existence of swelling minerals, alteration and weathering, will determine their mechanic characteristics and behaviour [9]. Soils are tested and classified according to international systems that allow a first approach to the expected performance when submitted to loads. Other factors such as the geological history, geological processes, seismicity, the presence of faults, the discontinuities network and the identification of paleo-mass movements are of utmost importance in assessing a site for construction. The rock mass properties and tests will be covered in another chapter, and therefore, this text will focus on soils study and tests. Conversely, the role of engineering geology mapping in geotechnical practice is detailed in another chapter.

The site investigation aims at: assessing the suitability for engineering and building works; obtaining the parameters that affect the design and construction; predicting geological and geotechnical hazards; guarantying security for the workers and the populations; reducing uncertainties from data collection, and adding value to the project (e.g., [7, 10]). The collection of further data must be decided by weighting the cost and the improvement in the knowledge, which is about 20% benefit per phase [11], and its contribution to the performance of the structure.

For some special structures, steps must be added, while some stages are unnecessary for others and are omitted. The sequence of the different investigation steps may also need to be adapted, although always following a logical and well-structured program [12]. For special large structures such as power plants, dams, tunnels, bridges, railways and highways, there might also be the need to choose alternative places and solutions from the geotechnical viewpoint. The geological and geotechnical survey is developed parallel to the design phases, extending in some cases to the construction phase and, in some special structures, eventually until the closure phase if the monitoring of the work shows that unexpected conditions are observed [13, 14]. The project phases of geotechnical investigation and site characterisation activities are presented in Fig. 1. The flowchart summarises the main engineering geology activities commonly carried out according to an ongoing cycle from observation to evaluation and analysis [15].

1.2   Scope of the Investigation

To assess the suitability of a site, a desk study of all the maps and publications about the location and its vicinity is mandatory. It must be followed by a walkover that allows for drafting the preliminary engineering geological map of the area. This map is successively modified by iteration as more data are acquired, effort and time are put into the analysis, aiming at obtaining an interpretative 3D model. In this concern, substantial work was developed by the Commission C25 of the International Association for Engineering Geology and the Environment, as summarised in [17]. The authors highlight the need to include the surficial geology and the bedrock, the hydrogeological conditions, the tectonic relevant features, and the geotechnical parameters. In addition, information is added during the drilling and in situ surveys to generate a geotechnical model that includes mathematical and physical analysis [17]. Figure 2 is a summary of the activities to be carried out during site investigation.

As the project phases evolve, the site characterisation is more and more intensive, detailed and expensive. It is based on increasingly sophisticated techniques and methods, simultaneously reducing uncertainty and financial risk. Parameters are obtained for each geotechnical unit, such as deformation, the potential to settlement, seismic response, seepage and stability [18, 19]. Scales also vary from 1:10,000 available maps at the conceptual stage to 1:10 in logs and 1:100 in vertical geological cross-sections (e.g., [17, 20]).

The information obtained in each phase will define the plan's scope and the objectives of the work to develop in the following phase, namely the coordinates of placement of the surface and subsurface exploration methods, the depth to be reached and the techniques to use. A wide range of methods can be applied in the subsurface investigation. The engineering geologist is responsible for selecting the appropriate methods for testing and sampling depending on the type and magnitude of the load,

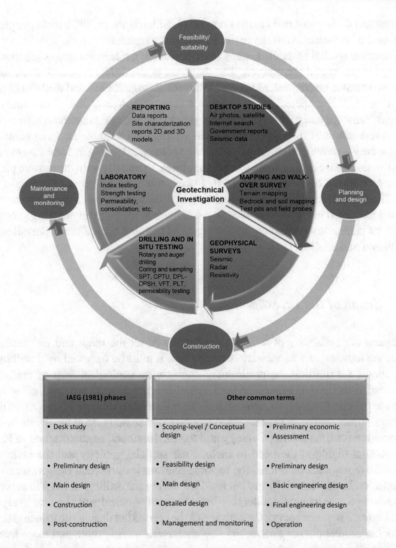

Fig. 1 Phases of the project and the methodology of geotechnical investigation and site characterisation activities focusing on the feasibility and design phases (adapted from [7, 13, 16])

the sensitivity of the structure and the geological complexity of the ground [7]. For example, in EN 1997-1 and 2 [4, 5], three categories of works are defined and guidelines offered regarding the intensity of site investigation and the methods to apply, their depth and spacing. Regarding costs, BRE Digest 322 [21] recommends that expenditure on the ground investigation is a minimum of 0.2% of the total project cost, but Hytiris et al. [22] found out that an adequate site investigation for low-rise

Fig. 2 Methodology of site investigation (adapted from [17])

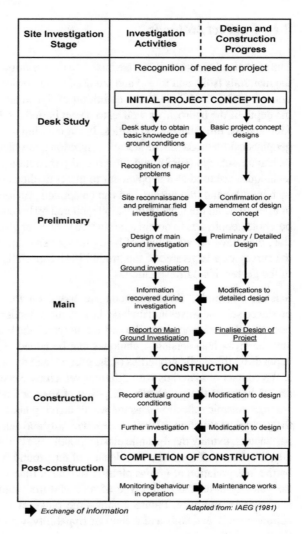

Site Investigation Stage	Investigation Activities	Design and Construction Progress
	Recognition of need for project	
	INITIAL PROJECT CONCEPTION	
Desk Study	Desk study to obtain basic knowledge of ground conditions	Basic project concept designs
	Recognition of major problems	
Preliminary	Site reconnaissance and preliminar field investigations	Confirmation or amendment of design concept
	Design of main ground investigation	Preliminary / Detailed Design
	Ground investigation	
Main	Information recovered during investigation	Modifications to detailed design
	Report on Main Ground Investigation	Finalise Design of Project
	CONSTRUCTION	
Construction	Record actual ground conditions	Modification to design
	Further investigation	Modification to design
	COMPLETION OF CONSTRUCTION	
Post-construction	Monitoring behaviour in operation	Maintenance works

➡ *Exchange of information*   *Adapted from: IAEG (1981)*

buildings should be a minimum of 0.42% of the cost of the project. Anyway, the site investigation costs are always a very low percentage of the project.

2 Methods of Site Investigation

The subsurface exploration methods can be divided into various ways. The most common division is into indirect and direct methods and, for in situ testing, according to the parameters determined (e.g., [10, 18, 19]). These principles are adopted in the present work.

2.1 Indirect Methods

The so-called indirect methods are based on the variation of geophysics properties of the materials (soils and rocks) and are used to determine the thickness of the superficial deposits, contribute to the definition of the weathering profiles, and evaluate the depth of the bedrock, as well as to detect buried artefacts and voids, such as old foundations, mineshafts and cavities. In engineering geology, the most well-known geophysical methods are the seismic refraction and reflection surveys, the electrical resistivity surveys (ERS), and the ground penetration radar (GPR). These survey techniques are used as a supplement to direct methods of investigation, carried out by boreholes and trial pitting, and can be applied previously to these direct methods or after, to detail the local information gathered by boreholes (e.g., by using acoustic borehole geophysics to obtain quantitative values for ground stiffness) [23]. In addition, they have the advantage of being inexpensive and quick to perform, allowing the survey of a large area of the ground [24], especially relevant in the first phases of the geotechnical investigation.

Seismic Methods. These methods are based on the propagation of artificially produced seismic waves and include the seismic reflection and the seismic refraction, both usually performed from the surface, to which the borehole geophysical logging and the cross-hole seismic techniques can be added. The velocity of propagation depends on the elastic properties of the ground, and the wave is refracted or reflected at the contact of materials with different velocities. Seismic refraction uses the first arrival times at the geophones, disposed at regular intervals along an alignment, whereas seismic reflection uses the waves arriving later. The method aims to define the geological structure, evaluate the excavability and determine dynamic mechanical parameters, namely the dynamic elastic moduli, the Poisson coefficient and, eventually, the shear stiffness, when the density of the ground is known. Geology is defined by the interpretation of graphs plotting the travel time vs distance of geophones, as the velocity is related to the type of rock and the weathering grade, the porosity, and cavities' presence. Figure 3 shows the range of wave velocities for the most common rocks and soils and a chart of rippability, which is relevant for evaluating the excavability of the ground.

Seismic reflection is more common in the investigation at depth, such as in oil exploration. This is because it has the advantage of representing multiple horizons graphically with a single shot. In offshore works, the seismic survey is also used for bathymetric measurements and the definition of seabed morphology.

Electrical Resistivity Methods. These methods, especially the Continuous Vertical Electrical Sounding (CVES), the 2-D resistivity pseudo-sections and the tomography assess variations in lithology and the definition of major structures. In addition, they contribute to locating groundwater level and underground salty water, identifying fluctuations in groundwater quality, and correlating with parameters such as porosity and degree of saturation [10, 24]. The results obtained of apparent resistivity characterise the material affected by the passage of the electric current as a whole, and the

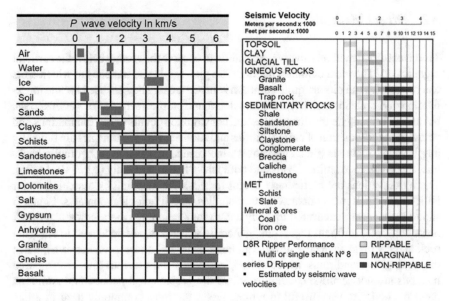

Fig. 3 Velocity of propagation of longitudinal waves in common rocks and soils (**left**) (adapted from [10]) and an example of a rippability chart (**right**) (after [25])

depth analysed depends on the distance between electrodes. Tomography is obtained by moving the set-up laterally, resulting in a model of real resistivities after applying an inversion process. Figure 4 shows a tomography and representative values of the apparent resistivity of common geological materials.

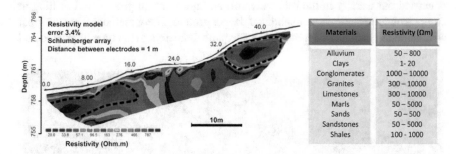

Fig. 4 Example of an apparent resistivity model and values of resistivity characteristic of common rocks and soils (adapted from [10])

2.2 Direct Methods

The visual direct methods involve excavation to observe the ground below the surficial organic soil. The most common methods are pits, shafts and trenches to which adits can be added. Trial pits are quite common in urban areas. They are used for site investigation and collection of large disturbed and undisturbed samples of geomaterials, but also to investigate the dimensions and construction details of old foundations and define the exact position of buried utilities and services. Pits and trenches are typically dug by hand methods (pick and shovel), mechanical excavator, or hydraulic backhoe, and allow to examine the ground both laterally and vertically (e.g., [10, 14, 26]. These excavations are particularly useful for geologic mapping by assessing major fault traces, rock mass characterisation studies, and lithological contacts. Figure 5 shows the trenches excavated to study a dam foundation in a metasedimentary rock mass with intense fracturing. Adits, generally expensive, are dedicated to studying rock masses for special structures (e.g., large dams and underground caverns, when the geological conditions are complex), and when the previous site investigation methods are not accurate enough to obtain the geotechnical parameters. Although these methods are very useful in remote areas, the most commonly used in urban sites are boreholes and in situ tests. Boreholes are a direct method for ground investigation, while they demand that empirical interpretation of the ground between each borehole is carried out, contributing to uncertainties in the engineering geologic and geotechnical models.

Boring and drilling are perforation methods of small diameter used to study the underground and obtain information on the geological, geotechnical and geochemical characteristics. They are used for ground investigation for buildings, roads, bridges, power plants, tunnels, dams, marine structures and wind farms. Boring is carried out usually in the relatively soft and uncemented ground and drilling in competent/cemented/overconsolidated deeper ground. These methods have the great advantage to reach large depths, using adequate diameters to the type of ground and

Fig. 5 Tranches excavated to investigate a dam site (Portugal) (**left**) and a detail of a fault containing clay gauge observed in a trench wall (**right**)

the study's specific aim. They can be destructive (e.g., auger drilling and bottom-hole hammer) but, for geotechnical site investigation for large structures, engineering geologists elect as preferential the core drilling with continuous sampling.

Drilling and boreholes must extend to the deepest substratum affected by the structure load [27]. Drilling methods are used to collect different types of samples, from disturbed samples, as in auger, wash boring, and percussion drilling, to undisturbed samples (cores), collected with sophisticated samplers [11]. One of the most commonly used perforation methods in soils is the continuous-flight auger with a hollow stem, which is screwed into the ground by rotation and acts as a casing. Penetration in strong soils/gravel layers can be difficult, but samples can be collected or in situ tests performed inside the auger. Thus, they are suitable for soft to firm cohesive soils.

In rotary drilling [28], a hole is made by a rotary action combined with downward force to grind away the material at the bottom of the hole. The drilling fluid, commonly water, mud or foam, is pumped down to the bit through hollow drill rods, aiming at: lubricating and cooling the bit; clean the borehole by flushing the drill cuttings to the surface; stabilising the borehole; provide hydrostatic pressure [14]. For soils in which the hole walls tend to collapse, usually, a water solution of bentonite, a thixotropic clay, is used. The mud also contributes to sealing off the water flow into the shaft from the permeable water-bearing strata.

Other destructive methods, which produce loose cutting, are rock roller bits or open hole drilling, down-the-hole hammers (DTH) and water jets. The advantage of percussion drilling is that the method produces boreholes quickly and cheaply compared with core drilling. The advancement occurs due to alternatively lifting and dropping a heavy tool [24], and a range of bits can be used. The drill cuttings brought to the surface in the flushing medium can only indicate the ground conditions being encountered. Although it is difficult to correlate the cutting samples to the exact depth [14], the methods can be helpful for rapid advancement (e.g., for field testing or instrument installation). As already highlighted, for engineering purposes, conventional core drilling is preferred. It allows the establishment of the nature of the materials, defining the depth and characteristics of the strata and the rock mass structure, and obtaining samples for laboratory testing (Fig. 6).

In core drilling, the drilling rig rotates a string of rods, and a downward force is applied hydraulically. At the end of the string, an annular diamond or tungsten carbide coated bit, fixed to the outer rotating tube of a core barrel, cuts the core. In the interior of the drilling tool, there is a stationary core barrel attached to the rig containing a core catching device to retain the core sample [18]. The core barrels can be single-tube, but preferably in site investigation, double-tube for rocks or triple-tube containing detachable liners and retractable shoe for decomposed materials and soils should be used [14], isolating the soil or rock core from the drilling fluid. The core is brought to the surface inside the sampler for examination and laboratory testing. There are several sizes, but larger core diameters produce, usually, greater recovery. Boreholes also allow the performance of in situ tests at a depth of interest and the installation of monitoring equipment such as inclinometers, extensometers and piezometers, with importance during the site investigation, construction and

Fig. 6 Types of samples and core drilling bits: **a** cuttings obtained by destructive methods; **b** drill core; **c** opposing hemispheres of the split-spoon sampler open, exposing the recovered clay sample of an SPT test; **d** tungsten carbide bits for destructive drilling; **e** drilling bits for core sampling, diamond-impregnated and tungsten carbide tipped cutting shoe

maintenance phases. The depth and spacing of the boreholes depend on the size and sensitivity of the project and the complexity and characteristics of the ground. In the samples collected from core drilling in rock masses, information such as lithology and its variability, weathering grade, fracture spacing, Rock Quality Designation (RQD) and total core recovery (TCR) can be obtained, providing important geotechnical parameters for the characterisation of the rock mass [18]. Table 1 summarises the relevant parameters to be evaluated in coherent and incoherent soils and rock masses.

2.3 In Situ Tests

The majority of in situ tests were created at the beginning of the twentieth century. Since then, they have evolved dramatically to comply with the development of information technologies and the challenges of larger and more complex constructions (e.g., pipeline foundations and wind farms, as presented in [29, 30]). Table 2 summarises the main parameters that can be obtained by the most common in situ tests.

Table 1 Parameters to evaluate in different soil types and rock masses (adapted from [17])

Cohesive soils (clay)	Granular soils (sand, silt or gravel)	Rock mass
Layering	Layering	Mineralogy, lithology and structure
Grain size distribution	Grain size distribution	Water absorption
Water content	Water content (silt)	In situ permeability
Total unit weight	Maximum and minimum density	Unit weight of the intact rock
Atterberg (plastic and liquid) limits	Relative density	Unconfined compression strength
Indicative shear strength (miniature vane, torvane, pocket penetrometer, fall cone, UU, etc.)	Drained angle of shearing resistance	Discontinuities characteristics
Remoulded shear strength	Soil stress history and over-consolidation ratio	RQD
Sensitivity	Angularity	Fracturing
Soil stress history and over-consolidation ratio	Carbonate content	Weathering grade
Organic material content	Organic material content	

Several publications elaborate on the most common site characterisation methods (e.g., [11, 14, 18, 31–34]), and several standards and scientific papers discuss the different types of tests, their advantages and limitations (e.g., [35]). In EN 1997-2 [5], useful information about the most common methods, empirical correlations, equations and examples are provided, based on relevant literature on each method. The most common methods for soils are listed according to the purpose, i.e. to the parameters obtained, namely penetration resistance, strength and/or compressibility and permeability [18]. In situ tests are particularly important for soils in which recovering samples is complex. The performance of each method and the data obtained depend on the type of soil, which is roughly divided into coherent (clay) and incoherent (granular). The nature of the soil influences the in situ drainage conditions, and usually, the tests in clays are done rapidly to consider undrained conditions. In consequence, tests carried out in clay soils provide the undrained shear strength, and in granular soils, the (drained-) peak effective angle of friction is obtained.

Penetration Resistance. Penetration resistance is usually obtained by two methods: dynamic hammering (e.g., standard penetration test, SPT, and dynamic probing, DP); static penetration by pressing the equipment into the soil (e.g., cone penetration test, CPT). These methods were established many decades ago, and there has been a continuous improvement. However, the procedures include some assumptions and corrections and, under some ground conditions, the joint employment of SPT and CPT together has the greatest potential for geotechnical engineering (e.g., [36]). The penetration resistance is obtained using empirical approaches to test soils' strength

Table 2 Applicability of the most common in situ tests in soils (adapted from [11]; further information in [3, 18, 32])

Test	Parameters											
	Establish vertical profile	Relative density D_r	Angle of friction φ'	Undrained shear strength S_u	Pore pressure u	Stress history OCR and K_0	Young modulus E, G'	Compressibility index C_c	Coefficient of consolidation c_v	Coefficient of permeability k	Stress–strain curve σ–ε	Liquefaction resistance
Seismic refraction	C						B					B
Electrical resistivity	B	B	B	C				C				
Standard penetration test (SPT)	B	B	C	C				C				A
Cone Dynamic (DPL/M/H/SH)	A	B	C	C		C						
Cone Electrical piezocone (CPTU)	A	B	B	B	A	A	B	B	A	B	B	A
Cone Mechanical (CPTM)	A	B	C	B		C	B	C				A
Cone Seismic (SCPTU)	C	C					A				B	B
Flat plate dilatometer (DMT)	A	B	C	B		B	B	C			C	B
Plate load tests (PLT)	C	B	B	C		B	A	B	C	C	B	B
Pressuremeter Ménard (MDM)	B	C	B	B·		C	B	B			C	C
Self-boring pressuremeter (SBP)	B	A	A	A	A	A	A	A	A	B	A	A
Shear vane test (SVT)	C			A		B						

(continued)

Table 2 (continued)

Test	Parameters											
	Establish vertical profile	Relative density D_r	Angle of friction φ'	Undrained shear strength S_u	Pore pressure u	Stress history OCR and K_0	Young modulus E, G'	Compressibility index C_c	Coefficient of onsolidation c_v	Coefficient of permeability k	Stress–strain curve σ-ε	Liquefaction resistance
Borehole permeability					A				B	A		

OCR Overconsolidation ratio; K_0 Ratio between horizontal and vertical effective stresses for OC soils
Code A = most applicable; B = may be used; C = least applicable

at various depths [37]. Penetrometers are divided into dynamic, driven by blows of drop weight, and static pushed hydraulically into the soil. For the first group, the most well-known is the Standard Penetration Test (SPT) and the Dynamic Probing Light/Medium/High/Super High (DPL/DPM/DPH/DPSH), the former using a bore-hole, the last being independent of boreholes. The most common static penetrometer is the Cone Penetration Test (CPT), measuring the interstitial pressure (undrained, CPTU). Examples of empirical correlations and analytical methods can be found in [5].

Standard Penetration Test. SPT aims to determine granular soil's strength and defor-mation properties, but data can be obtained for other soil types [5]. It is simple and inexpensive, performed discontinuously every 1.5 m of depth along with the execution of a borehole, using a split barrel sampler driven into the ground by blows produced by dropping a 63.5 kg hammer through 762 mm. At the depth defined for the test, the split barrel sampler is driven in three steps of 150 + 150 + 150 mm of penetration length, and the number of blows needed for each 150 mm is registered. The test results consider the second and the third steps to overcome the volume of soil disturbed by the borehole operations (300 mm in total). The sum of blows (N) corresponds to the penetration resistance. At the end of the test, the split spoon is pulled from the hole, and a small disturbed sample is stored in an airtight container. Although these features correspond to the original SPT, the test is not totally stan-dardised, and the results will depend on the test equipment, the standard followed (e.g., [38, 39]), the disturbance during the perforation, the cleanliness of the bottom hole, the stability of the borehole wall, the type of soil and its age (e.g., [5, 10, 18]). For example, the resistance of sand to penetration is higher for longer geological periods of consolidation of the soil, increasing the blow counts.

The presence of groundwater can affect severely the performance of sands, which become loosened. The test is particularly interesting when it is impossible to collect good quality borehole samples in sand, silts or sandy clay. However, it is also common to perform SPT alternating with core sampling. The limitations of the test have been discussed since the 1960s in several publications (e.g., [36, 40, 41]). Some corrections were introduced to the N obtained, as explained in [4] and [39], namely regarding the actual energy delivered by the drive-weight apparatus, the borehole diameter, the length of the rope, the sampling method and the pressure of soils at rest, which itself depends on the effective overburden pressure. For granular soils, $(N_1)_{60}$ represents the number of blows from the SPT corrected to energy losses and normalised for effective pressure and atmospheric pressure (e.g., [42], and other references in [43]). The result N_{60} reflects the number of blows needed with a machine delivering 60% of the theoretical energy. Empirical correlations allow to obtain a rough approximation of the relative density and peak drained friction angle for sand, the Young modulus of elasticity for sandy soils, the undrained cohesion and the overconsolidated ratio for clay, the liquefaction potential of sand, bearing capacity and the settlement of foundations for granular soils ([43] and references therein). Information can only be quantitative for well-known local conditions where the results are correlated with other more reliable tests. The correlations between SPT and soil properties are empirical, based on international databases.

Dynamic Probing test. The dynamic Probing test (DP) method is outlined in, for example, EN ISO 22476–2 [44]. It aims at determining a soil profile of the resistance of soil or soft rock to the dynamic penetration of a solid cone. Correlations can be used for fine soil to obtain the soil's strength and deformation properties and determine the depth of very dense ground layers. A metal cone attached to several rods is driven into the ground by dropping a hammer of a given mass from a certain falling height and counting the blows. Depending on the hammer's mass, a distance of penetration is defined and the blows counted, which provides a continuous record along with the depth. As the method does not allow sampling, the results of the DP must be complemented by sampling obtained from other investigation methods, namely drilling [5]. Apparatus features and the specifications regarding the penetration depth for different masses of the hammer can be found in the standards: dynamic probing light (DPL), medium (DPM), heavy (DPH) and super-heavy (DPSH).

Cone Penetration Test. CPT/CPTU is a very popular and quite precise test method, and correlations have been established with SPT. CPT is used to characterise soft and compressible soils, such as weak clays [43], although some equipment can be used in stiff to hard clays and dense sands. The static cone test consists of a 60° cone, and a surface sleeve continuously pushed into the ground, at a standard rate of penetration, and the resistance offered by the cone and sleeve is measured by load cells located just behind the tapered cone [36, 45–47], Fig. 7. It is a continuous test, with measurements made at every 0.20 m, allowing that the subsoil profile is obtained in much more detail than in the SPT, including when thin granular layers are found within soft cohesive soils. However, the results are not absolute and should be calibrated using the vane shear test or triaxial laboratory tests. Besides the mechanical CPT, there are electric cone penetrometers (e.g., [48]) with pore pressure measurements (piezocone, CPTU) and the seismic cone penetrometer (SCPT), measuring the shear wave velocity in depth. During the test, the tip resistance (q_c) and the sleeve friction (f_s) are recorded, simultaneously, as in the electric cone, or separately, in the mechanical method, which ratio is called the friction ratio (R_f, %), Fig. 7.

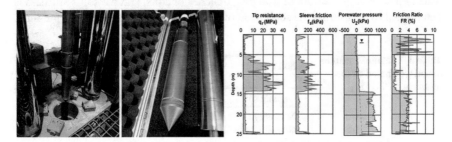

Fig. 7 CPTU test and equipment (**left**) (courtesy of Geocontrole) and example of charts of results (**right**) (adapted from [26])

The measurement of pore pressure (u), made by using a pore pressure transducer usually located where cone and sleeve meet, is a great advantage of the CPTU method, allowing dissipation tests and determining the coefficient of consolidation for clays. According to the equations published, the pore pressure is used to correct the values of q_c and f_s, (e.g., [49]). The porous filter for pore pressure measurement might be located either at three different positions: at the apex or mid-face of the cone tip (u_1), at the shoulder (u_2), behind the cone sleeve (u_3). The electrically operated devices also register soil resistivity, ground vibration, core inclination, gamma-ray backscatter and pressuremeter values, depending on the type of device [18]. The method provides correlations with design parameters, depending on the equipment used, namely relative density and friction angle for sands, undrained shear strength, sensitivity, overconsolidation ratio (OCR) and compressibility for clayey soils. The results allow to calculate the ultimate bearing capacity of shallow and deep foundations, correlate with unit weight and permeability. However, borehole sampling should be done for the correct identification of the type of soil. In [43], a summary is made on the parameters obtained, the correlations and the equations proposed by different authors. CPT can also assess liquefaction, with some advantages compared to SPT, as presented in [50], namely the quality and repeatability and the detection of the variability of soil deposits. This test is quick and inexpensive when comparing with those that need boreholes. A deep push PCPT system was recently created for offshore site investigation, obtaining continuous sub-sea profiling, with options such as the spherical ball penetrometer and the T-bar.

Charts have been created by different authors correlating the CPT data with the type of soil (e.g., [32, 51–53]), based on the USCS (Unified Soil Classification System). Table 3 contains information about a rough classification of the soils based on the different parameters registered during the CPTU tests.

Strength and Deformability Parameters. Strength and deformability parameters are obtained by several methods, and there are established correlations with the penetrometer tests.

Plate Load In Situ *Test*. Plate Load in situ Test (PLT) is a very accurate method to characterise the stiffness and the bearing capacity of soils [24, 54]. The plate is usually placed at the bottom of trial pits, trenches or adits, and the soil's stiffness is determined by measuring the settlement of the plate subjected to a defined pressure.

Soil type	Cone resistance	Friction ratio	Excess pore pressure
Organic soil	Low	Very high	Low
Normally consolidated clay	Low	High	High
Sand	High	Low	Zero
Gravel	Very high	Low	Zero

Table 3 Diagnosis features of soil type based on the results of CPTU [18]

The test also allows to validate the quality of soils during compaction and to estimate foundation settlements. It is performed following different standards (e.g., [3, 37, 55, 56]). The load is applied in increments by a hydraulic jack to the steel plate, varying between 150 and 600 mm in diameter and equipped with pressure gauges or load cells, and each increment is maintained until the penetration of the plate has ceased. The time needed will depend upon the soil type and its permeability. The depth of influence is less than 1.5 times the plate diameter, which corresponds to 0.2 of the applied load [54]. The stiffness is calculated considering the vertical displacement of the plate measured in three dial gauges, the plate diameter, the Poisson coefficient, and the contact stress assumed to be uniformly distributed and obtained as the ratio of the applied load by the plate area. The pressure to be applied depends on the working load. The required pressure in compaction control tests should be three times the allowable bearing capacity used in the design and 1.5 times the working load to determine stiffness [54].

Pressuremeter Tests. Pressuremeter Tests (PMT) are divided into four different types [5]: pre-bored pressuremeters (PBP), e.g. the flexible dilatometer test (FDT) and the Ménard pressuremeter (MPM); the self-boring pressuremeter (SBP); and the full-displacement soil pressuremeter (FDP). In addition, some methods require that a borehole is made specifically for the test (as in MPM) whilst others are pushed into the ground independently from boreholes (SBP and FDP). These methods aim at determining the pressuremeter modulus and the limit pressure or the stiffness and strength parameters.

The selection of the self-boring pressuremeter or the pre-bored pressuremeter depends on the ground type (e.g., in very stiff soil, preboring is needed, and the second method is used, whilst in soft soils, the SBP is adequate). As a result, some improved pressuremeters have been created, some of them also applied in soft and hard rocks. The great advantage of the SBP is the application in undisturbed soil, requiring expert operators.

Preboring Ménard Pressuremeter. Preboring Ménard Presuremeter (MPM) can be performed as in ASTM D4719 [57] or EN ISO 22476-4 [58]. The test aims at measuring the in situ deformation of soil and soft rock produced by the radial expansion of a borehole section by using an expandable cylindrical, flexible membrane under pressure produced by water, compressed air or gas [4, 5, 14]. It is required that a high-quality borehole is open, the Poisson ratio of the soil is known, and the isotropic and elastic behaviour of the soil is assumed. Briaud [59] summarises the method and refers to the applicability of the test. The device is inserted into a pre-formed borehole at a predetermined depth to read the pressure and the expansion until a maximum expansion of the device is reached. The pressure is applied in equal increments, and volumetric readings are registered at time intervals after applying each pressure increment [59]. The results depend on the borehole's quality, and the diameter and limit values are established for the borehole and the probe. The friction angle is also obtained, although the method seems to overestimate the parameter for deep tests (>12 m depth) and underestimate for tests close to the surface [60].

The Self-Boring Penetrometer. The Self-Boring Penetrometer (SBP) [61] was developed to reduce the soil disturbance due to the perforation of the borehole. The probe is inserted hydraulically as the internal cutter rotates, and fluid flushes the debris to the surface through the hollow centre of the probe. There are different versions of SBP, namely the Cambridge pressuremeter [62] and the French Pressiomètre Autoforeur PAF [63]. The method has been developed since the 1970s to evaluate the initial state, deformation and strength characteristics of soil (e.g., [64]).

The Flat Plate Dilatometer. The Flat Plate Dilatometer, also known as the Marchetti dilatometer test (DMT), is a self-boring pressuremeter that has been used in soil investigations for decades, and several publications explain the functioning, advantages and limitations of the test (e.g., [65–67]). It can be performed according to ASTM D6635-15 [68] or EN ISO 22476-11 [69], and guidelines were published as a Report of the ISSMGE ([70] and references therein). The equipment consists of a thick steel blade having an expandable circular steel membrane on one of the faces and is equipped with pressure gauges, a valve for gas pressure control, vent valves, and an audio-visual signal (Fig. 8). The blade and rods are pushed hydraulically into the ground, and the membrane inflated to obtain the lift-off pressure. The blade is then pushed to the next test depth, usually in increments of 0.20 m. An update is presented in Marchetti et al. [71], focusing on the importance of the stress history for estimating the settlement and the liquefaction resistance of the soil and the combination of DMT and CPT to assess OCR in sands. The authors discuss the applicability of the seismic dilatometer (SDMT), a combination of the DMT with a seismic module to measure the shear wave velocity every 0.5 m of depth, providing the evaluation of the liquefaction potential.

Vane Shear Test. Vane Shear Test (VST) method covers the in situ determination of the undrained shear strength and the sensitivity of weak intact cohesive soils, including clays, silts and glacial clays. It is more relevant when combined with CPT. With four delicate rectangular blades, a cruciform vane is pushed into de soil with a protective shoe to the required depth. The vane is then advanced a short distance (0.5 m) ahead of the shoe into the undisturbed soil to perform the test. The blade is

Fig. 8 Flat dilatometer and measuring equipment (courtesy of Geocontrole)

turned with a torque of sufficient magnitude to shear the soil (e.g., [72, 73]) at a rate defined in each standard (e.g., 6°/min). The length (H) is double of the width (D) of the blade, and different sizes can be used (e.g., 150×75 mm^2 blade for soils up to shear strength of 50 kPa, 100×50 mm^2 blade shall be used for soils of shear strength between 50 and 75 kPa [37]). The test is restricted to uniform, fully saturated soil, usually in clay, and is not adequate when there are thin layers of sand, coarse particles or dense silt. Many factors influence the results of the VST, namely the type of soil, the strength anisotropy, the rate of rotation, the time elapsed between the insertion of the vane into the soil and the test and the possible disturbance due to the insertion of the blades [18]. Usually, pre-boring is adequate to reach the desired layer. The maximum resistance offered by the soil to the vane rotation is noted and, to obtain the remoulded shear strength, the vane is further turned by ten complete rotations [14].

Okkels and Andersen [74] suggested an alteration of the VST test to become useful in all types of soils, designated by M-VST, fast multi-soil vane shear test. In traditional VST, the vane is rotated between 2 and 12°/min to produce failure in 0.5–3 min. The Danish "deep" VST system is more robust and has different dimensions, sharpened edges of the blades, and is used to be applied in overconsolidated soils for the strength of 10 to 700 kPa. The blades are 3 mm in thickness, and the shaft is 20 mm in diameter to allow the use to firm, gravelly and stony fine soils. The blades have rounded corners to minimise the stress effects at the corners during rotation. The torque is produced to a uniform rate of 360°/min, and the failure occurs in 6 s. The test is performed during boring with soil core sampling, which allows knowing the type of soil tested. This rapid test, with 1 min waiting time, complies with the conclusions in [75], who concludes that by using a long time before shearing, as in the traditional VST, the increase in strength due to consolidation could be more significant than the reduction in strength caused by the vane insertion, resulting in an overestimation of the strength.

Permeability In Situ Tests. In situ water tests are the most accurate to assess the permeability of the ground as they cover a much larger volume of soil than any laboratory test [10, 14, 18]. Permeability of soils depends on several factors, namely the particle size and shape, the structure of the soils, the void ratio, the degree of compaction and the degree of saturation. The permeability coefficient can be obtained using different methods that can be grouped as pumping-out and pumping-in tests. Standards such as [76–78] can be followed. Depending on the type of soil, two types of permeability tests are carried out: Constant Head test (CH) for more permeable soils, and Falling Head test (FH), for soils of low permeability [79]. Lefranc test is one of the most common tests for soils, and it is performed by generating a water flow with a constant or variable water head into a cavity of a given shape, named lantern, at the bottom of a borehole. After filling the borehole with water, and verifying that all air has been expelled, the flow rate needed to keep the water level constant (constant head test, CH), or the velocity of drop or change in the water level (falling and rising head tests, FH) are measured (e.g., [80, 81]). The variable rising-head test includes the third method in the Canadian standard (CAN/BNQ 2501–135 [82]). In the CH test, the flow rate is measured every 5 min so that the water level at the top of the borehole is kept constant for 45 min. If there is a very high intake of water, the measurements are carried out every minute during the first 20 min and every 5 min thereafter up to a total of 45 min [10]. The variable-head (VH) consists of filling the drill casing and measuring the variation in elevation from a set reference level with time. Different methods of interpretation have also been discussed [83].

3   Example of Challenging Structures

Marine investigation for pipeline foundations, wind farms, oil platforms, bridges and tunnels is one of the most challenging, as summarised in [29, 30], not only due to the difficult environmental load conditions, tides, waves, winds, currents, icebergs and ice sheets but also in consequence of the depth of the seabed and the geological complexity and risks associated, namely earthquakes and seabed landslides. Two modes for the geotechnical investigations in offshore environments are common: the seabed mode, when the sampler or in situ tester is placed directly on the seabed, and the drilling mode, using platforms or vessels [83]. Several field tests can be performed, most of them listed in this work. The geophysical survey allows obtaining the seabed bathymetry using echo-sounding. It may be used in association with sophisticated equipment such as ROV (remotely operated vehicles) and AUV (autonomous under-water vehicles) or geoBAS survey (geophysical burial assessment survey) to provide quantitative information of the soil below seabed [30, 84]. The most common in situ tests include CPTU, vane shear tests, standard penetration tests, dilatometer, and permeability tests [14, 85]. Disturbed samples can be obtained by quick methods such as the grab sampler, but it only reaches 0.5 m depth. Other relatively surficial corers are: the gravity corer, an open barrel of 3 m in length; the vibrocorer, which can penetrate 3 to 6 m in the seabed (further details in [86]); and the box corer. For deep foundations and other offshore demanding structures, high-quality samples are mandatory, such as using a deep water sampler (DWS) or a driven deep penetration piston corer as the one developed by [87], which can collect soft soil samplers with over 95% recovery of 110 mm diameter samples.

Some interesting examples of offshore site investigation can be found in topical publications (e.g., [88, 89]) and guidelines (e.g., [84, 86]). The report from ISSMGE [86] is exhaustive and presents different structures and the site investigation methods most adequate for each situation. Methods slightly different from the most common can be applied as CPT with additional sensors such as thermal conductive probe, electrical conductivity cone, seismic cone and natural gamma and dilatometer [86].

Lunne et al. [88] studied a site for hydrocarbon in deepwater (3000 m) with 30 m thick, soft clays below the seabed. Due to the difficulty in collecting high-quality soil samples, in situ test methods were selected, although adapted to the adverse conditions, namely: CPTU with 1000–1500 mm^2 tip area and logging interval of 20 mm; vane field tests with 40–65 mm diameter and height double of the diameter; and additional laboratory tests. The authors refer to the decrease in the accuracy of

CPTU for deep water, especially in soft soils, and suggest the use of T-bar (TBT) and ball penetration test (BPT) data, both with a projected area of 10,000 mm^2 and remarkably accurate for very soft soils [90] and ball penetrometers, commonly used in offshore site investigation. In deep water, these methods are also adequate for determining strain rate dependency of soil strength, soil stratigraphy, and consolidation parameters by varying the penetration rate during a penetration test [91]. Some future improvements for in situ techniques and equipment are also suggested, namely incorporating additional sensors and pore pressure sensors for the full-flow penetrometers (e.g., [92]).

One of the most recent large works developed is the Fehmarnbelt link, connecting Denmark to Germany, in which site investigation took place first in 1995/96 and later between 2008 and 2014. Kammer et al. [93], the report GDR 00.1-001 [94], and Morrison et al. [95] present the geotechnical investigation carried on for the Fehmarnbelt Fixed Link, when two options were considered, namely an immersed tunnel or a cable-stayed bridge. The study involved: geophysical offshore and onshore surveys, comprising marine shallow seismic investigations, marine side-scan investigations, marine magnetic measurements and bathymetric measurements, onshore reflection seismic surveys and Continuous Vertical Electrical Sounding (CVES), down-hole geophysical borehole logging; onshore and offshore geotechnical type borings for different quality of sampling, seabed CPTUs; Advanced Laboratory Testing; Large Scale Testing including mainly a phased offshore trial excavation with advanced instrumentation, different kinds of plate load testing, pile installation and tension load testing as well as onshore installation and load testing of ground anchors, all in clays of Palaeogene origin; establishment of an overall geological model and detailed description of the ground conditions, boring with continuous sampling with high quality core recovery to 100 m depth, and large scale in situ tests such as CPTU, in a total of 71 tests in depths of 25 m and 50–100 m. The interpretation of the seismic surveys provided the seismic ground model allowing the selection of locations for drilling in a total length of 3675 m. The large scale trial excavation (construction tests) in the Palaeogene clay aimed at investigating the behaviour when subjected to a below seabed level excavation. It included bored and driven pile tests, instrumented trial excavation, CPTU testing, plate load tests and block sampling from the base of the trial excavation, multibeam surveys of the trial excavation area and ground anchors tests. The site investigation program allowed the definition of the complex lithology sequence, the tectonics and the geotechnical parameters needed, first to decide between a tunnel and a bridge, as well as to define the parameters for the design of this very complex structure.

4   Final Remarks

From current buildings to special structures such as power plants, dams, deep cavities (e.g., for compressed air storage, CO_2 sequestration, storage of dangerous substances, large tunnels) and marine structures, the site investigation is a cost-effective compo-nent of any engineering design, concurring to ensuring the required performance of the structure during the design lifetime. It is undertaken in progressive phases, which extent and sequence depend on the geological complexity of the site and the type and dimensions of the structure. It constitutes a multidisciplinary approach in which different players are engaged to obtain relevant information about the construction site. The methods are applied in an iterative mode, from a large area survey using less expensive methods to a design phase for which reliable methods for obtaining geotechnical parameters are crucial. Although most methods were created decades ago, improvements have been carried out to respond to increasingly demanding structures and less favourable locations, taking advantage of the development of technology and scientific knowledge. The development of novel methods and the improvement of the existing ones occur on a daily basis, with scientists, academia, consultants and contractors willing to overcome the challenging situations faced by a demanding and sustainable construction industry.

Acknowledgements This work was funded by the Portuguese Fundação para a Ciência e Tecnologia (FCT) I.P./MCTES through national funds (PIDDAC) – UIDB/50019/2020. Several projects partially supported H. I. Chaminé: LABCARGA|ISEP re-equipment program (IPP-ISEP|PAD'2007/08), and Centre GeoBioTec|UA (UID/GEO/04035/2020). Special thanks to L. Freitas (LABCARGA|ISEP) for the careful final editing of the figures. We also recognise the company Geocontrole for kindly sharing some of the equipment images. Our gratitude to the colleague editors, especially J. Neves, for the challenging invitation and full support throughout the editorial stage. This chapter is dedicated to J. M. Cotelo Neiva, R. Oliveira and A. Costa Pereira, outstanding geol-ogists and pioneers of engineering geology in Portugal, as well as to the hydrogeologist J. Martins Carvalho.

References

1. Goldsworthy, J.S., Jaksa, M.B., Kaggwa, W.S., Fenton, G.A., Griffiths, D. V, Poulos, H.G.: Cost of foundation failures due to limited site investigations. In: The International Conference on Structural and Foundation Failures, pp. 2–4. Singapore (2004)
2. ICE: Institution of Civil Engineers: Inadequate Site Investigation. Thomas-Telford, London, England (1991)
3. BS 5930:2010: Code of practice for site investigations. Br. Stand. Inst. 1–192 (2010)
4. EN 1997-1: Eurocode 7: Geotechnical design - Part 1: General rules. In: CEN, Brussels, Belgium (2011)
5. EN 1997-2: Eurocode 7: Geotechnical design - Part 2: Ground investigation and testing. In: CEN, Brussels, Belgium (2007)
6. Hatheway, A.W., Kanaori, Y., Cheema, T., Griffiths, J., Promma, K.: 10th Annual Report on the International Status of Engineering Geology—Year 2004–2005; Encompassing hydrogeology, environmental geology and the applied geosciences. Eng. Geol. **81** (2005)
7. APEGBC Professional Practice Guidelines: Site characterisation for dam foundations in BC (2016)
8. Matula, M., Holzer, R.: Engineering typology of rock masses. In: Proceeding of the Felsmekanik Kolloquium, Grundlagen und Andwendung der Felsmekanik, pp. 107–121, Karlsruhe, Germany (1978)

9. Palmstrom, A.: Rock Masses as Construction Materials. Ph.D. Thesis. Oslo University (1995)
10. González de Vallejo, L.I.F.M.: Geological engineering. CRC Press, Taylor-Francis Group, Boca Raton (2011)
11. Look, B.G.: Handbook of Geotechnical Investigation and Design Tables, Second Edition. CRC Press (2014)
12. TR-01-29: Site investigations. Investigation methods and general execution programme. Svensk Kärnbränslehantering AB (2001)
13. Oliveira, R.: Engineering geological investigations of rock masses for civil engineering projects and mining operations. Memória LNEC **693**, 1–28 (1987)
14. GEO-2: Hong Kong, Geoguide 2- Guide to Site Investigation. Geotechnical Engineering Office, Civil Engineering and Development Department, HKSAR Government (2017)
15. McLelland, C.: Nature of science and the scientific method. Geol. Soc. Am. 1–11 (2006)
16. IAEG: Engineering geological maps, A guide to their preparation (Earth Sciences Series, no 15). Int. J. Rock Mech. Min. Sci. Geomech. Abstr. **14**, 5–6 (1977)
17. Parry, S., Baynes, F.J., Culshaw, M.G., Eggers, M., Keaton, J.F., Lentfer, K., Novotny, J., Paul, D.: Engineering geological models: An introduction: IAEG commission 25. Bull. Eng. Geol. Environ. **73**, 689–706 (2014)
18. Clayton, C.R.I., Matthews, M.C., Simons, N.E.: Site Investigation. Halstead Press, New York (1995)
19. Matos Fernandes, M.: Analysis and design of geotechnical structures. CRC Press (2020)
20. de Freitas, M.H.: Future developments for ground models. Q. J. Eng. Geol. Hydrogeol. **54** (2020). https://doi.org/10.1144/qjegh2020-034
21. BRE 322: Site Investigation for Low-Rise Building. Procurement. Building Research Establishment, UK (1987)
22. Hytiris, N., Stott, R., McInnes, K.: The importance of site investigation in the construction industry: A lesson to be taught to every graduate civil and structural engineer. World Trans. Eng. Technol. Educ. **12** (2014)
23. Butler, D.K., Curro, J.R.: Crosshole seismic testing—procedures and pitfalls. Geophysics. **46** (1981)
24. Patel, A.: Geotechnical investigation. In: Geotechnical Investigations and Improvement of Ground Conditions. Elsevier (2019)
25. HGI: Caterpillar Handbook of Ripping. https://www.hgiworld.com/services/engineering/rip pability-analysis/
26. NZGS: Earthquake geotechnical engineering practice. Module 2: Geotechnical investigations for earthquake engineering. New Zealand Geotechnical Society Inc and Ministry of Business Innovation & Employment (MBIE) (2016)
27. Hunt, R.E.: Geotechnical Engineering Investigation Handbook. CRC Press (2005)
28. BDA: Manual for rotary drilling. British Drilling Association, London (2021)
29. Danson, E. (ed): Geotechnical and Geophysical Investigations for Offshore and Nearshore Developments. In: Technical Committee 1, International Society for Soil Mechanics and Geotechnical Engineering ISSMGE, p. 101. Technical Committee 1, International Society for Soil Mechanics and Geotechnical Engineering (2005)
30. Dean, E.T.R.: Offshore geotechnical engineering Principles and practice. Thomas Telford (2010)
31. Robertson, P.K.: Guide to In-situ Testing. Gregg Drilling & Testing Inc. (2006)
32. Robertson, P.K.: Interpretation of In-situ Tests—Some Insights. Proc. Fourth Int. Conf. Site Charact. 1–22 (2012)
33. Mayne, P.W., Barry, R.C., DeJong, J.: Subsurface Investigations (Geotechnical Site Characterization). F. Hydrogeol. (2011)
34. Mayne, P.W.: Quandary in geomaterial characterisation : New versus the old. Shaking Found. Geo-engineering Educ. 15–26 (2012)
35. Wroth, C.P.: The interpretation of in situ soil tests. Geotechnique **34**, 449–489 (1984)
36. Rogers, J.D.: Subsurface exploration using the standard penetration test and the cone penetrometer test. Environ. Eng. Geosci. **12**, 161–179 (2006)

37. BS 1377-9:1990: Methods of test for soils for civil engineering purposes—Part 9: In-situ tests. Br. Stand. Inst. (1990)
38. ASTM D1586-11: Standard test method for standard penetration test (SPT) and split-barrel sampling of soils. ASTM Stand. ASTM Int. West Conshohocken, Penn. US. (2008)
39. EN ISO 22476-3: Geotechnical investigation and testing—Field testing—Part 3: Standard penetration test. In: CEN, Brussels, Belgium.
40. Fletcher, G.F.A.: Standard penetration test: Its uses and abuses. J. Soil Mech. Found. Div. **91**, 67–75 (1965)
41. Skempton, A.W.: Standard penetration test procedures and the effects in sands of overburden pressure, relative density, particle size, ageing and overconsolidation. Geotechnique **36**, 425–447 (1986)
42. Liao, S.S.C., Whitman, R.V.: Overburden correction factors for SPT in sand. J. Geotech. Eng. **112**, 373–377 (1986)
43. Ameratunga, J., Sivakugan, N., Das, B.M.: Correlations of Soil and Rock Properties in Geotechnical Engineering. Springer India, New Delhi (2016)
44. EN ISO 22476-2: Geotechnical investigation and testing Field testing—Port 2: Dynamic probing. In: CEN, Brussels, Belgium
45. ASTM D3441-16: Standard Test Method for Mechanical Cone Penetration Testing of Soils. ASTM Stand. ASTM Int. West Conshohocken, Penn. US. (2016)
46. ASTM D5778-07: Standard test method for electronic friction cone and piezocone penetration testing of soils. ASTM Stand. ASTM Int. West Conshohocken, Penn
47. EN ISO 22476-1: Geotechnical investigation and testing—Field testing—Part 1: Electrical cone and piezocone penetration tests. In: CEN, Brussels, Belgium
48. Robertson, P.K., Campanella, R.G.: Guidelines for Use and Interpretation of the Electric Cone Penetration Test. Hogentogler & Company Inc, 3rd ed (1986)
49. Lunne, T., Powell, J.J.M., Robertson, P.K.: Cone Penetration Testing in Geotechnical Practice. CRC Press (2002)
50. Youd, T.L., Idriss, I.M.: Liquefaction resistance of soils: Summary report from the 1996 NCEER and 1998 NCEER/NSF workshops on evaluation of liquefaction resistance of soils. J. Geotech. Geoenvironmental Eng. **127**, 297–313 (2001)
51. Robertson, P.K.: In situ testing and its application to foundation engineering. Can. Geotech. J. **23**, 573–594 (1986)
52. Robertson, P.K.: Soil classification using the cone penetration test. Can. Geotech. J. **27**, 151–158 (1990)
53. Olsen, R.S., Mitchell, J.: CPT Stress normalisation and prediction of soil classification, pp. 257–262. Int. Symp. Cone Penetration Testing, CPT (1995)
54. Barnard, H., Heymann, G.: The effect of bedding errors on the accuracy of plate load tests. J. South African Inst. Civ. Eng. **57**, 67–76 (2015)
55. ASTM D1196M-12: Standard Test Method for Nonrepetitive Static Plate Load Tests of Soils and Flexible Pavement Components, for Use in Evaluation and Design of. ASTM Stand. ASTM Int. West Conshohocken, Penn. US.
56. EN ISO 22476-13: Geotechnical Investigation and testing—Field testing—Part 13: Plate loading test. In: CEN. Brussels, Belgium
57. ASTM D4719-07: Standard Test Methods for Prebored Pressuremeter Testing in Soils. ASTM Stand. ASTM Int. West Conshohocken, Penn. U.S. (2007)
58. EN ISO 22476-4: Geotechnical investigation and testing—Field testing—Part 4: Menard pressuremeter test. In: CEN. Brussels, Belgium
59. Briaud, J.L.: Ménard lecture: The pressuremeter test: Expanding its use. In: 18th International Conference Soil Mechanices Geotechnical Engineering Challenges Innovations Geotechnics. ICSMGE 2013, vol. 1, 107–126 (2013)
60. Monnet, J.: Numerical analysis for the interpretation of the pressuremeter test in granular soil. In: 18éme Congrés Français de Mécanique. Grenoble (2007)
61. EN ISO 22476-6: Geotechnical investigation and testing Field testing—Part 6: Self-boring pressuremeter test. In: CEN, Brussels, Belgium.

62. Wroth, C.P.. H.I.M.O.: An instrument for the in situ measurement of the properties of soft clays. Technical Report Soils TR13. University of Cambridge (1973)
63. Baguelin, F., Jezeguel, J.F., Shields D.H.: The Pressuremeter and foundation engineering. Trans. Tech. Public. (1978)
64. Wang, K., Xu, G., Wang, J., Wang, C.: Self-boring in situ shear pressuremeter testing of clay from Dalian Bay. China. Soils Found. **58**, 1212–1227 (2018)
65. Marchetti, S.: In situ tests by flat dilatometer. J. Geotech. Eng. Div. ASCE. **106**, 299–321 (1980)
66. Marchetti, S.: The flat dilatometer: Design applications. In: 3rd Geotechnical Engineering Conferences, Keynote Lecture, pp. 421–448 (1997)
67. Monaco, P., Marchetti, S., Totani, G., Calabrese, M.: Sand liquefiability assessment by Flat Dilatometer Test (DMT) Évaluation de la susceptibilité à la liquéfaction des sables par l'essai de dilatomètre (DMT). In: Proceedings of the 16th International Conference on Soil Mechanics and Geotechnical Engineering: Geotechnology in Harmony with the Global Environment, pp. 2693–2697 (2005)
68. ASTM D6635-15: Standard Test Method for Performing the Flat Plate Dilatometer. ASTM Stand. ASTM Int. West Conshohocken, Penn.
69. EN ISO 22476-11: Geotechnical investigation and testing—Field testing—Part 11: Flat dilatometer test. . In: CEN, Brussels, Belgium.
70. Marchetti, S., Monaco, P., T.G. & C.M.: The flat dilatometer test (DMT) in soil investigations—A Report by the ISSMGE Committee TC16. In: Proceeding of the 2nd International Conference on the Flat Dilatometer (2001)
71. Marchetti, S.: Some 2015 Updates to the TC16 DMT Report 2001. In: 3rd International Conference Flat Dilatom, pp. 43–65 (2015)
72. ASTM D2573-08: Standard test method for field vane shear test in cohesive soil. ASTM Stand. ASTM Int. West Conshohocken, Penn
73. EN ISO 22476-9: Geotechnical investigation and testing—Field testing—Part 9: Field vane test. . In: CEN, Brussels, Belgium
74. Andersen, J.D., Okkels, N., Andersen, J.D.: Introduction of a fast multi-soil test to field vane standards. In: Proceedings of the XVII ECSMGE-2019, pp. 1–8 (2019)
75. Chandler, R.: The In-Situ measurement of the undrained shear strength of clays using the field vane. In: Vane Shear Strength Testing in Soils: Field and Laboratory Studies. pp. 13–13–32. ASTM International (2009)
76. ASTM D2434-19: Standard Test Method for Permeability of Granular Soils (Constant Head). ASTM Stand. ASTM Int. West Conshohocken, Penn.
77. ASTM D4044: Standard test method (field procedure) for instantaneous change in head (slug) tests for determining hydraulic properties of aquifers. ASTM Stand. ASTM Int. West Conshohocken, Penn.
78. ISO 22282-2: Geotechnical investigation and testing—Geohydraulic testing—Part 2: Water permeability tests in a borehole using open systems. Geneva, Switzerland: ISO (2014)
79. Hvorslev, M.J.: Time lag and soil permeability in ground-water observations. Bull. **36**, 53 (1951)
80. Cassan, M.: Les essais de perméabilité sur site dans la reconnaissance des sols. Presses des Ponts, Paris (2005)
81. Monnet, J.: In Situ Tests in Geotechnical Engineering. Wiley (2015)
82. CAN/BNQ2501-135: Determination of Permeability by the Lefranc Method. National Standard of Canada, Ottawa, ON (2014)
83. Randolph, M., Cassidy, M., Gourvenec, S., Erbrich, C.: Challenges of offshore geotechnical engineering. In: Proceedings of the 16th International Conference on Soil Mechanics and Geotechnical Engineering: Geotechnology in Harmony with the Global Environment. pp. 123–176 (2005)
84. SUT: Offshore Site Investigation and Geotechnics Group (OSIG) of the Society for Underwater Technology (SUT): Guidance Notes on Site Investigations for Offshore Renewable Energy Projects (2005)

85. Fung, A.K.L., Foott, R., Cheung, R.K.H., Koutsoftas, D.C.: Practical conclusions from the geotechnical studies on offshore reclamation for the proposed Chek Lap Kok Airport. Hong Kong Eng. **12** (1984)
86. Danson, E. (ed.): Geotechnical and geophysical investigations for offshore and nearshore developments. Tech. Comm. 1, Int. Soc. Soil Mech. Geotech. Eng. ISSMGE (2005)
87. Lunne, T., Tjelta, T.I., Walta, A., Barwise, A.: Design and testing out of deepwater seabed sampler. In: Offshore Technology Conference, Houston, Texas. Society of Petroleum Engineers (SPE) (2008)
88. Lunne, T., Andersen, K.H., Low, H.E., Randolph, M.F., Sjursen, M.: Guidelines for offshore in situ testing and interpretation in deepwater soft clays. Can. Geotech. J. **48**, 543–556 (2011). https://doi.org/10.1139/t10-088
89. Merritt, A.S., Schroeder, F.C., Jardine, R.J., Stuyts, B., Cathie, D., Cleverly, W.: Development of pile design methodology for an offshore wind farm in the north sea. Offshore Site Investig. Geotech. 2012 Integr. Technol. - Present Futur. OSIG 2012. 439–447 (2012)
90. Norsok Standard G-CR-001: Marine soil investigations. NTS (Norwegian Technology Standards Institution) (1996)
91. Boggess, R., Robertson, P.: CPT for soft sediments and deepwater investigations. In: Proceeding of CPT'10, 2nd International Symposium on Cone Penetration Testing. pp. 1–17. Society of Petroleum Engineers (SPE), Huntingdon Beach, California (2011)
92. Kelleher, P.J., Randolph, M.F.: Seabed geotechnical characterisation with a ball penetrometer deployed from the portable remotely operated drill. In: Frontiers in Offshore Geotechnics, ISFOG 2005—Proceedings of the 1st International Symposium on Frontiers in Offshore Geotechnics, pp. 365–371 (2005)
93. Kammer, J., Frederiksen, J.K., Hansen, G.L., Hammami, R., Morrison, P., Mortensen, N., Skjellerup, P.: Fehmarnbelt fixed link—Geotechnical investigations. In: 12th Baltic Sea Geotechnical Conference (2012)
94. GDR 00.1-001: Ground investigation report. Rambøll Arup Joint Venture (2014)
95. Morrison, P., Kammer, J., Hammami, R., Frederiksen, J.K., Hansen, G.L., Humpheson, C.: Fehmarnbelt fixed link—Trial excavation. In: Geotechnical Engineering for Infrastructure and Development—Proceedings of the XVI European Conference on Soil Mechanics and Geotechnical Engineering, ECSMGE 2015, pp. 3029–3034. ICE Publishing (2015)

Laboratory and Field Testing of Rock Masses for Civil Engineering Infrastructures

José Muralha(ID) **and Luís Lamas**(ID)

Abstract Several types of relevant civil engineering infrastructures, such as the foundations of large buildings, bridges and dams, rock slopes, tunnels and caverns, encompass construction of structures on or in rock masses. Rock masses, specifically those within a few hundreds of meters from the surface where civil infrastructures are implanted, being composed of intact rock and discontinuities (e.g., faults, joints, schistosity and bedding planes), often behave as discontinuum media, with the latter determining their behaviour. The assessment of rock mass properties and conditions is crucial for the design of rock engineering structures, and for assuring safety during their life-time exploration. Since the development of rock mechanics as a distinct engineering discipline in the 1950s and early 1960s, the importance of laboratory rock testing emerged. Additionally, the recognition that tests on small size specimens could not be representative of the behaviour of the rock mass led to the emergence and development of specific in situ tests, where comparatively large rock mass volumes are tested in order to estimate engineering properties suitable for design. This chapter presents laboratory and in situ tests currently used to estimate the relevant parameters required to model the behaviour of rock mass—a naturally occurring material with unknown in situ stresses—at a scale compatible with the dimensions of engineering infrastructures.

Keywords Rock mass · Intact rock · Discontinuity · Laboratory testing · In situ testing

J. Muralha (✉) · L. Lamas
LNEC, National Laboratory for Civil Engineering, Lisbon, Portugal
e-mail: jmuralha@lnec.pt

L. Lamas
e-mail: llamas@lnec.pt

© The Author(s), under exclusive license to Springer Nature Switzerland AG 2023   55
C. Chastre et al. (eds.), *Advances on Testing and Experimentation in Civil Engineering*, Springer Tracts in Civil Engineering,
https://doi.org/10.1007/978-3-031-05875-2_3

1   Introduction

Rock mechanics is a discipline that applies mechanics principles to rocks and is used to design and monitor structures built on or in rock masses, such as dams, large bridges and buildings, natural and excavated slopes, tunnels, caverns, hydroelectric schemes, nuclear repositories, or mines.

Throughout this chapter, the terms rock mechanics and rock engineering will be used in a sense as defined by the International Society of Rock Mechanics and Rock Engineering (ISRM): "*The field of rock mechanics and rock engineering includes all studies of the physical, mechanical, hydraulic, thermal, chemical and dynamic behaviour of rocks and rock masses, and engineering works in rock masses, using appropriate knowledge of geology*". As a consequence, rock mechanics is generally taken to include rock engineering, though occasionally both terms may be used separately, since rock mechanics is the key for dealing with many problems met in rock engineering projects.

As opposed to common man-made materials used in engineering projects, such as steel or concrete, rocks and rock masses are historical materials that during geological times have gone through quite long history of natural phenomena, being acted on chemically, thermally and mechanically, and undergoing deformation, fracture and weathering. Even at a smaller scale, intact rock is a bonded or cemented aggregate of grains, generally individual crystals or amorphous particles from different minerals, but rarely do not include inter or intragranular cracks. At a rock engineering scale, rock mechanics deals with rock masses, which are media where discontinuities, anisotropy and heterogeneity are nearly always present requiring particular approaches.

Recognition that rock masses are particular media not covered by continuous mechanics led to the seminal reply by Leopold Müller to the question "*Do we know the strength of rock?*". Müller replied: "*For rock (specimens) tested in the laboratory, yes. For a rock mass, no.*" [1]. Though engineering properties of rocks were already being studied all around the world, it commonly acknowledged that Rock Mechanics emerged as an independent discipline at that time [2].

Regrettably, the beginning and the early development of rock mechanics is also related to the occurrence of three catastrophic events: the failure of the foundation of the Malpasset concrete arch dam, in December 1959 (Fig. 1, left) [3, 4], the collapse of the coal mine pillars at Coalbrook, in January 1960 [5], and the landslide of the left bank of Vajont dam reservoir, in October 1963 (Fig. 1, right) [6, 7]. These serious accidents led to understanding that discontinuities, regardless of their origin, play a significant role in the behaviour of rock masses as their reduced shear strength may convert a sound rock masses into a crumbling block system for stresses acting along particular orientations. They also triggered much debate, new research and promoted the development of new tests and methodologies to assess rock mass properties.

In the 1950s, construction of large concrete dams and underground caverns and tunnels for hydroelectric schemes were seeing a notable expansion worldwide.

Fig. 1 Malpasset dam failure (**left**), and Mont Toc (Vajont) landslide (**right**)

Though rock mechanics tests already were an important component in the investigations that supported the design of these structures, the improvement of existent rock testing methods and the development of new experimental testing methods and techniques was also related to the emergence rock mechanics as an autonomous discipline within the geomechanics framework, encompassing a distinct body of knowledge. At that time, novel in situ testing methods started being developed at the Portuguese National Laboratory for Civil Engineering (LNEC), under the leadership of Prof. Manuel Rocha [8]. This chapter will make reference to the authors' experience in this subject, while mentioning relevant technological updates and alluding to other worth mentioning testing techniques.

2 The Relevance of Testing in Rock Mechanics

It is accepted that any structural engineering design comprises some kind of modelling of the physical, mechanical, or hydraulic behaviour of the components involved in the construction. In the design of structures to be built on or in rock masses—a natural, discontinuous, heterogeneous, anisotropic, often highly variable material—the behavioural models depend critically on the input parameters, namely their deformability, strength, permeability and boundary conditions (i.e. natural in situ stresses).

Current developments in computing capabilities, that have allowed the proliferation and availability of numerical analyses, have led to more and more elaborated models, as well as to an increasing demand of a better understanding of the mechanisms occurring in rock masses, once they are disturbed by natural actions or by new man-made structures. These requirements make the need of rock testing an always current topic.

In rock engineering, the behaviour of rocks and rock masses concerns mainly the following properties: deformability and strength, and how they vary with the direction and magnitude of the loads, permeability, susceptibility to weathering, and the natural in situ stresses acting on them before construction starts.

Before the 1960s, researchers and civil and mining engineers working in rock were developing independently their own tests and methods to assess rock mass properties. It is not surprising that early efforts of the ISRM were the establishment of a common to all terminology and the standardization of the different testing techniques and procedures that were used to determine rock and rock mass properties. This second task led to the creation of the Commission on Standardisation of Laboratory and Field Tests (now the ISRM Commission on Testing Methods) that has been working until today ever since 1966 [9]. Though its mandate was to go ahead with the development of test standards, documents published by the commission were not issued as standards but rather as Suggested Methods. It is a term that was carefully chosen, since Suggested Methods do not intend to be testing standards, but documents where practitioners that have not been involved with a particular subject can find guidance, explanations and recommended (not strictly mandatory) procedures [10, 11]. Many ISRM Suggested Methods deal with tests that are not (or were not at the time of publication) available as test standards. Description of rock mechanics tests presented in this chapter derives from ISRM Suggested Methods and other applicable standards, such as ASTM, EN and ISO.

Very often, rock masses include many discontinuities so that they have a blocky structure. The three-dimensional basic elements of these structures are the elementary blocks, without visible macroscopic fractures basically, made of more or less massive, intact rock. Discontinuities are two-dimensional geologic features that occur in rock masses in a large diversity of forms, and their classification is not straightforward. The most conventional differentiation considers simply joints and faults. In general terms, faults are considered to be fractures in rock continuity along which an identifiable shear displacement of the adjacent faces has occurred, usually resulting from rock mass movements occurring over geologic times. Opposed to faults, joints are fractures within the rock that do not exhibit shear displacements between their surfaces. Joints are caused by fractures of the rock body as a result of tensile stresses induced by geologic events such as the folding of rock masses, shrinkage of a rock body due to a temperature decrease or the reduction of stresses caused by the erosion of overbearing rock layers.

Geometrically, both can be considered as approximately plane surfaces, currently defined in Geology by a pair of angles (strike and dip, or dip and dip direction), though some folding often occurs, mainly in the case of larger discontinuities. Usually, joints display a smaller extent or persistence and they occur in an ordered manner: joints with approximately parallel orientations form a joint set. The evaluation of the geometric characteristics of the discontinuities (orientation, intensity, spacing and persistence), and also other descriptive parameters (roughness, aperture, wall strength and filling), is usually performed during geotechnical surveys, and they will not be addressed in this chapter.

Whether they are joints or faults, discontinuities are responsible for not allowing rock masses, at the scale of rock engineering projects, to comply with the basic assumptions of solid mechanics of continuous, homogeneous, isotropic and linear elastic media (known as CHILE media): first of all, they turn rock masses into discontinuous and inhomogeneous media, and additionally their occurrence defines preferential directions that make several characteristics and properties display anisotropic, non-reversible and non-linear elastic behaviours (known as DIANE media).

A basic approach could lead to consider the behaviour of rock masses as the result of some kind of sum of the behaviour of their components: intact rock plus discontinuities. As a consequence, the assessment of rock mass properties could be reached by sampling and testing rock and discontinuities separately in the laboratory and extending the aggregate of the results to the field scale. The other approach would be to evaluate the rock mass properties performing in situ tests involving a tested volume large enough to be considered representative of the rock mass, being the representative elementary volume (REV) the minimum volume of rock mass that encompasses the relevant features of any larger volume. The notion of size effect in the scope of materials testing refers to the variation of a certain property with the size of homothetic samples. In rock mechanics testing, the term "scale effects" is often used in a broader sense denoting not just the difference between sample sizes, but including also the consideration of greater rock volumes that comprise discontinuities and the upscaling to the dimension of the engineering project. Results of laboratory and in situ tests are thus affected by both the chosen testing locations and the volumes representativeness involved in the tests, particularly their relationship with the engineering work that is being considered. Figure 2 shows a schematic representation of scale effects as it is interpreted in rock mechanics.

Fig. 2 Schematic representation of scale effects in rock masses [12]

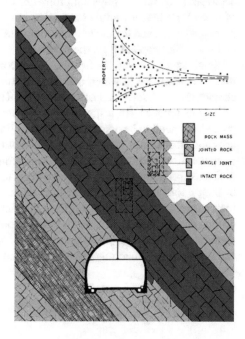

The development of in situ testing methods and techniques specifically dedicated to the geotechnical characterization of rock masses derives from the need to address the issues related with scale effects. In early years, it was also responsible for the recognition of rock mechanics as a distinctive scientific discipline within the geotechnical sphere.

Rock has been used by mankind as a building material and for other purposes since early years. Records of the first mechanical testing of materials are attributed to Leonardo da Vinci ca. 1500, and the first documented rock mechanics experimental study, performed by Gautier around 1770, referred to a testing machine with a lever system that was used to measure the compressive strength of specimens for the pillars of Sainte Genevieve Church in Paris [11].

There are several possible ways to classify the different types of rock tests, none of which being fully satisfactory. In this chapter testing techniques were simply divided into laboratory and field tests. However, even this simple distinction is not undisputable, as several tests can be performed in the field using portable laboratory equipment.

Another informative subdivision is to classify the tests according to their purpose. On the one hand, design tests are those that are used to provide a quantitative measure of given rock or rock mass characteristics, such as the deformability modulus or the shear strength. On the other hand, index tests are simple testing techniques used to give indications about a given characteristic. Since they are generally inexpensive, they can offer important sets of data and thus provide useful estimators for characterization of several physical properties of rock [2]. Another relevant advantage of index tests is that useful correlations have already been established, such as the point load index and the unconfined compressive strength, or they can be specifically defined in the scope of a given project.

Some sandy, clayey, carbonate, or evaporitic geomaterials, referred to as soft rocks, are sensitive to water, and display crumbling, foliated, slaking or expansive characteristics. Additionally, they are difficult to sample, requiring special cutting and drilling techniques, for instance without water, and testing equipment and standard procedures need to be adapted considering limits for specimen deformations. Index tests and correlations may play an important role in overcoming such issues.

In the subsequent sections, the most preponderant tests used for the estimation of rock mass properties usually included in geotechnical characterization for the design of major civil engineering infrastructures are described. Tests used for assessing rock hardness or abrasivity and their interaction with the wear and capabilities of drilling and cutting equipment, tests carried out to characterize rock as a construction material (aggregates or ornamental stones), and tests specially devised for mining and petroleum engineering, are not addressed in the chapter.

3 Laboratory Tests

3.1 Uniaxial Compression

Regarding deformability, many intact rocks show an almost linear elastic behaviour under loadings lesser than 40–50% of their strength, which can be described by two elastic constants, Young's or elasticity modulus E and Poisson ratio v in the isotropic case, or by five or more depending on the anisotropic degree. These parameters are determined in uniaxial compression tests of cylindrical rock specimens taken from borehole cores or of prismatic or cylindrical specimens cut from rock blocks. The same specimens can be also used to determine the uniaxial or unconfined compressive strength (UCS) of the intact rock.

Specimens diameter or side should not be less than 54 mm, or at least greater than 10 times the rock grain size. Specimens should have a height to diameter, or side, ratio of 2.5–3.0. Flat ends and perpendicularity of the specimens should be ensured by an adequate specimen preparation [10, 13, 14].

To determine the elastic constants or simply to control the test, the axial and diametric or lateral strains are measured using strain gauges applied directly on the specimen's faces or displacement transducers coupled to the specimen with specially designed devices (Fig. 3). Standard procedures specify that measuring devices should be placed close to the mid-height of the specimen, and they should average at least two strain measurements. The measuring length of the gauges or devices should be at least ten times the rock grain size. The test is carried out in a loading device to

Fig. 3 Uniaxial compression test specimens with electric strains gauges (**left**) and displacement measuring devices (**right**)

consistently apply load at a required stress or strain rate. It is pointed out that stress-controlled tests may lead to explosive failure of the specimens, due to the brittle behavior of hard rocks, and only strain-controlled devices can capture the behavior of the specimens close to and after failure occurs. This requirement leads to the use of stiff servo-controlled testing systems with displacement or strain control to perform these tests [15].

Tests are performed by applying the axial load continuously at a pre-defined stress or strain rate until failure occurs or a predetermined amount of strain is achieved. The stress or strain rates should be selected in order that failure is reached in a test time between 2 and 15 min.

Young's modulus of the specimen, defined as the ratio between a certain axial stress change and the axial strain produced by it, can be calculated using several methods: tangent modulus measured at a fixed percentage of the compressive strength (usually 50%), average modulus of a linear portion of the axial stress axial strain curve, or secant modulus up to a fixed percentage of compressive strength. Figure 4 shows an example of a graph from a uniaxial compression test with the latter calculation.

In some cases, it is preferable to apply two or three loading–unloading cycles at a given stress rate up to an axial stress in accordance with project design requirements, use them to calculate the elastic constants, and then apply a strain-controlled loading cycle until failure.

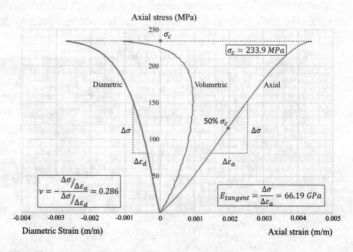

Fig. 4 Graph with the results of a uniaxial compression test

3.2   Triaxial Compression

Assessment of rock strength is necessary for the rational design of underground structures, such as caverns and tunnels. In engineering, the relationship establishing the stress condition by which ultimate strength is reached is referred to as a "failure criterion". They are often expressed as a function of the major principal stresses that rocks can sustain for given values of the other two principal stresses. The Mohr–Coulomb and Hoek–Brown are the most frequently used failure criteria, but both incorporate only the major σ_1 and minor σ_3 principal stresses, and the effect of the intermediate stress is not considered.

Parameters for failure criteria can be determined empirically or from laboratory tests, aiming at characterizing strength and deformation behaviour under stress conditions simulating, as close as possible, those encountered in situ [16]. However, most laboratory tests are conducted on cylindrical specimens subjected to uniform confining pressure, reproducing only a particular field condition where intermediate and minor principal stresses are equal ($\sigma_2 = \sigma_3$). Triaxial tests have been widely used for the study of mechanical characteristics of rocks because of equipment simplicity and convenient specimen preparation and testing procedures.

The main difference between triaxial and uniaxial compression tests lies in the fact that the specimen is inserted in a triaxial cell. Inside this cell, a confining pressure is applied to the specimen by a hydraulic fluid inside the cell, usually oil, that is kept from penetrating into the rock pores by a flexible membrane (Fig. 5, left) [17]. The confining pressure is controlled by a hydraulic system that has to be able to keep it constant during the whole test, taking into account that changes in the specimen's volume resulting from stress changes will affect the oil pressure inside the triaxial cell. The axial stress is applied by a loading device with steel platens of prescribed

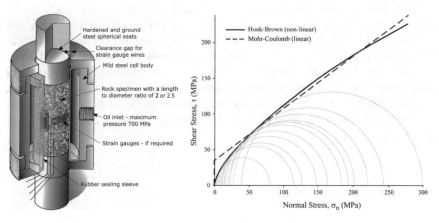

Fig. 5 Cut-away view of a triaxial cell (**left**) [17], and graph with results of a set of triaxial tests and the resulting envelopes for the Mohr–Coulomb and Hoek–Brown failure criteria (**right**) [11]

hardness. It is possible to measure axial and diametric strains using electrical strain gauges applied to the rock surface or displacement transducers inside the cell.

The most frequent test procedure starts with inserting the specimen in the triaxial cell and applying a confining pressure. The cell is then placed in the loading device that will continuously increase the axial load until failure and peak load are obtained. Performing a series of tests with different confining pressures on specimens sampled from the same rock lithology or horizon, test results, σ_1–σ_3 pairs, allow calculating the parameters of the considered failure criteria [18]. Figure 5 (right) shows a graphical representation, in the shear stress—normal stress plane, of the failure envelopes obtained from triaxial tests results.

3.3 Diametral Compression

The diametral compression test, also referred to as Brazilian, or Brazil test or splitting test, is an indirect tensile test intended to estimate the tensile strength of intact rock. It was first developed in 1943, while studying the correlation between compressive strength and flexural tensile strength [19].

By definition, the tensile strength of intact rock should be obtained from the direct tensile test. However, direct tensile test preparation is difficult for routine applications, since it is problematic to attach a cylindrical rock specimen to the jaws of a testing machine. The Brazilian test soon presented itself as an attractive alternative because it is much simpler and inexpensive. Furthermore, rock mechanics design usually deals with complicated stress fields, including various combinations of compressive and tensile stress fields, and testing across different diametrical directions allows determining variations in tensile strength for anisotropic rocks.

This test involves compressing a cylindrical specimen along diametrically opposed longitudinal thin surfaces of a cylindrical specimen using a common load system [10, 18]. Under the action of such load, tensile stresses develop perpendicularly to the loaded diameter and as load is steadily increased the specimen breaks. The load is transmitted to the specimen by steel jaws with cylindrical loading surfaces with larger radius than the specimen's radius until failure. The specimens are right circular cylinders with a height equal to the radius (disks). Figure 6 shows the loading and the stresses occurring along the loaded diameter (left) [20], and a specimen being tested (right) [21].

3.4 Elastic Wave Velocity

The propagation of artificially generated elastic waves through a rock medium can be used to assess the elastic properties of rocks. It is a common non-destructive method that measures the velocities of compressional V_P and shear V_S waves, and, given the

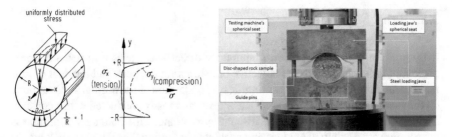

Fig. 6 Schematic representation of the loading and of the stresses along the loaded diameter in a diametrical compression test [20] (**left**), and picture of a specimen being tested [21] (**right**)

bulk density, allows estimating the dynamic Young's modulus and Poisson's ratio of intact rock.

Laboratory wave velocity measurements are usually performed on cylindrical rock specimens prepared for other strength tests, namely uniaxial and triaxial compression tests. The equipment includes an ultrasonic pulse wave generator, a transmitter and a receiver that are coupled to the flat end surfaces of the specimen with a bonding product to improve acoustic transmissivity (Fig. 7). Travelling time of the waves is measured by an oscilloscope, enabling to calculate V_P and V_S, given the length of the rock specimen is also measured [22]. If the mass density of the rock specimen is determined, the Young's modulus and Poisson coefficient can also be calculated. These values are usually referred to as dynamic parameters.

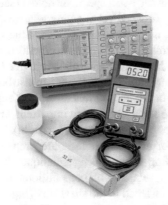

Fig. 7 Ultrasonic velocity test equipment

3.5 Joint Shear

It is common practice to perform laboratory direct shear tests on relatively small discontinuity samples with the objective of estimating the peak and residual or ultimate shear strength of rock discontinuities, as a function of the normal stress applied on the sheared plane [11, 23]. Direct shear tests are mostly conducted with a constant normal load (CNL) applied to the discontinuity plane. This boundary condition is appropriate for a group of engineering problems involving the sliding of rock blocks near the ground surface (e.g., rock slope stability and surface excavation stability). However, when dilation of a discontinuity is constrained during sliding (e.g., around an underground excavation), the normal stress on the sliding surface may change as shear displacement occurs. For this class of problems, constant normal stiffness (CNS) shear tests are more appropriate for determining joint shear strength.

Under CNL conditions, shear strength determination usually includes the application of several different magnitudes of normal stresses on multiple samples from the same joint to determine its shear strength. Alternatively, in cases where it is not possible to sample a representative number of specimens, the same specimen can be tested repeatedly under different constant normal loading conditions. For a single rock joint, at least three, but preferably five, different normal stresses should be used. To minimize the influence of damage and wear, each consecutive shear stage is performed with an increasingly higher normal stress. Usually, multi-stage shear tests are not practical under CNS conditions.

Commonly, direct shear testing machines include a relatively stiff frame against which the loading devices can act, a stiff specimen holder (shear box) in which the two halves of the joint are firmly fastened yet allowing relative and shear displacements, loading devices to apply the normal and shear loads to the specimen, and devices to measure both shear and normal loads and displacements (Fig. 8, left).

The applied normal and shear forces are usually provided by actuators (hydraulic, pneumatic, or gear driven), and cantilever systems can also be used to apply a constant normal load for CNL tests under low normal stresses. Keeping the normal load or stiffness constant during the shear test is very important, and it is usually achieved by servo-controlled close-loop systems (Fig. 8, top right).

Rock joint specimens for direct shear tests are prepared from rock blocks or drilled core samples containing the joint using techniques that minimize disturbance. Usually, specimens are encapsulated with cementitious mortar or similar material, allowing them to be tightly fastened in the shear box (Fig. 8, bottom right).

Specimen sizes depend on the dimension of the shear box, and usually their length along the shear direction ranges between 100 and 200 mm and does not exceed around 400 mm. Length of the specimens should cover the main roughness features of the rock joint, but frequently low frequency waviness is not tested.

Results of rock joint shear tests are presented as plots with the shear stress versus shear displacement graphs. Using these graphs and the records of the measured stresses and displacements, the peak and residual shear strength of each rock joint can be determined. Then, these values are used to calculate the strength parameters of

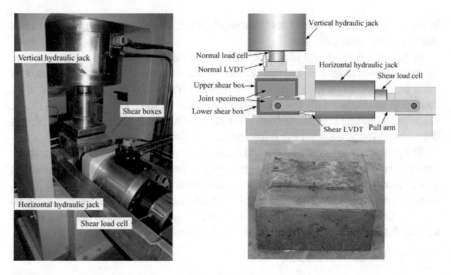

Fig. 8 Rock joint shear test equipment (**left**), schematic representation of the loading frame (**top right**) [24], and encapsulated half of a joint specimen (**bottom right**)

a prescribed failure criterion. Figure 9 shows the plots of the shear stress versus shear displacement graphs of a multi-stage rock joint shear test and the respective peak and residual shear strengths, that allow calculating the parameters of the relevant strength envelope.

Despite the non-linear strength envelope usually obtained for peak shear strength, results of rock joint shear tests are often modelled by the linear Mohr–Coulomb criterion, thus allowing to calculate the friction angle and the apparent cohesion.

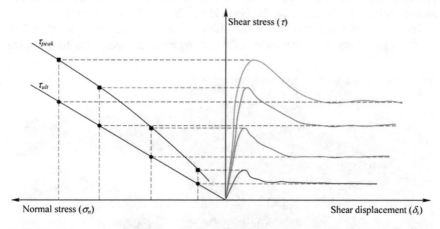

Fig. 9 Plots of the shear stress versus shear displacement graphs of a multi-stage rock joint shear test and the respective peak and residual shear strengths [11]

Particular care should be paid not to extrapolate below the value of the lowest normal stress applied during the test.

In the case of rough or non-planar joints, a non-linear shear strength envelope may be more representative of the test results. In these cases, it is possible to consider other well-established failure criteria, calculate the respective parameters, and deliver them also as results of the tests (e.g., the i value of Patton bilinear criterion [25], or the joint roughness coefficient (JRC), the joint wall compressive strength (JCS) and the residual friction angle (ϕ_r) values of Barton-Bandis criterion [26, 27].

The procedure for joint shear tests described in this section is not intended to cover direct shear tests of intact rock or other types of natural or artificial discontinuities that display tensile strength, such as rock–concrete interfaces or concrete lift joints. However, if the testing equipment holds certain capabilities, namely regarding its loading devices and servo-controlled system, it can be adapted to perform similar tests to determine the shear strength of bonded interfaces.

3.6 Tilt and Pull Tests

Several rock joint shear strength criteria require performing tilt or pull tests to determine some of their intrinsic parameters, being the most prominent the Barton-Bandis model [28, 29]. Tilt tests or pull tests are carried out to assess the basic friction angle (ϕ_b) and the JRC value [30].

Tilt tests are related with the concept of angle of repose of a solid body on an inclined surface. They are carried out by means of simple apparatuses essentially consisting of a rigid plane, which can be rotated around an axis (Fig. 10). A rock joint or a rock surface is placed horizontally on this plane, with the bottom half prevented from moving. The plane is then rotated until the upper part of the joint or surface moves. At this moment, the dip angle of the plane is the friction angle.

In the case of rough rock joints, the tilting angles reach values higher than 70°, generating high stress concentration at the rotating toe of the specimen. To minimize

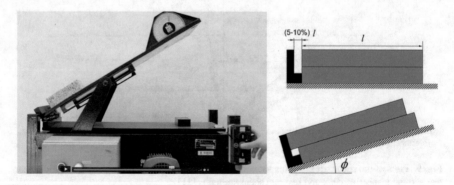

Fig. 10 Tilt test equipment (**left**), and schematic representation (**right**)

Fig. 11 Pull test equipment

this effect, specimens should have a length to height ratio of the upper block in excess of 4, and pull tests a preferable alternative. Figure 11 shows a pull test apparatus featuring a hard plastic block pulled over roller bearings, that pushes the upper half of the joint sample without any kind of overturning caused by the pull force if it is not parallel to the joint mean surface. The pull force is increased until shear displacement occurs and, given the weight of the upper half of the specimen, the friction angle is easily determined [31].

3.7 Index Tests

Fundamental tests directly measure an intrinsic rock property, such as the compressive strength, while, on the other hand, index tests are simple, cheap and can be performed quickly, but may not determine an intrinsic property. The point load and the Schmidt hammer rebound tests are the best-known examples. Consequently, it is good practice to perform many index tests and calibrate them against fewer fundamental tests, but still with statistical significance according to the property variability.

Point load test. This test method is performed to determine the point load strength index of rock specimens, which is used as an index for strength classification of rock materials or in correlations with the unconfined compressive strength. Since uniaxial compression tests are comparatively more time-consuming and expensive than point load tests, the latter can be used to make timely and more informed decisions during the exploration phases and more efficient and cost-effective selection of samples for more precise and expensive laboratory tests.

Rock specimens for point load tests may be in the form of rock cores (the diametral and axial tests), cut blocks (the block test), or irregular lumps (the irregular lump test), with diameter values D between 35 and 80 mm (Fig. 12). Tests can be performed in either the field or in the laboratory, because the testing machine is portable and little or no specimen preparation is required [10, 32].

Fig. 12 Point load test equipment with rock specimens and respective size requirements [33]

The result of a single test is the size-corrected Point Load Strength Index $I_{s(50)}$, defined as the value that would be measured in a diametral test with D equal to 50 mm. For a sample of the same rock type several tests should be performed, and the mean $I_{s(50)}$ value is to be calculated after deleting the two highest and the two lowest values as the average of remaining results, for test batches with 10 or more valid tests.

Schmidt hammer rebound. The Schmidt impact hammer is a light, portable apparatus consisting of a spring-loaded piston that transfers its energy as it is released and impacts on a rock surface. Part of this energy is recovered depending on the hardness of the impacted rock. The result of each test is the rebound value R. Though intended to provide a measure of rock hardness, R is most frequently used as an index in rock mechanics practice for estimating rock and joint wall strength, as well as rock excavability and drillability [11].

Though it is a very simple and quick determination, many factors can affect the results. Firstly, as the impact area and released energy are very small, the Schmidt hammer tests only affect a thin band of a few millimetres or centimetres of rock. If the rock specimen is not securely fastened, energy will be dissipated returning a false result. As a consequence, special core specimen holders with V-shaped steel cradles are often used to test cylindrical rock cores, and a large number of impacts should be averaged to render the R value (Fig. 13, left). Moreover, tests should be performed by experienced personnel in order to assure the quality of the result produced by this test method [34].

Schmidt hammer rebound can also be used in the field on rock exposures (Fig. 13, right). As rock faces occur with any given orientation, corrections for reducing the rebound value when the hammer is not used vertically downwards are required.

Fig. 13 Schmidt hammer and core holder (**left**), and used in the field (**right**)

4 Field Tests

4.1 In Situ Stresses

Several authors present descriptions, limitations and fields of application of existing in situ stress measurement methods [35, 36]. For the design of underground structures in civil engineering projects, they are usually classified as methods based on hydraulic fracturing, methods based on complete stress release, and methods based on partial stress release. Methods based on the observation of the rock mass behaviour are less frequently used.

Overcoring and hydraulic fracturing tests are used when the zones of interest can only be reached with boreholes. In most cases, they are performed during the geotechnical survey stage. Flat jack tests require direct access to rock mass surfaces, so they are usually carried out when excavation reaches regions near the underground works. Often their results are used to confirm previous stress field estimates.

Tests for determination of the in situ stresses in rock masses for the design of underground structures are usually scarce in numbers, due to cost and time constraints, they have limitations inherent to their nature, and their results are only valid in the exact locations where they are executed. Owing to these factors, characterization of the in situ stress field in the rock mass at the location of the underground infrastructure often requires a global model for the interpretation of results from all the tests.

Global interpretation methodologies start by establishing a set of assumptions regarding the stress field in the rock mass. For instance, it is common to consider that the vertical and horizontal stresses increase linearly with depth, since the stresses are, in a large proportion, due to the weight of the overlaying ground. Then, three-dimensional numerical models are used to calculate the stresses at the locations where the stress measurements were performed, and an inverse methodology is applied to estimate the in situ stress field that better reproduces the test results [37, 38].

Overcoring tests. Overcoring tests use a complete or partial stress release method allowing to obtain the stress tensor components at a given location in a borehole.

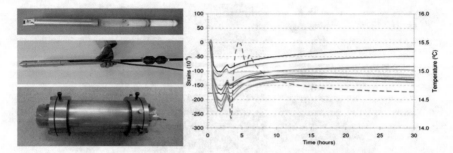

Fig. 14 STT (**top left**), USBM (**middle left**), biaxial test chamber (**bottom left**), and typical strains measured during STT cell overcoring (**right**)

CSIRO and LNEC's STT triaxial cells, and the Borre probe allow determining all six stress components from a single test, while with USBM and doorstopper deformation gauges only the three stress components in a plane can be obtained [10].

STT stress cells are 2-mm-thick epoxy resin hollow cylinders with embedded strain gauges. Test starts by cementing the cell inside a 37-mm-diameter borehole. Then the in situ stresses are released by overcoring with a larger diameter. Strains are measured during overcoring until temperature stabilizes by an in-built data logger. Stresses are calculated using the rock elastic constants obtained in a biaxial test of the recovered core with the cemented cell. Figure 14 presents a STT cell (top left) with the data logger, a biaxial test chamber (bottom left) and a diagram with the typical evolution of the measured strains and temperature during the overcoring process (right).

Flat jack method. The flat jack method is based on partial stress release. LNEC's SFJ (small flat jack) test consists in cutting a 10-mm slot in a rock surface, with a 600-mm- diameter circular disk saw, where a flat jack is inserted. Pressure is applied by the flat jack until deformation caused by opening of the slot is restored. With each flat jack, a single stress component is obtained. Usually, at a given location, several tests in slots with different orientations are performed (Fig. 15) [37, 38].

Hydraulic tests. Two types of hydraulic tests can be performed for the determination of in situ stresses: hydraulic fracturing (HF) and hydraulic tests on pre-existing fractures (HPTF) [10, 39]. HF tests induce a fracture in the rock by applying water pressure in a borehole section isolated by packers, enabling to estimate the minimum horizontal stress. In HTPF tests, water pressure is applied in a borehole comprising an isolated existing fracture whose opening allows to determine the stress component perpendicular to the fracture plane. Figure 16 shows the hydraulic fracturing equipment being inserted in a borehole (top left), an electrical image of a tested fracture (bottom left) and a scheme with the general setup for the hydraulic fracturing tests (right).

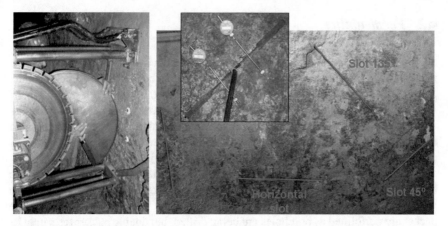

Fig. 15 Flat jack being inserted in the slot, array of slots and instrumentation

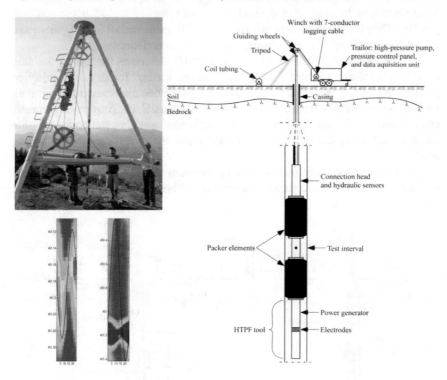

Fig. 16 Hydraulic fracturing equipment (**top left**), electric image of a tested fracture (**bottom left**) and hydraulic fracturing test setup (**right**)

4.2 Permeability

Seepage in rock masses occurs mainly through conductive discontinuities and, for most civil engineering purposes, crystalline rocks can be considered impermeable. This is why reference to permeability tests appears here in the field tests section.

The most commonly used in situ test to estimate permeability in rock engineering works is the Lugeon test, which is also known as "packer test" or "water pressure test". It was designed by Maurice Lugeon in 1933 as a means of assessing rock mass permeability and the need for grouting at dam sites [40].

The Lugeon test is a stepwise, constant head permeability test performed in a borehole section isolated by one or two packers, whether the isolated section is located at the end of the borehole or not, respectively. Lugeon tests with a single packer are performed as boreholes are being drilled, but double packer tests may be performed after the borehole is concluded (Fig. 17). The injection section length has to be adapted to the jointing of the rock mass, but values of 3 and 5 m are common practice. They are standard tests usually included in geotechnical investigations and in rock mass drainage and grouting curtains in dam foundations.

Test results are expressed as Lugeon units (LU) defined as the loss of one litre of water per minute, per metre of the borehole test section, for an excess injection pressure of 1 MPa measured at the middle of the test section. Estimation of equivalent rock mass permeability from Lugeon tests is controversial, but conversion formulas can be used to calculate the permeability coefficient assuming stationary pressure and flow, and steady-state transmission of water from the borehole to the surrounding medium.

Standard Lugeon tests include several pressure stages, usually five to nine, between a minimum and a maximum pressure. When five pressures are used, the

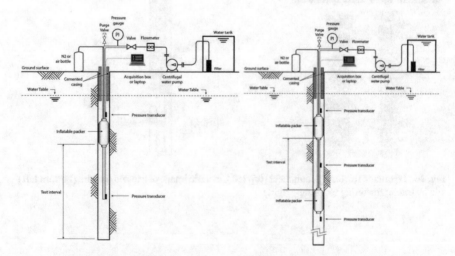

Fig. 17 Scheme of Lugeon tests with a single packer (**left**) and double packers (**right**) [41]

first pressure stage in performed at the minimum pressure, the second at an intermediate value, the third at the maximum pressure, the fourth again at the intermediate value, and the last again at the minimum value. If nine pressure stages are considered, a similar increasing–decreasing sequence is carried out, but with three intermediate pressures. The maximum pressure, which should not exceed 1.0 MPa, is defined taking into account several factors, such as the objective of the test, the depth of the test section and the need to assure that hydraulic fracturing of the rock mass does not occur. After steady flow is reached, each pressure stage lasts 10 min.

A Lugeon value is calculated for each one of these pressure stages, and test interpretation follows from the analysis of the LU values versus pressure plots. Different evolution trends of these graphs during the increasing–decreasing pressure allow to define if flow in the injected rock mass section can be considered laminar or turbulent, or if wash-out or void filling occurred, or even if hydraulic fracturing was reached.

Particular projects may require the execution of particular permeability tests, such as pressure drop test, in which water is injected into a borehole section up until a given pressure is reached and then water injection is stopped and pressure drop (or build-up) is measured, or the constant head Lefranc-type tests used in the case of high permeability environments.

4.3 Deformability

Rock mass deformability plays an important role in the design of several types of structures, because their behaviour depends on the displacements undergone by the rock mass. This is the case of concrete dams, large bridge foundations, underground caverns and tunnel linings. For the design of these important types of structures, it is not adequate to characterize the rock mass deformability by only using laboratory tests on intact rock specimens and extrapolating their results to the rock mass based on geomechanical classifications. For these structures, in situ deformability tests such as borehole expansion tests, plate loading tests or flat jack tests, are required.

Borehole expansion tests. Several types of borehole expansion tests are available to evaluate rock mass deformability, but they involve relatively small rock mass volumes around 0.1 m^3, which are seldom a representative elementary volume (REV). A major advantage is that they are not expensive, as they are performed in boreholes that are generally used for other purposes in the scope of geotechnical investigation of the rock masses, and it is possible to carry out a significant number of tests and use these results for zoning the rock mass deformability at a given site.

Borehole expansion tests can be performed with borehole jacks, also known as stiff dilatometers, that apply a unidirectional pressure over two diametrically opposed sectors of a borehole wall. As an alternative, dilatometers are probes that apply a uniform radial pressure via a flexible rubber membrane pressed against the borehole walls by a fluid. Some of this second type of apparatuses, derived from soil pressuremeters, measure the rock mass deformation indirectly by recording the volume change of the probe, while others, like LNEC's BHD dilatometer, measure directly the diametric displacements with displacement transducers contacting the borehole wall [10, 42].

Dilatometer tests are carried out after the probe is installed at the desired borehole depth and an initial low pressure is applied so that the flexible membrane expands and contacts the walls. Tests usually follow a loading programme including several loading–unloading pressure cycles with increasing peak values and prescribed pressure stages at which pressure is maintained for 1–2 min, displacements stabilize and data (pressure and displacements or probe volume) are measured. Gradual pressure increase and cautious monitoring of the displacements is required, since the applied radial pressure induce tensile stresses that, if in excess, may cause rock fracturing [43, 44].

Test results are plotted as stress versus displacement curves and the deformation modulus can be calculated assuming that the rock mass is isotropic, elastic and linear-elastic. In Fig. 18 a BHD dilatometer probe (left), and the full standard equipment (probe, winch, positioning rod, water pump and read-out unit) (right) are displayed.

Plate loading tests. Plate loading tests are widespread in situ deformability tests, but in some cases they do not provide satisfactory results, because the rock mass in the tested zone is often disturbed by the excavation. They consist of applying pressure via steel loading plates, about 1 m in diameter, to a rock surface in an exploratory adit or test chamber, and calculating the rock mass deformation modulus from the measured deformation [10, 42].

Most frequently double tests are performed on opposite walls at the same location, as one surface is used as reaction for the other. Accordingly, the loaded surfaces have to be coplanar, and any unevenness has to be compensated with cement mortar. The loads are applied by hydraulic jacks, and displacements are usually measured at the

Fig. 18 BHD dilatometer probe (**left**) and full equipment (**right**)

Fig. 19 Plate loading tests in an adit (**left**) [45], and in a tunnel (**right**) [46]

steel plates and at the rock surface around it with displacement transducers, and on occasions also inside the tested rock with extensometer rods.

Relations between the pressure changes and induced displacements of the rock mass allow calculating an equivalent rock mass deformation modulus. Figure 19 shows two plate loading test set-ups, a vertical test in an exploratory adit (left) and a slightly inclined test in a tunnel (right).

On occasions when rock masses are relatively competent, high pressures are required to produce appraisable displacements, which may be hazardous given the precarious stability conditions of the set-ups. In other cases, if the load surfaces are not adequately chosen and prepared, loads may be applied to disturbed rock mass in the excavation damaged zone.

Large flat jack tests. To avoid the shortcomings of plate load tests, large flat jack tests (LFJ) are preferably used, as they also allow testing relatively large volumes of rock mass, of a few cubic meters, while determining the deformability in less disturbed zones of the rock mass [10, 47].

LFJ tests consist in cutting a thin slot in the rock mass, by means of a disk saw, and inserting a flat jack that is then pressurized in order to load the slot walls while measuring the rock mass deformation with several displacement transducers. In order to obtain a mean value of the modulus of deformability in large rock volumes, as well as information about the rock mass heterogeneity, a group of two co-planar contiguous slots is usually cut for each test.

The equipment for cutting the slots includes a machine, with a 1000 mm diameter diamond disk saw mounted at the end of a rig that houses the system that transmits the rotating movement to the disk. A central 168 mm diameter hole with a depth of 1.10 m is previously drilled by the same machine, in order to allow the introduction

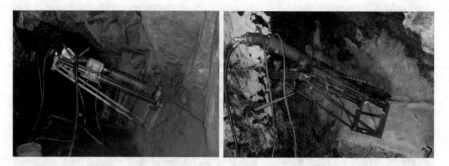

Fig. 20 Plate loading tests in an adit (**left**), and in a tunnel (**right**) [47]

of the disk supporting column. The disk saw cuts 1.50 m deep slots (Fig. 20). Once a slot is cut, a flat jack is introduced and, after the central hole is filled with cement mortar, the jack is ready to be filled with hydraulic oil and pressurized. Usually, as tests are carried out with two flat jacks side by side, this procedure is repeated for the second jack. Each flat jack consists of two steel sheets less than 1 mm-thick, welded around the edges. Inside the flat jack, four transducers measure the opening and closure displacements of the slot. The flat jacks are then inflated to adjust to the surface of the slots and a low initial pressure, usually of about 0.05 MPa, is applied (Fig. 21).

A LFJ test comprises at least three loading and unloading cycles reaching increasing maximum pressures. Displacements are measured by the four transducers in each flat jack and, in some cases, by transducers mounted on the rock surface across the slot. The raw test results are the pressure versus displacement curves obtained in the test.

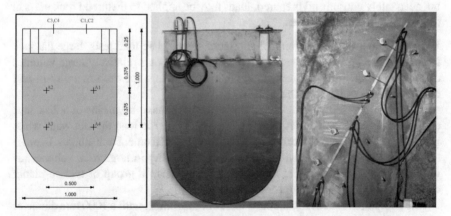

Fig. 21 Schematic representation of a large flat jack (**left**), a large flat jack (**centre**), and s LFJ test set-up with two jacks on a vertical wall of an adit (**right**) [47]

Pressure applied to the rock mass by the flat jacks in the cut slots causes tensile stresses to develop at the edge of the cut slot. As a test is carried out, pressures and stresses increase and if the combination of rock mass tensile strength and in situ stresses is exceeded, which is common, a tension crack will develop around the slot. Though it might not be visible at the surface, it will be noticed in the pressure versus displacement curves as they will show a decrease in the deformability. This conclusion is used to outline the continuation of the test and establish the maximum pressures of the following cycles. It is also used in the model for interpretation of LFJ test results to calculate the rock mass deformability modulus, which is based on the theory of elasticity for homogeneous, isotropic and linear elastic bodies, and takes into account the possible development of the tension crack.

4.4 Shear Strength

Best shear strength estimates are obtained from in situ direct shear tests as they inherently account for any possible scale effect. However, due to the duration and high cost of such tests, they are solely performed in special cases, to assess the shear strength of particular interfaces in the rock mass relevant for design, such faults and joints with thick fillings, veins or weathered bands, bedding or interlayer planes, and concrete-rock contacts.

In situ direct shear tests can be performed underground in exploratory adits and test chambers on discontinuities with any orientation, or at the surface. The walls and roof of the adit or tests chambers provide the reactions for the normal and shear forces, often reaching 4 MN.

Preparation for an in situ shear test is very complex and time consuming. First, after defining the test location and the shear direction, a rock block with the discontinuity, around 1 m² in area and 0.5 m in height, is cut using disk saws or drilling overlapping boreholes. Then, the block is encapsulated with reinforced concrete or a steel frame. Concrete blocks are built on the roof or sidewalls of the adit for reaction of the normal and shear forces. All these operations have to be executed ensuring that the discontinuity does not move and that all filling materials are not disturbed (Fig. 22 left). In situ direct shear tests performed at the ground surface, anchored concrete and steel structures are required to provide reaction blocks for both normal and shear forces (Fig. 22 right).

Sometimes, the direction of the shear jacks is inclined in relation to the discontinuity plane, but acting through its centroid. The forces are applied using hydraulic jacks, either cylinders or flat jacks. Load cells can be used, but usually stresses are calculated from the pressure of the jacks, considering the area of the discontinuity and its inclination: Transducers are used to measure normal and shear displacements and if environmental conditions at the testing site allow, a data acquisition system may be used.

Owing to the high costs of these tests, they are typically performed as multi-stage shear tests under several, usually five, increasing normal stresses. Tests start

Fig. 22 Example of in situ direct shear tests on the floor of an adit (**left**) [48], and schematic representation of a test at the surface (**right**) [49]

by applying the lowest normal stress until stabilization of the normal displacements is reached. Normal stress should be applied at a slow rate in order to allow excess pore pressures in the filling material to dissipate. Then, shearing at a constant shear displacement rate (0.1–0.5 mm/min) is initiated and continues until the shear displacement progresses under an approximately constant shear stress. If the shear force is inclined, as it is increased, it produces an increase of the normal load that needs to be continuously compensated as shear displacement goes on. After this first stage, the shear stress is slightly decreased and the normal stress is increased to the value established for the second stage, and a similar sequence follows. Completion of the test happens after several stages under increasing normal stresses are performed.

Results of a test are plotted as shear stress versus shear displacement curve that allows defining the shear strength for each normal stress, and subsequently enables plotting these values and the calculation of the strength parameters of the tested discontinuity, for instance the friction angle and the apparent cohesion.

5 Concluding Remarks

In the scope of large civil engineering projects, the laboratory and field tests carried out for the characterization of rock masses can be seen as small pieces of a large puzzle aimed at providing fundamental elements about the rock mass for the design. The values of the parameters that characterize the rock mass properties, which will be used in design, shall result from an expert and cautious judgement of the whole range of values obtained from the testing program. Once integrated in a safety verification procedure with adequate safety requirements, they will provide an important basis for assuring safety during the construction and exploration stages of the project.

The laboratory and field tests presented in this chapter were considered as the most relevant and commonly performed for the characterization of rock masses. It has to be recognized that even a simple enumeration of all currently available methods and techniques is an unfeasible task, and that the biased selection that was inevitably necessary reflects the experience of the authors. Furthermore, all descriptions of testing equipment, methods and procedures included in this chapter had to be seriously shortened to a minimum, but still allowing to fully comprehend the underlying basic principles of the tests, their objectives and results, and their benefits and shortcomings. Detailed description of equipment, in particular of the measuring devices, was intentionally excluded given their continuous advancements.

Referencing had to be considerably abbreviated also. References in this chapter have to be understood as starting points for wider searches. It was sought to provide the source references for each test and they often comprise the ISRM Suggested method and the corresponding ASTM standard.

References

1. Fairhurst, C.: The formation of rock mechanics. In: Hudson, J.A., Lamas, L. (eds.) ISRM 50th Anniversary Commemorative Book. ISRM, Lisbon (2012)
2. Brown, E.T.: The First 50 Years of the ISRM. In: Hudson, J.A., Lamas, L. (eds.) ISRM 50th Anniversary Commemorative Book. ISRM, Lisbon (2012)
3. Bernaix, J.: Étude Géotechnique de la Roche de Malpasset. Dunod, Paris (1967)
4. Londe, P.: The Malpasset dam failure. Eng. Geol. **24**(1-4), 295–239 (1987)
5. Bryan, A., Bryan, J.G., Fouché, J.: Some problems of strata control and support in pillar workings. Min. Eng. **123**(41), 238–266 (1964)
6. Müller, L.: The rock slide in the Vajont Valley. Rock Mech. Eng. Geol. **2**, 148–212 (1964)
7. Müller, L.: New considerations of the Vajont slide. Rock Mech. Eng. Geol. **6**(1), 1–91 (1968)
8. Rocha, M.: Analysis and design of the foundations of arch dams. In: International Symposium on Rock Mechanics Applied to Dam Foundations, General Report—Theme III, Rio de Janeiro, Brazil (1974)
9. Brown, E.T (ed.).: Rock Characterization, Testing and Monitoring. Pergamon Press, Oxford (1981)
10. Ulusay, R., Hudson, J.A. (eds): The Complete ISRM Suggested Methods for Rock Characterization, Testing and Monitoring: 1974–2006. ISRM Commission on Testing Methods (2017)
11. Ulusay, R. (ed.): The ISRM Suggested Methods for Rock Characterization, Testing and Monitoring: 2007–2014. Springer, Cham (2015)
12. Pinto da Cunha, A. (ed.): Scale Effects in Rock Masses 93. Taylor & Francis. Lisbon, Portugal (1993)
13. ASTM: D7012-14 Standard Test Methods for Compressive Strength and Elastic Moduli of Intact Rock Core Specimens under Varying States of Stress and Temperatures. ASTM Int. West Conshohocken (2014)
14. ISRM: Suggested methods for determining the uniaxial compressive strength and deformability of rock materials. In: Video prepared by the Seoul National University (South Korea), www.isrm.net (2018)
15. Hudson, J.A., Crouch, S.L., Fairhurst, C.: Soft, stiff and servo-controlled testing machines: A review with reference to rock failure. Eng. Geol. **6**, 155–189 (1972)
16. Hoek, E., Brown, E.T.: Empirical strength criterion for rock masses. J. Geotech. Engng Div. ASCE 106(GT9), 1013–1035 (1980)
17. Hoek, E.: Practical Rock Engineering. https://www.rocscience.com (2017)
18. ASTM: D3967-16 Standard Test Method for Splitting Tensile Strength of Intact Rock Core Specimens. ASTM International, West Conshohocken (2016)

19. Carneiro, F.L.L.B.: A new method to determine the tensile strength of concrete. In: Proceedings of the 5th Meeting of the Brazilian Association for Technical Rules, Section 3, 126–129, in Portuguese (1943)
20. Wittke, W.: Rock Mechanics. Theory and Applications with Case Histories. Springer, Berlin (1990)
21. ISRM: Suggested method for determining tensile strength of rock materials. In: Video prepared by the Seoul National University (South Korea), www.isrm.net (2018)
22. ASTM: D2845-08 Standard Test Method for Laboratory Determination of Pulse Velocities and Ultrasonic Elastic Constants of Rock (withdrawn 2017). ASTM International, West Conshohocken (2008)
23. ASTM: D5607-16 Standard Test Method for Performing Laboratory Direct Shear Strength Tests of Rock Specimens Under Constant Normal Force. ASTM International, West Conshohocken (2008)
24. Zhang, X., et al.: Laboratory investigation on shear behavior of rock joints and a new peak shear strength criterion. Rock Mech. Rock Eng. **49**, 3495–3512 (2016)
25. Patton, F.D: Multiple modes of shear failure in rock. In: 1st Congress of the ISRM, vol. 1, pp. 509–513. Lisbon, Portugal (1966)
26. Barton, N.: Review of a new shear strength criterion for rock joints. Eng. Geol. **7**, 287–332 (1973)
27. Barton, N.R., Choubey, V.: The shear strength of rock joints in theory and practice. Rock Mech. **10**, 1–54 (1977)
28. ISRM: ISRM Suggested Method for Determining the Basic Friction Angle of Planar Rock Surfaces by Means of Tilt Tests. Rock Mechanics and Rock Engineering, **51**, 3853–3859 (2018)
29. ISRM: Suggested Method for Determining the Basic Friction Angle of Planar Rock Surfaces by Means of Tilt Tests. In: Video prepared by Universidad de Vigo (Spain), www.isrm.net (2019)
30. Pérez-Rey, I., Alejano, L., Muralha, J.: Experimental study of factors controlling tilt-test results performed on saw-cut rock joints. Geotech. Test. J. **42**(2), 307–330 (2019)
31. Muralha, J.: Rock joint shear tests. Methods, results and relevance for design. In: Proceedings of Eurock 2012 Symposium. Stockholm, Sweden (2012)
32. ASTM: D5731-16 Standard Test Method for Determination of the Point Load Strength Index of Rock and Application to Rock Strength Classifications. ASTM International, West Conshohocken (2016)
33. ISRM: Suggested method for determining point load strength. In: Video prepared by the Seoul National University (South Korea), www.isrm.net (2016)
34. ASTM: D5873-14 Standard Test Method for Determination of Rock Hardness by Rebound Hammer Method (2014)
35. Cornet, F.H.: Stress in rock and rock masses. In: Hudson, J. (ed.) Comprehensive Rock Engineering, vol. 3. Pergamon Press, Oxford (1993)
36. Ljunggren, C., Chang, Y., Janson, T., Christianson, R.: An overview of rock stress measurement methods. Int. J. Rock Mech. Min. Sci. **40**, 975–989 (2003)
37. Lamas, L., Muralha, J., Figueiredo, B.: Application of a global interpretation model for assessment of the stress field for engineering purposes. ISRM News J. **13**, (2010)
38. Lamas, L., Espada, M., Figueiredo, B., Muralha, J.: Stress measurements for underground powerhouses—three recent cases. In: AfriRocks 2017 Proceedings, Cape Town, South Africa (2017)
39. Figueiredo, B.: Integration of in situ stress measurements in a non-elastic rock mass. Ph.D. thesis, University of Strasbourg, France (2013)
40. Lugeon, M.: Barrages et Géologie. Dunod, Paris (1933)
41. ISRM: Suggested Method for the Lugeon Test. Rock Mech. Rock Eng. **52**, 4155–4174 (2019)
42. Wittke, W.: Rock Mechanics Based on an Anisotropic Jointed Rock Model. Ernst & Sohn, Berlin (2014)

43. Muralha, J.: L'emploi du dilatomètre du LNEC dans la caractérisation geomécanique des massifs rocheux. In: International Symposium 50 Years of Pressure Meters. LCPC, Paris. (2005)
44. ISO: EN ISO 22476-5 Geotechnical investigation and testing—Field testing—Part 5: Flexible dilatometer test (2009)
45. Rezaei, M., Ghafoori, M., Ajalloeian, R.: Comparison between the *In Situ* tests' data and empirical equations for estimation of deformation modulus of rock mass. Geosci. Res. **1**(1), 47–59 (2016)
46. Solexperts website, https://www.solexperts.com/en/geotechnics/services/plate-loading-tests, Last accessed 31 Dec 2020
47. Figueiredo, B., Bernardo, F., Lamas, L., Muralha, J.: Interpretation of rock mass deformability measurements using large flat jack tests. In: 12th International Congress of the ISRM. Beijing (2011)
48. Sanei, M., et al.: Shear strength of discontinuities in sedimentary rock masses based on direct shear tests. Int. J. Rock Mech. Min. Sci. **75**, 119–131 (2015)
49. Barla, G., Robotti, F., Vai, L.: Revisiting large size direct shear testing of rock mass foundations. In: 6th International Conference on Dam Engineering. LNEC, Lisbon, Portugal (2011)

Testing and Monitoring of Earth Structures

Rafaela Cardoso⬤, Anna Ramon-Tarragona⬤, Sérgio Lourenço⬤, João Mendes⬤, Marco Caruso⬤, and Cristina Jommi⬤

Abstract Monitoring structural behavior of earth structures during construction and in service is a common practice done for safety reasons, consolidation control and maintenance needs. Several are the techniques available for measuring displacements, water pressures and total stresses, not only in these geotechnical structures but also at their foundations. Materials testing has been used for calibrating models for structural design and behavior prediction, and these models can be validated with instrumentation data as well. Relatively recent investigation on the behavior of these materials considering their degree of saturation focuses on monitoring the evolution of water content or suction as function of soil-atmosphere interaction, necessary to predict cyclic and/or accumulated displacements, and has huge potential to predict

R. Cardoso (✉)
CERIS, Department of Civil Engineering, Architecture and Georesources, Instituto Superior Técnico, University of Lisbon, Lisbon, Portugal
e-mail: rafaela.cardoso@tecnico.ulisboa.pt

A. Ramon-Tarragona
Division of Geotechnical Engineering and Geosciences, Department of Civil and Environmental Engineering (DECA), Universitat Politècnica de Catalunya, Barcelona, Spain
e-mail: anna.ramon@upc.edu

S. Lourenço
Department of Civil Engineering, The University of Hong Kong, Hong Kong, China
e-mail: lourenco@hku.hk

J. Mendes
Department of Mechanical and Construction Engineering, Faculty of Engineering and Environment, Northumbria University, Newcastle, UK
e-mail: joao.mendes@northumbria.ac.uk

M. Caruso · C. Jommi
Materials Testing Laboratory, Politecnico Di Milano, Milan, Italy
e-mail: marco.caruso@polimi.it

C. Jommi
Department Geoscience and Engineering, Delft University of Technology, Delft, The Netherlands
e-mail: c.jommi@tudelft.nl

© The Author(s), under exclusive license to Springer Nature Switzerland AG 2023
C. Chastre et al. (eds.), *Advances on Testing and Experimentation in Civil Engineering*, Springer Tracts in Civil Engineering,
https://doi.org/10.1007/978-3-031-05875-2_4

the impact of climate changes on the performance of existing geotechnical structures. This new need justifies the investment on developing sensors able to be used for in situ monitoring of water in the soils, such as those presented here. Testing and monitoring becomes even more important nowadays when, for sustainability purposes, traditional construction materials are replaced by other geo-materials with unknown behavior and long-term performance (mainly accumulated displacements). Existing experimental protocols and monitoring equipment are used for such cases, however new techniques must be developed to deal with particular behaviors. Three case studies are presented and discussion is made on monitoring equipment used and how monitored data helped understanding the behaviors observed.

Keywords Suction · Climate changes · Non-traditional geo-materials · Monitoring · Accumulated displacements · Chemical properties · Mineralogy

1   Instrumentation of Embankments and Their Foundations

Earth structures such as dykes, dams and road embankments have been built since more than 4000 years and construction techniques have been updated as function of technological development and human needs. Nowadays, monitoring structural behavior during construction and in service is a common practice for safety reasons, consolidation control, maintenance needs, etc. Instrumentation data is also necessary to confirm design assumptions, and also for research purposes. Several are the techniques available for measuring displacements, water pressures and total stresses (Table 1), not only in the geotechnical structures but also in their foundations.

The instruments used for the different applications are being updated along time taking advantage of technological development, improving their accuracy, response time and maintenance needs. Materials testing has been used for calibrating models used in structural design and behavior prediction, and then these models can be validated with instrumentation data as well. Nevertheless, for the most usual earth structures and almost for practical cases, the monitored parameters have remained practically unchanged the last century and their use in many practical cases has been

Table 1 Summary of typical instruments on embankments

Parameter			Equipment
Displacements	Internal	Horizontal	Inclinometers (electric, magnetic)
		Vertical	Settlement plates
			Extensometers
	External	Topographic targets, …	
Total stress		Loading cells	
Pore pressure		Piezometers (electric, hydraulic, magnetic, etc.)	
		Pore pressure cells	

extensively reported. For this reason, this work is not about standard instrumentation or standard applications of such known instruments, but on how the instruments may be used for particular needs illustrated by three case-studies.

Design mechanical and hydraulic properties of the geotechnical materials refer to saturated conditions, although compacted materials are in unsaturated state and their degree of saturation experience cyclic changes due to the exposition to climate actions. Changes on the degree of saturation or water content significantly affect these properties and can be responsible for cyclic and/or accumulated (irreversible) volume changes, compromising structural performance. Relatively recent investigation on the behavior of these materials considering their degree of saturation focuses on monitoring the evolution of water content (or suction) in earth structures, which is necessary to predict such cyclic behavior. In addition, this knowledge has huge potential to be used on predicting the impact of climate changes on the performance of existing geotechnical structures (in the structure itself and also considering changes on water table levels at the foundation). This new need justifies the investment on developing sensors able to be used for in situ monitoring of water presence in the soils in unsaturated states (soil suction sensors), and for this reason a small introduction to existing sensors and working principles is presented here.

Testing and monitoring or earth structures becomes even more important nowadays when, for sustainability purposes, traditional construction materials are replaced by other geo-materials which behavior and long-term performance (mainly accumulated displacements) are unknown. Existing experimental protocols and monitoring equipment are used for such cases, however new techniques must be developed to deal with the particular behavior of these non-standard materials. Some case-studies are presented here to deal with different situations: (i) when non-standard materials for embankment construction are used, such as light weight aggregates, construction residues, lime or cement treated materials, etc.; (ii) when displacements caused by unexpected chemical reactions occur, causing structural damages and forcing the adoption of repair interventions; (iii) when the effects on earth structures of permanent changes in water table levels at their foundations must be quantified. The cases presented here are examples in which standard instrumentation was installed to monitor parameters related with the particular nature of such structures. Monitored data presented helped to investigate instability phenomena and to find solutions for the problems observed.

2 Monitoring Soil-Atmosphere Interaction

2.1 Introduction

Soil structures, such as embankments and excavations, are typically built in an unsaturated state, above the water table and exposed to the atmosphere. Vegetation also plays a critical role, and it often has opposite effects, by providing root reinforcement and contributing to stability or by enhancing drying by evapotranspiration. The ground is subjected to continuous wetting and drying cycles. Under extreme weather events, through excessive rainfall and prolonged droughts, slope instability (creep and slides) and shrinkage (desiccation cracks) endanger the structures leading to significant economic losses. Therefore, approaches to prevent or mitigate the negative effects of soil-atmosphere interactions in soil structures are based on (i) classic engineering (ground stabilization, retaining structures), (ii) new approaches based on bioengineering and (iii) monitoring or sensing of structures. This chapter will focus on the latter.

At the pore scale, soils are composed of three distinct phases: particles, water, and air, forming water menisci in-between soil particles with a negative water pressure. Suction, the difference between pore air and pore water pressure across the menisci, controls soil behavior, by increasing the soil strength and compressibility, controlling the soil volumetric behavior (swelling and shrinkage) including the soil water retention (relation between soil water content and suction). The response of soil structures to atmosphere interactions and its effects can thus be accessed through the monitoring of suction.

This section presents a review of methods to measure suction in the field, with a focus on methods that provide a continuous measurement and that can be installed permanently in the field. At first, indirect methods that rely on the measurement of water content will be presented, followed by methods that measure the pore water pressure, mostly conventional tensiometers and high capacity tensiometers. Finally, field examples will be provided of the monitoring of pore water pressure with high capacity tensiometers, including their sensitivity to soil-atmosphere interactions (speed of response to rainfall events).

2.2   Measurement of Soil Suction

Over the past decades, various techniques have been developed for measuring the different soil suction components (total, matric and osmotic) using direct and indirect methods. Direct methods, as the name suggests, are methods in which the measured value is directly related to the soil suction. Whereas indirect methods are methods in which the soil suction is inferred from the equilibrium between the suction in the soil and a known medium. Although many techniques have been developed for measuring different soil suction components in the laboratory (Total [1–3], Matric [4–9] and Osmotic [10]) and most have been adapted to field use, only a few have been adapted for the continuous measurement of soil suction on a permanent field installation basis. Below is a brief description of different methods that have been adapted for continuous measurement of soil matric suction in the field that are commercially available.

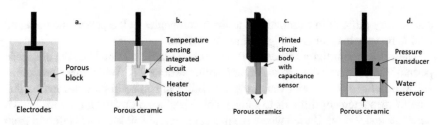

Fig. 1 Schematics of soil matric suction sensors (not to scale): **a** Electrical conductivity [5]; **b** Thermal conductivity [6]; **c** Dielectric permittivity [7]; and **d** High capacity tensiometer [12]

Indirect methods. Indirect methods for measuring soil matric suction include electrical conductivity, thermal conductivity, and dielectric permittivity.

Electrical conductivity sensors infer soil matric suction from the relation that the degree of saturation has with the suction and resistivity of the soil. Because this relation is different for different soil types [5] the measurement of soil suction is, instead, inferred from the electrical impedance on a porous block with known characteristics. The most typical electrical conductivity sensors are the Gypsum Blocks, as shown schematically in Fig. 1a. These sensors are composed of two concentric electric electrodes installed in a permeable block. Each sensor is required to undergo calibration where the electrical impedance is directly related to the water content in the block that, in turn, is indirectly related to soil matric suction. The measurement of soil matric suction is achieved when an equilibrium between the suction in the soil with the suction in the block is reached. These sensors are simple to use and easy to install in the field and can measure soil suctions typically in the range of 30–1500 kPa. However, they are known to underperform when matric suction values are higher than 300 kPa, the measurement is affected by the salt concentration in the soil pore water and the initial equilibration time can be significant (up to two weeks).

Thermal conductivity sensors infer soil matric suction from the water flux into the sensor until suction reaches and equilibrium between the porous medium of the sensor and the soil. Much like the electrical conductivity sensors, a known porous medium must be used due to the variability in the thermal conductivity of soils. The typical schematic for a thermal conductivity sensor is shown in Fig. 1b. These sensors are composed of a temperature sensing integrated circuit and a heater resistor encased in a ceramic porous medium with known porosity. The measurement of soil suction is made indirectly by determining the water content by heating the porous block with the heater resistor and by measuring the temperature increment during heating. These sensors must be calibrated before use for a reliable measurement of soil suction, different readings must be obtained at different degrees of saturation of the porous block varying from fully dry to fully saturated and, if possible, at know values of suctions using, for example, the pressure plate technique [11]. These sensors are relatively easy to use and to install in the field, measurement is not affected by salt concentrations and have a measuring range of 10–1500 kPa. However, these sensors can be very difficult to calibrate due to the heterogenies in the porous medium and tend to be very inaccurate at high values of suction.

Dielectric permittivity sensors (DPS) infer soil matric suction from the dielectric permittivity properties of air, water, and solid particles [7]. While these properties are well defined for air and water, the variability in soil solid particles makes this type of measurement difficult to implement. Thus, a porous medium with known pore size distribution is used where the measurement of soil suction is inferred from the water content in the porous medium when it is in equilibrium with the soil. The typical schematic for a DPS is shown in Fig. 1c. These sensors are composed by a capacitance sensor placed between two ceramics discs with known properties. The measurement of soil suction is inferred from the water content in the porous ceramic discs derived from the dielectric permittivity of the amount of air, water, and solid particles in the ceramic discs. Because the dielectric permittivity is highly dependent on the amount of water an accurate and wide range of the measurement is possible. Factory calibration of these sensors are based on the water retention curve of the ceramic discs and is directly imbued in the sensor circuit board. The measurement is made by submerging the whole sensor (ceramic discs and circuit board) into the soil until equilibrium between the soil and ceramic discs is reached. These sensors are very easy to use and install in the field, have a measuring range of 9–2000 kPa and the factory calibration is reasonably accurate due to the close control on the porosity and pore size dimensions of the ceramic filter during fabrication.

All indirect methods that rely on measuring the water content of a porous medium have the same limitation regarding field installations, the porous medium undergoes hysteresis during wetting and drying cycles that can result in under and overestimation of soil suction in dynamic climatic environments during drying and wetting cycles, respectively.

Direct methods. Direct methods for measuring soil matric suction include conventional tensiometers (CT) and high capacity tensiometers (HCT). Both methods use the same measuring principle where the measurement of soil matric suction is made directly from the hydraulic equilibrium of the pore water in the soil with a pressure transducer or pressure gauge. These sensors are composed of a high air entry value (AEV) porous ceramic filter with known largest pore size, a water reservoir, and a pressure transducer/gauge as shown schematically in Fig. 1d. The main difference between these two methods is in the measuring range, CTs have a measuring range of 100 kPa [8], whereas HCTs have a measuring range of > 15,000 kPa [12]. The difference in the measuring range is due to the smaller largest pore size within the porous ceramic filter used in HCTs which is directly related to the AEV of the ceramic filter and, in turn, related to the measuring range of the sensor [13]. Measurement of soil suction is only possible if the water reservoir and ceramic filter are fully saturated. This is achieved by filling the sensor existing pore space (pores in ceramic filter and water reservoir) with deaired water under pressure that should be above the AEV of the ceramic filter using a pressure controller. Calibration of different sensors is required as the voltage output of the pressure transducer must be converted to pressure values. Calibration is performed on the positive pressure range that is linearly

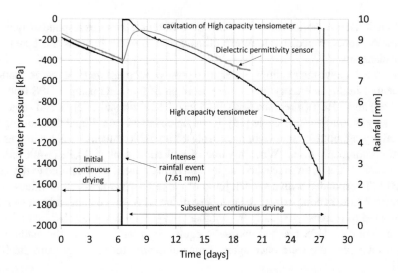

Fig. 2 Comparison of pore-water pressure readings of a DPS versus a HCT during drying-wetting–drying cycle, after [13]

extrapolated to the negative pressure range. The measurement of soil matric suction is then possible, and it can be achieved by placing the CT or HCT in intimate contact with the soil. During measurement, water from the water ceramic filter flows into the soil until hydraulic equilibrium between sensor and soil is reached.

CTs and HCTs, if properly saturated, can react very quickly to changes in suction in the soil, are not affected by hysteresis due to wetting and drying cycles, can be used as a typical piezometer as it measures both positive and negative pore-water pressures, and are the only techniques available that measures soil suction directly. However, these sensors have some important limitations, such as cavitation. Cavitation is the result of tension breakdown within the water reservoir inside the CT or HCT when water is under tension. This is easily identifiable in HCTs by a sudden jump in the reading to values close to -100 kPa, as shown in Fig. 2. When cavitation occurs, the sensor is unable to measure soil suction and must be re-saturated. Thus, to continuously use these types of sensors in the field a more complex installation and maintenance than any of the indirect methods mentioned before is required.

2.3 Soil Suction Monitoring in Soil Structures

Soil structures are typically exposed to dynamic climates where they undergo through multiple wetting and drying cycles influencing the pore-water pressure in the soil. During a dry spell, while evaporation occurs, the suction in the soil will tend to increase; while, during rainfall events, as water infiltrates into the soil, soil suction will tend to reduce. Thus, when choosing which methods to deploy for monitoring soil suction it is important to understand how different techniques respond to dynamic climatic events; especially if future climate and climatic patterns are taken into consideration.

Figure 2 shows the readings of negative pore-water pressure (soil matric suction) measured in the laboratory on a large specimen of fine sand using a DPS (indirect method) and a HCT (direct method), both installed at the same depth, and subjected to drying-wetting–drying cycle [13]. It can be observed from Fig. 2 that, during the initial drying, both sensors were measuring comparable values of soil suction as the sand continuously dried. When a value of soil suction of 400 kPa (−400 kPa of pore-water pressure) was reached, an intense rainfall event was imposed to the sand. The HCT responded with a sudden jump of the pore-water pressure to values close to zero. While, in comparison, the response of the DPS was considerably slower that continued to increase gradually for another 2 days well into the subsequent drying stage. After the DPS reached hydraulic equilibrium with the soil, the readings of pore-water pressure were once again comparable with the readings from the HCT [13].

The observed behavior of the DPS in Fig. 2 can be explained by the measuring principle used in this and other indirect methods of inferring soil suction from the measurement of the water content of a porous medium. During continuous measurement when the soil is drying, and while the sensor and soil are in equilibrium (initial drying in Fig. 2), desaturation of the pores will occur as water gradually flows from the sensor into the soil. However, when the flow is quickly reversed (intense rainfall event in Fig. 2) the hydraulic equilibrium is lost as the intake of water by the porous medium is not sufficient to maintain it. This is because the water flux is controlled by the hydraulic conductivity and the shape of the pores (responsible for the hysteretic water retention behavior) inside the porous medium. The reason why HCTs do not show this behavior is because the porous ceramic filter is required to be fully saturated during measurement. When significant desaturation occurs in the porous ceramic filter the HCTs cavitate and will require to undergo re-saturation.

Thus, when considering which method to employ for monitoring soil suction in the field a mixed approach of direct methods, that have an immediate response to suction changes but can cavitate, coupled with indirect methods should be considered.

3   Some Case-Studies

3.1   Embankment Built with Compacted Marls Sensitive to Water

Description of the case-study. Several embankments from A10 Motorway, in Portugal, were built with fragments of marls in order to reuse the excavated material

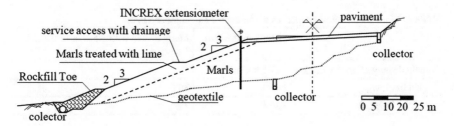

Fig. 3 Schematic profile of embankment AT1 from A10 Motorway (adapted from [14])

from the site. Marls are soft rocks which exhibit evolutive behaviour due to exposition to atmospheric actions (wetting–drying cycles). Crack opening and/or loss of bonding occur on wetting, with negative impact on the strength and compressibility of the material.

Relatively large fragments remained after the compaction processes and their further physical degradation or swelling deformations can have strong effect on global behaviour. For this reason, in the embankment investigated the construction procedure was adapted to add more water than what is usually done in road embankments, and the marls used to build the shoulders were treated with lime. The construction procedure adopted intended to reduce the size of the fragments minimizing the impact on overall behaviour of their eventual volume change. Lime addition would reduce the swelling potential of the marls.

Due to the particular behaviour of this non-traditional material, the vertical deformations and the evolution of the water content in embankment AT1 (simplified profile at Fig. 3) were measured during the construction and in the following two years. The instruments installed during the construction were (i) extensometers, to measure vertical deformations due to wetting and drying, and (ii) electrical resistive sensors to measure changes in water content. Traditional instruments such as inclinometers (operating in the gutters of the extensometers) and topographic targets were also installed, but they will not be discussed here.

Embankment AT1 was selected because it is very high (about 10 m at pavement axis) and was installed in a slope. The material treated with lime was at the shoulders in a layer with 5 m thickness (zoned profile in Fig. 3). Compaction was done in the wet side of optimum (interval $[w_{opt}, w_{opt+2\%}]$, energy equivalent to modified compaction effort and minimum of 95% of relative compaction), and using a vibratory sheepstoot roller to promote fragments breakage. Drainage systems were installed in the foundation and at the pavement levels (Fig. 3) to prevent water access, as it is usually done in embankments.

Overall description. The marls used in the construction of AT1 are named as Abadia marls (upper Jurassic) having moderate swelling potential. The main minerals present are carbonates (calcite), quartz, mica, dolomite, feldspar, clay minerals (almost no smectite), expansive minerals such as chlorite and gypsum, and a very small percentage of organic matter (0–2%) [14]. Laboratory tests were performed on

Table 2 Main geotechnical properties of the treated and untreated marls

	Marls (untreated)	Marls treated with 3.5% of lime (after 1 day)
Weight density of solid particles	27.5 kN/m³	
Liquid limit, w_L	49%	36%
Plasticity index, PI	25%	7%

samples with different weathering degrees to find the main geotechnical properties summarized in Table 2. The treatment with lime (3.5% in weight) reduced the plasticity of the marls (due to the reduction of the w_L thus on PI), improving the workability of the material and changing the classification of the fine fraction from CL to ML. The effect of this treatment on the hydraulic and mechanical properties of the marls was also investigated (details in Cardoso and Maranha das Neves [14] and Cardoso et al. [15]).

Instrumentation. The instruments installed are (i) magnetic extensometers, INCREX [16], on a PCV gutter, to measure vertical displacements, and (ii) electrical resistive sensors, ECH₂O [17], for measuring water content. Two vertical sections were instrumented, each having one gutter for measuring displacements (both vertical and horizontal can be done in the same gutter) and seven water content sensors (five in the marls and two in the treated marls, Fig. 4) distributed 1.5 m along height. To avoid interferences between the different instruments, the sensors were installed at 3 m distance (in longitudinal direction) from the INCREX gutters.

INCREX are INCRemental EXtensometers [16] and measure small vertical deformations with large precision. The system required to measure the magnetic field allows a precision higher than the one obtained when standard measuring systems are used ($\pm$0.02 mm for INCREX, and $\pm$ 0.5 mm for more general measurement equipment). Operation is based on the measurement of relative vertical displacements of magnetic rings spaced exactly one meter in the installation (Fig. 5). The rings are installed outside the gutter and are free to slide along it, but must be connected to the

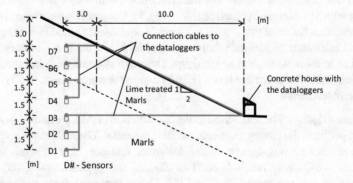

Fig. 4 Profile instrumented with sensors ECH₂O

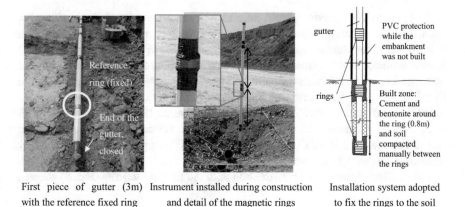

First piece of gutter (3m)   Instrument installed during construction   Installation system adopted
with the reference fixed ring   and detail of the magnetic rings   to fix the rings to the soil

Fig. 5 Gutters and magnetic rings of INCREX instruments

soil to displace with it. This connection was done by creating a thin cement-bentonite layer in the ring zone and a compressible layer made of compacted material between these zones, as illustrated in Fig. 5. Unfortunately, the two gutters were broken during the construction and readings were interrupted. Nevertheless, it was possible to measure the vertical displacements during almost the construction period. They were similar to those from topography readings, showing that the installation system adopted was efficient. In addition, these readings allowed calibrating a finite element model of the embankments used to estimate long-term displacements [18].

Sensors ECH_2O (Fig. 6) measure water content and changes in water content, related to suction changes through the water retention curve. These sensors must be in contact with soil and measure its electrical resistivity, which changes in function of changes in water content. The electrical wires had to be protected during sensors installation and the construction of the embankment, to be connected to the dataloggers, also shown in Fig. 6. The dataloggers were stored in a locked case build for that purpose in the access way.

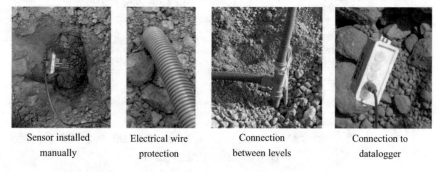

Sensor installed   Electrical wire   Connection   Connection to
   manually   protection   between levels   datalogger

Fig. 6 Installation of the resistive sensors ECH_2O for measuring water content

Soil resistivity depends on void ratio as well, and therefore sensors calibration had to be done in the laboratory. This calibration consists in the relationship between voltage and water content, so the sensors were installed in samples compacted with known different water contents for the same dry volumetric weight and voltage was measured. The relationship between voltage and water content is presented in Fig. 7 for each dry volumetric weight, where it can be seen that the slope found was independent from dry volumetric weight (average value 19.9). The voltage for null water content increases with the dry unit weight, so the real dry volumetric weight had to be measured in situ to select the proper calibration curve. It was done by measuring the voltage and water content of the soil when each sensor was installed.

Data collected during the construction with the ECH_2O sensors (Fig. 8) showed

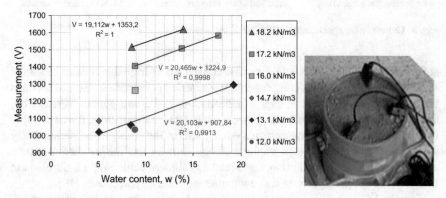

Fig. 7 Calibration of sensors ECH_2O

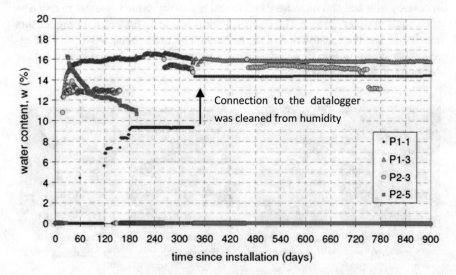

Fig. 8 Evolution of water content with depth and time (after [18])

some malfunction of the sensors, explained by humidity attack in their connection to the dataloggers. Nevertheless, data available allowed identifying an initial period where water content of the soil surrounding the sensor equalized the value of the layer where it was installed (around 15%) after this period. It was possible to detect the homogenization in depth of the water content (between 12 and 16%). No oscillations were detected during the service period monitored, which somehow was expected due to the depth where the sensors were installed. Nevertheless, this result indicates that the drainage system was efficient to prevent water access to the embankment, and therefore the swelling nature of the marls appear not to compromise the performance of the embankment during service.

Final considerations. The instrumentation of AT1 was conceived to monitor the expansive behaviour of the marls. Vertical deformations and displacements were recorded, as a direct measurement of this behaviour, but also its causes were investigated by introducing the water content sensors. This would allowed a more informed decision in case of problems.

Creative solutions had to be implemented to install the instruments used, in particular the INCREX. The location of the dataloggers necessary for recording ECH$_2$O measurements also requested the construction of a small locked case. Humidity problems caused the malfunctioning of some sensor's connection to the dataloggers. This is a typical problem in long term monitoring.

The need of complementary experimental tests is highlighted when using nontraditional materials. In this case, the sensors had to be calibrated before installation through laboratory testing. Experimental tests were also performed to characterize the hydraulic and mechanical behaviour of the materials, information used in other works.

Unfortunately, the gutters were broken, but no relevant displacements were detected by topographical instruments also installed in AT1. This was in accordance with the lack of significant oscillations on the values recorded for the water content along the exploitation, monitored by the resistive sensors. This showed that water access to the interior of the embankment was not significant, indicating the effectiveness of the drainage system installed.

3.2   Massive Expansions in Compacted Embankments Due to Sulphate Attack

Introduction. Deformations and movements in embankments and fills may result from chemical processes that involve crystal growth. For instance, sulphate attack is at the origin of surface heave developed in cement- and lime-treated soils that contain sulphates or are in contact with a source of sulphated water. This attack induces a decrease in soil strength and has damaged a number of compacted road bases and sub-bases [19–22]). The soil stabilization commonly concerns thin layers of soil.

However, massive sulphate attack can also develop when the treatment encompasses larger masses of soil. This is the case described in this section.

Pallaressos embankments performance. Pallaressos embankments belong to the high-speed rail connection between Barcelona and Madrid in Spain. Their construction finished in 2004. The embankments give access to a bridge 196 m long, Pallaressos Bridge. Alonso and Ramon [23] detail structural characteristics of the bridge. The embankments have a maximum height of 18 m in the vicinity of bridge abutments and its thickness decreases gradually at farther distances from abutments.

The abutment structures are founded directly on a Tertiary hard marl formation. Figure 9 presents the internal design of the embankments. The transition wedge built in the proximity of the abutment guarantees a smooth transition from the compacted fill to the rigid structures of the abutment and bridge. The design specifications of the embankments defined the construction of the wedge closest to the abutment with a cement-soil mixture As-built data obtained during construction showed that the compaction was in general on the dry side of optimum. The origin of the compacted material of the embankments core was a previous excavation in a nearby Tertiary natural formation of gypsiferous claystone with limestone and sandstone interstratified layers. The extreme expansive problems that suffered Pont de Candí Bridge [24, 25] and Lilla tunnel [26, 27] also involved this geologic formation.

Railway administration detected heave development of the tracks located above the embankments, in the vicinity of abutments, a short time after the end of construction of the embankments. Periodical rail leveling revealed a maximum vertical displacement rate ranging from 4.0 to 4.5 mm/month at the beginning of 2006.

The measured heave rates did not decrease and a grid of jet-grouting columns 1.5 m in diameter was carried out in each embankment (Fig. 10) in October 2006. The distribution and length of the columns resulted in an enlarged transition zone in both embankments. The reinforcement treated the central part of the embankments along an approximate length of 30 m. The application of the treatment was more intense in the proximity of abutments. The implementation of the reinforcement was

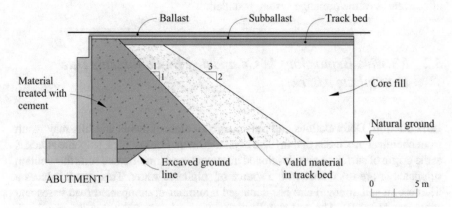

Fig. 9 Longitudinal section of pallaressos embankment

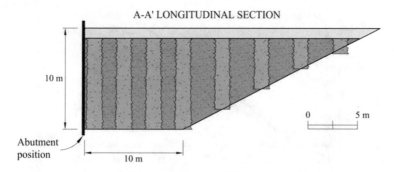

Fig. 10 Longitudinal cross section of the jet-grouting treatment applied at both embankments

justified by the weakness and disintegration easiness of the cement-treated soil in the embankment material, and also by the expected poor wedge design observed from boreholes drilled through the two embankments.

Despite of this treatment solution the embankment heave continued. In addition, heave rates measured after the treatment accelerated to values up to 6.5 mm/month in some positions.

Embankments monitoring. An monitoring campaign was launched to investigate the extension and origin of expansions. The full operation of the railway line posed challenging difficulties for the field investigations.

A monitoring of the surface of embankments by means of topographic marks identified the sustained development of vertical and horizontal surface displacements. The surveillance indicated relevant horizontal movements in the transverse direction (perpendicular to rail tracks). A topographic mark installed 10 m away from the abutment structure of one embankment showed an accumulated horizontal transverse displacement of 150 mm and an accumulated heave of 59 mm during the first 18 months of monitoring. Surface topographic control provided also longitudinal horizontal displacements. However, the values were significantly lower than the displacements measured in the other two directions. Maximum values of 22 mm towards the structure of the abutment developed during the same period along the first 10 m from the position of the abutment.

The distribution of transverse horizontal movement and heave measured at both embankment surfaces followed the pattern indicated in Fig. 11. The maximum displacement in both embankments occurred at a position 10–13 m far from the abutments and no development of significant movements occurs at distances more than 30 m away from the abutments. Topographic monitoring outside the embankments, on the natural ground surface, did not detect any displacement.

Recorded heave rate was not constant in time. In fact, rainfall events seemed to have an influence on the development of heave. The comparison between measured heave and precipitation indicated that heave rates increased immediately after significant rainfall events.

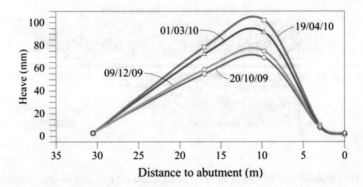

Fig. 11 Distribution of surface heave along the distance from abutment position. Initial topographic measurement: 26 May 2008

Several high precision (±0.003 mm/m) vertical continuous extensometers (sliding micrometers [28]) installed in boreholes allowed investigating the vertical strains of the embankments along depth. The extensometers crossed the whole embankment and reached the natural stratum under the embankments. The instruments were installed at different distances from the abutments. Records showed that vertical expansions develop along the first 8 – 10 m and smaller compressive deformations occur within the deeper part of the embankments over time (Fig. 12). Micrometers

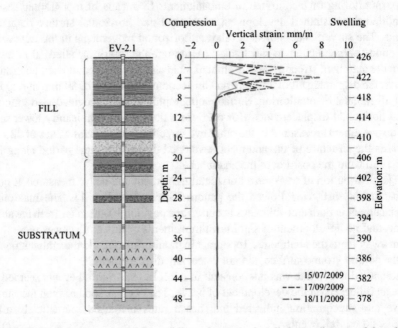

Fig. 12 Vertical strains measured in a sliding micrometer installed at a distance of 8 m from abutment. Initial recording: 20/05/2009

installed 40 m far from abutments measured only a small compression. The pattern of strain records remained invariable over time. The upper expanding layer did not enlarge downwards.

The integral of deformation measured along depth in each micrometer was consistent with the surface vertical displacement in topographic records. This verification is important because it shows that the extensometer length covered the whole "active" expansive zone that originates the heave observed at surface level.

Inclinometer measurements identified the development of horizontal movements in depth along mainly the first 8–10 m of boreholes. Inclinometers recorded a horizontal movement about 9–12 mm in the transverse direction in 2.5 months of monitoring. The horizontal measurements indicated that the embankments were swelling also in a lateral direction.

The combination of the recorded three-dimensional expansion of the embankments and the reduced deformation measured in longitudinal direction suggested that the internal swelling may result in the development of high horizontal loads against the abutments and, therefore, against the bridge structure. In fact, an examination of the structure showed the existence of spalling damage and fissures at the contact between the abutment and bridge structural elements, a displacement of the abutment structure towards the deck of the bridge and relevant damage in the drainage and communications conduits. In addition, forces generated from confined displacements were acting against both sides of the bridge, which agrees with a bending-induced cracking pattern observed at the lower part of the pillars.

The reduced longitudinal strain also pointed out to the risk of occurrence of a passive failure on the upper part of the embankments due to the development of strong passive stresses at the upper level of the embankment, in the longitudinal direction. A finite element analysis simulated well the expansion and heave measured in the embankment and verified the development of a dangerous state of passive stresses at the upper 8–10 m of the embankment. The model estimated a total thrust of 2.32 MN/m against bridge abutment.

Laboratory testing. The material recovered from different depths in the boreholes drilled for the monitoring campaign was investigated in detail. Test results suggest an overall heterogeneous distribution of properties along the depth of the embankments. Identification tests showed that the fine fraction of the compacted fill is a low-plasticity clay and that water content within the embankment does not reach the plastic limit. Heterogeneity was also observed in the wide ranges of the grain size fraction contents found in different samples: fines (7–69.3%), sand (10.5–35.1%); and gravel contents (18.6–62.1%). Some boulders larger than 100 mm were scattered within the compacted material. Figure 13a) shows the aspect of the material at the upper meters of the embankments at the exposed surface excavated during the underpinning works described later. The presence of gravels and boulders immersed in the reddish clay matrix is appreciated in the photo. Also, two columns of jet-grouting treatment and some masses with cement-treated soil can be observed. A detail of a cement-soil mixture recovered from the cut surface is shown in Fig. 13b.

(a)   (b)

Fig. 13 **a** Upper meters of the compacted fill of one of the embankments. Exposed surface excavated during the underpinning works; **b** detail of a cement-soil mixture found at the cut surface

An important outcome of the laboratory tests was the distribution of soluble sulphates in the soil along depth. Contents range from 2.0 to 2.5% in the upper 8 m and drop to values lower than 0.5% at increasing depths. Presumably, the existence of two different source areas for the material used for the construction of the embankment may explain these findings.

The heterogeneity in the compacted fill pose difficulties to investigate in the laboratory the reasons for the expansions observed in the field. Large free swelling tests on compacted samples approximated the embankment conditions. The preparation of representative samples involved two steps: firstly, the homogenization of the material recovered in boring lengths of 1.20 m and, secondly, the compaction to the standard Proctor energy with a water content of 10%. Four representative samples were prepared to test the five upper meters of one embankment. The compacted unloaded samples, 160 mm in height and 120 mm in diameter, remained partially submerged in water and at a constant temperature of 8 °C during the tests.

The samples developed a significant long-term swelling that cannot be explained by a possible mechanism of clay minerals hydration (Fig. 14a). Expansion evolved

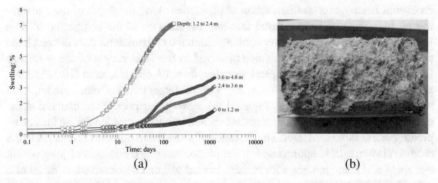

(a)   (b)

Fig. 14 **a** Vertical swelling strains measured in free swelling tests (boring S-2.1B); **b** Sample aspect after test dismantling (1.20–2.40 m deep)

throughout 11 months with no signs of stabilization. Figure 14b shows the aspect of the sample exhibiting larger expansion (built from material recovered at depths ranging 1.20–2.40 m). The mineralogical investigation of the sample by means of scanning electron microscopy (SEM) observations and X-ray diffraction analyses revealed thaumasite $(Ca_6[Si(OH)_6]_2(CO_3)_2(SO_4)_2 \cdot 24H_2O)$ and ettringite $(Ca_6[Al(OH)_6]_2(SO_4)_3 \cdot 26H_2O)$.

Samples of poorly cemented soil–cement mixtures or pure cement grout, recovered from different locations in the embankment and which had a wet-muddy consistency (Fig. 14b), also showed the presence of thaumasite and ettringite in its mineralogical composition. The analysis identified calcite, dolomite, gypsum, illite, kaolinite and quartz in the clay matrix.

Interpretation and final remarks. The mineralogical composition pointed out to the origin of expansions. Sulphate attack in treated soils results in the precipitation of the minerals ettringite and thaumasite and produces the destruction of the strength of the cement paste and a relevant expansion starting from inside the material. Note the high proportion of water molecules in these chemical compositions.

Mohamed [29] describes the development of both minerals in lime-stabilized soils throughout a complex process. The highly basic environment created after the hydration of lime triggers the attack. High pH favors the dissolution of sulphate minerals (i.e., gypsum) and clay minerals. Then, the combination of aluminum released from clays, calcium from cement or lime and sulphates with water molecules results in the precipitation of ettringite. Thaumasite precipitates after the dissolution of calcite in the presence of carbonic acid in the pore water.

Monitoring records identified in a conclusive manner that volumetric expansions extend along the upper 8–10 m of the embankments. The content of sulphates in this region (2.5%) is high enough for the occurrence of sulphate attack. In fact, Puppala et al. [21] identified very low threshold values (0.3%) to trigger the attack. A chemical simulation of the coupled hydraulic and chemical processes taking place at soil–cement interface verified that sulphate attack can develop within the jet-grouting treated volume of the embankments [27]. The distribution of expansions and heave along the embankment axis agrees with the extension and intensity of jet-grouting treatment. The confinement offered on one side by the abutment structure and by the non-treated embankment at the other side also explains the heave profile.

The chemical composition of the embankments and the availability of water from rainfall set up a scenario for future unlimited development of expansions. The risks of an increase of the damage generated to the bridge and a development of a worse passive state of stress motivated the excavation of the upper 6 m of the embankments along the stretch affected by sulphate attack. The remedial measures also included the construction of a new structure, founded on piles on both sides of the embankment, to support the rail tracks.

Lessons learned. Topographic investigations and the measurements of movements and deformations by means of sliding micrometers and inclinometers carried out in the site of Pallaressos embankments suggested that a volumetric swelling affected both embankments. Expansion derived in the generation of a passive state of stress, menacing the rail tracks. Swelling also resulted in high thrusts applied against the structure of both abutments that were damaging the bridge. A massive active sulphate attack agreed with the observed swelling. The treatments containing cement that were implemented at transition wedges joined with the high contents of sulphate at the upper 8 m of the compacted fill set a dangerous scenario for the development of sulphate attack. The availability of the necessary chemical components for sulphate attack was unlimited in the compacted soil (alumina, silicates, sulphates, calcium and water) and suggested that the formation of thaumasite and ettringite and, consequently, strains in embankments would continue if no comforting measures were built.

3.3   Monitoring Infrastructures and Their Foundations

Infrastructures and their foundations at Mestre (Italy). The area around Venice is undergoing remediation and protection works, to address the environmental concerns coming from high water in the Adriatic Sea and pollution from the chemical industrial area sitting on the onshore at Marghera. Besides the mobile barriers in the lagoon, which started functioning in 2020, extensive remediation of the industrial area includes the construction of continuous sheet-piles, which are designed to protect the lagoon from the discharge of polluted water coming from the industrial area.

The design of the sheet-piles has been accompanied by an extensive study on their impact on groundwater hydrology. The sheet-piles will act as a hydraulic barrier, preventing direct groundwater flow into the lagoon. Pumping wells have been designed to capture the groundwater ahead of the old industrial area and compensate for the reduced discharge into the sea. The groundwater hydraulic gradients are locally changed due to these remediation works, and concerns arose about the local water mass balance and pore pressure distribution in the ground, which may affect the intense network of existing infrastructures, including roads, highways and railways.

Besides a hydro-geological study on the entire watershed, which addressed the study at the regional scale, the water mass balance at the local scale of the specific infrastructure was investigated by monitoring the hydraulic response in the upper unsaturated soil and using the monitoring data to calibrate and validate a comprehensive numerical model.

Soil classification. In spite of the extension of the area and local heterogeneities, the topsoil grain size distribution is well defined, and consists of silty and clayey materials, having different percentages of sand. Typical grain size distributions are plotted in Fig. 15b and the relevant geotechnical properties are given in Table 3.

The retention properties were investigated in the laboratory in a pressure plate extractor, which was modified to apply appropriate vertical stresses [30]. The data allowed drawing lower and upper bounds for the retention curves expected in the area (Fig. 15b). The unsaturated hydraulic conductivity was assumed to be represented

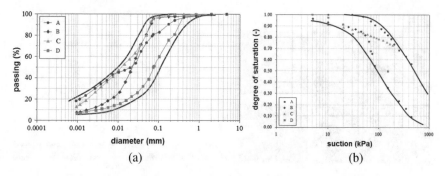

Fig. 15 Topsoil properties: **a** grading size distribution curves; **b** water retention curves

Table 3 Relevant geotechnical properties of the topsoils

Property	Topsoils
Specific gravity, Gs	2.74
Liquid limit, w_L, range	30–50%
Plasticity index, PI, range	10–25%
Average porosity	0.40
Saturated hydraulic conductivity, k	$10^{-6} \div 10^{-8}$ m/s

by the power function $k_r(S_r) = S_r^5$, which was chosen after careful and extensive assessment of the most common predictive functions [31].

Research questions. The laboratory and numerical investigation aimed at answering a number of research questions, including: (i) Quantifying the local net water recharge over one year, as a function of the depth of the water table; (ii) In case of exceptional design flood having 1 m height above the ground surface would affected the area, estimate the time needed for the subsoil and the embankments to get saturated; and (iii) given paradigmatic soil profiles, estimate the depth of the soil affected by water exchanges due to soil-atmosphere interaction. These questions were addressed with numerical models, implemented in the CODE_BRIGHT finite element platform [32]. The numerical model was calibrated and validated with the data from the field described and discussed as follows.

Choice of the instrumentation. The instrumentation had to be chosen in order to be effective in a variety of soils, from rather impervious clayey silts to more pervious sandy silts. Multiple monitoring depths were needed over a depth encompassing the unsaturated topsoil above the water table, located from −1 m to −4 m depth from the ground surface. Moreover, the monitoring period had to last over one year, with limited possibility for direct inspection and control.

Ideally, both water content and suction were of interest. However, given the previous constraints, it was decided to invest on monitoring water content changes, and use the retention curves to infer the changes in suctions, which are needed in the geotechnical analysis of the infrastructures and foundations. Among the available

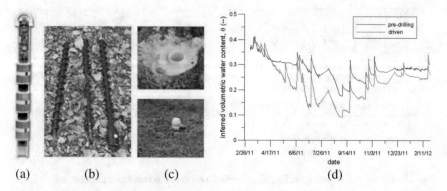

(a)   (b)   (c)   (d)

Fig. 16 Installation of the sensors: **a** sensor slide (from Sentek®); **b** drilling tools; **c** fluid filling and final configuration at the end of installation; **d** Comparison between two twin measurement installed with and without pre-drilling

possibilities, capacitance sensors were chosen, which could be installed at various depth interval on the same vertical section, and are less sensitive than comparable ones to the chemical composition of water, which is highly variable over the investigated area. The capacitance sensors are mounted on a plastic slide (Fig. 16a), inserted in an access tube, which was designed to be installed in the ground by pushing. However, the original design allowed to reach a depth of -1.0 to -1.5 depth only. To increase the installation depth of the sensor, pre-drilling was implemented, using specifically designed drilling tools (Fig. 16b). The pre-drilled hole was filled with a fluid mixture of kaolin and cement, to provide continuity between the sensors and the surrounding ground (Fig. 16c).

Calibration of the sensors. Typically, a reference calibration curve is provided for commercial sensors. One of the advantages of the capacitance sensors is that their calibration curve is rather insensitive to the type of soil, due to the limited contribution of the water chemistry at the input frequencies used in the measurement.

However, it was found that the installation procedures have a relevant influence on the measurement, as shown in Fig. 16d, where the results are reported from two twin sensors installed by driving and pre-drilling at a small distance from each other. The comparison in the figure shows that the kaolin-cement mixture at the interface between the access tube and the soil, which is chosen to keep saturated over a wide range of water content to guarantee optimal contact with the soil, partly shadows the water content changes in the surrounding ground, especially during the dry seasons. The calibration curves can be refined, by simulating the electric field in a numerical model, able to include the disturbance due to the different installation procedures. The numerical model adopted for the calibration refinement is shown in Fig. 17. Figure 18 presents the calculated electric fields in the soil accounting for installation disturbance (Fig. 18b), to be compared to the ideal field (Fig. 18a), aiming to evaluate the measurement dispersion. The analysis allowed to define calibration curves as a function of the installation procedure (Fig. 18c).

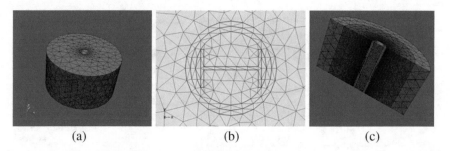

Fig. 17 Numerical model: **a** complete view; **b** top view, showing the instrumentation components; **c** cross section including variable properties to simulate installation effects

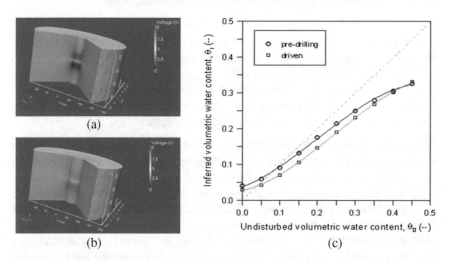

Fig. 18 Correction of the calibration curves to consider disturbance during installation: **a** numerical simulation of the electric field for ideal conditions; **b** numerical simulation of the electric field with disturbance from installation; **c** resulting calibration curves

Results. A number of vertical sections were investigated in the Mestre area. Among them, the data from a highway embankment and from the surrounding foundation soil are presented and compared in the following. Whenever possible, embankments are constructed with the soils locally available nearby, to reduce the construction and environmental costs. Thanks to the composition of the foundation soil, a good clayey and sandy silt, this was the construction choice for the tested embankment, which is part of the highway bypassing the area of Mestre in the E-W direction. Figure 19 presents selected results over two vertical sections nearby the embankment, showing the response of the natural soil profile to the climatic history. The soil profile presents a 50 cm cover, rich in organic matter, which clearly acts as a buffer over time for the deeper layers. The lower clayey silt has a hydraulic conductivity of about 10^{-7} m/s, which is low enough to avoid any significant change in the water content over time,

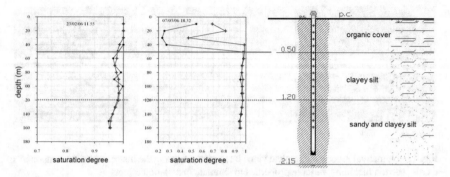

Fig. 19 Variation of the degree of saturation over time in the foundation soil

given the shallow position of the groundwater table. However, the behaviour of the same clayey silt used in the construction of the embankment is clearly different due to two concomitant effects. On the one hand, digging and compaction operations change the texture of the material, which results in an increased hydraulic conductivity. On the other hand, the absence of the protecting organic cover reduces the buffering effect and allows water exchanges to affect higher depths.

Final comments. the calibration of the numerical model on laboratory data showed to be reliable to simulate the response of the natural soil, hence of the foundation. When the same soil is used in compacted earth constructions, a wider pore size distribution is expected due to compaction of lumps, which typically cannot be reproduced in a standard laboratory investigations. Calibration of numerical models takes great advantage from the back-analysis of field data, as shown, e.g., in [33]. Finally, is worth mentioning that defects in sensors installation, like air pockets, would affect the measurement dramatically, as investigated and extensively discussed in a dedicated study [34]. Therefore, they must be avoided to get reliable data.

4 Conclusions

Monitoring soil suction, or the evolution of water content in earth structures, is necessary to predict deformations caused by wetting and drying cycles such as those from soil-atmosphere interaction, or by changes on water table levels at their foundations. This has huge potential to be used on predicting the impact of climate changes on the performance of existing geotechnical structures and is already a concern for all entities connected to infrastructures design, operation and maintenance. Novel working opportunities are expected for Geotechnical Engineers to deal with aspects related with unsaturated soils behavior.

As when traditional instruments are installed, the choice of the proper sensors to monitor water content or soil suction must consider their compatibility with the type of soil, the expected range of operation, accuracy and sensitiveness, response speed, calibration needs, installation and technical knowledge for their operation, and their durability and maintenance needs. For such sensors, the response to dynamic climatic events is very important if climatic patterns are taken into consideration.

The three case-studies presented are non-conventional cases where standard instruments were installed to help understanding problematic behaviors, expected or in progress. The lessons learned and some common features to all cases (also common to traditional earth structures), can be highlighted: (i) the choice of standard instruments is tailored for each case considering the expected behavior and its main causes; (ii) the displacements are the most relevant information to collect because they occur due to deformations, and therefore are liked to stress state by material mechanical constants; (iii) displacements allow identifying ongoing mechanisms and define adequate interventions to stop them; (iv) complementary instrumentation can also be installed to investigate the causes of the deformations, such as soil suction sensors; (v) creative solutions may have to be implemented to install the instruments and sensors, especially when they need to operate in different materials; (vi) disturbance affecting calibration must be considered; (vii) sensors need calibration, usually done through experimental tests in laboratory; (viii) laboratory tests can also be performed to characterize the hydraulic and mechanical behaviour of the materials, necessary to calibrate numerical models to simulate the behaviour observed; (ix) calibration of numerical models takes great advantage from the back-analysis of field data; (x) the behaviour predicted by the model can also help defining instrument location; (xi) instrumentation can be broken or malfunctioning, so it should be redundant by duplication or installing different instruments for cross information; (xii) monitored data also must be redundant and obtained by different type of instruments for validation.

To conclude, testing and monitoring earth structures requires instrumentation and sensors which are being continuously updated by technological advances and many new will be invented, always seeking for environment-friendly and sustainable solutions. New sensors and many interesting case studies are expected in the future.

References

1. Brown, R., Collins, J.: A screen-caged thermocouple psychrometer and calibration chamber for measurements of plant and soil water potential. Agron. J. **72**(5), 851–854 (1980)
2. Leong, C., Tripathy, S., Rahardjo, H.: Total suction measurement of unsaturated soils with a device using the chilled-mirror dew-point technique. Géotechnique **53**(2), 173–182 (2003)
3. Woodburn, J., Holden, J., Peter, P.: The transistor psychrometer: a new instrument for measuring soil suction. Unsaturated Soils 91–102 (1993)
4. Bulut, R., Lytton, L., Wray, W.: Soil suction measurements by filter paper. Geotech. Spec. Publ. **115**, 243–261 (2001)
5. Doussan, C., Ruy, S.: Prediction of unsaturated soil hydraulic conductivity with electrical conductivity. Water Resour. Res. **45**(10) (2009)
6. Phene, J., Rawlins, L., Hoffman, G.: Measuring soil matric potential in situ by sensing heat dissipation within a porous body: II. Exp. Results Soil Sci. Soc. Am. J. **35**(2), 225–229 (1971)
7. METER Group: TEROS21 Soil water potential sensor manual (2017)

8. Fredlund, D., Rahardjo, H., Fredlund, M.: Unsaturated Soil Mechanics in Engineering Practice. Wiley, Hoboken (2012)
9. Ridley, A., Burland, J.: A new instrument for the measurement of soil moisture suction. Géotechnique **43**(2), 321–324 (1993)
10. Manheim, F.: A hydraulic squeezer for obtaining interstitial water from consolidated and unconsolidated sediments. U.S. Geol. Surv. Prof. Pap. **550**, 256–261 (1966)
11. Hilf, J.: An Investigation of Pore Water Pressure in Compacted Cohesive Soils, Technical Memo No. 654, US Bureau of Reclamation, Denver, CO (1956)
12. Lourenço, S., Gallipoli, D., Toll, D., Augarde, C., Evans, F., Medero, G.: Calibrations of a high-suction tensiometer. Géotechnique **58**(8), 659–668 (2008)
13. Mendes, J., Gallipoli, D.: Comparison of high capacity tensiometer designs for long-term suction measurements. Phys. Chem. Earth **115**, 102831 (2020)
14. Cardoso, R., Maranha das Neves, E.: Hydro-mechanical characterization of lime-treated and untreated marls used in a motorway embankment. Eng. Geol. **133–134**, 76–84 (2012)
15. Cardoso, R. Maranha das Neves, E., Alonso, E.E.: Experimental behaviour of compacted marls, Géotechnique **62**(11), 999-1012 (2012)
16. ECH2O: Eco-Sensors ECH2O—soil Moisture Measurement. Decagon Devices, Inc., Pullman, USA (2004)
17. Interfels: Boart Longyear Interfels Catalogue- Geotechnical Instrumentation. Interfels (2004)
18. Cardoso, R., Maranha das Neves, E., Almeida Santos, P.: Suction changes during the construction of an embankment from A10 Motorway in Portugal. In: Proceeding of the International Conference on Unsaturated Soils, Barcelona, Spain (2010)
19. Sherwood, P.T.: Effect of sulphates on cement and lime stabilized soils. Highwaay Res. Bull. **353**, 98–107 (1962)
20. Mitchell, J.K., Dermatas, D.: Clay soil heave caused by lime-sulphate reactions. In: Walker Jr D.D., Hardy, T.B., Hoffman, D.C., Stanley, D.D. (eds.) Innovations and uses for lime, ASTM STP 1135, pp. 41–64. West Conshohocken, PA, USA: ATM Int (1992)
21. Puppala, A.J., Wattanasanticharoen, E., Punthutaecha, K.: Experimental evaluations of stabilisation methods for sulphate-rich expansive soils. Ground Improv. **7**(1), 25–35 (2003)
22. Rajasekaran, G.: Sulphate attack and ettringite formation in the lime and cement stabilized marine clays. Ocean Eng. **32**, 1133–1159 (2005)
23. Alonso, E.E., Ramon, A.: Massive sulfate attack to cement-treated railway embankments. Géotechnique **63**(10), 857–870 (2013)
24. Alonso, E.E., Ramon, A.: Heave of a railway bridge induced by gypsum crystal growth: field observations. Géotechnique **63**(9), 707–719 (2013)
25. Ramon, A., Alonso, E.E.: Heave of a railway bridge: Modelling gypsum crystal growth. Géotechnique **63**(9), 720–732 (2013)
26. Alonso, E.E., Berdugo, I.R., Ramon, A.: Extreme expansive phenomena in anhydritic-gypsiferous claystone: The case of Lilla tunnel. Géotechnique **63**(7), 584–612 (2013)
27. Ramon, A., Alonso, E., Olivella, S.: Hydro-chemo-mechanical modelling of tunnels in sulfated rocks. Géotechnique **67**(11), 968–982 (2017)
28. Kovári, K., Amstad, C.: A new method of measuring deformations in diaphragm walls and piles. Géotechnique **32**(4), 402–406 (1982)
29. Mohamed, A.M.O.: The role of clay minerals in marly soils on its stability. Eng. Geol. **57**(3–4), 193–203 (2000)
30. Caruso, M., Jommi, C.: Enhancement of a commercial pressure plate apparatus for soil water retention curves. In Mancuso, C., Jommi, C., D'Onza, F. (eds.) Unsaturated Soils: Research and Applications, pp. 63–70. Springer, Berlin (2012)
31. Caruso, M., Jommi, C.: An evaluation of indirect methods for the estimation of hydraulic properties of unsaturated soils. In: Proceedings of International Conference on Problematic Soils, pp. 25–27. Eastern Mediterranean University, Famagusta, N. Cyprus (2005)
32. Olivella, S. Gens, A., Carrera, J., Alonso E.E.: Numerical formulation for a simulator (CODE_BRIGHT) for the coupled analysis of saline media. Eng. Comput. **13**(7), 87–112 (1996)

33. Chao, C.-Y., Bakker, M., Jommi C.: Calibration of a simple 1D model for the hydraulic response of regional dykes in the Netherlands. In: Proceeding of the E-UNSAT 2020, E3S Web of Conferences, vol. 195, 01012, (2020)

34. Caruso, M., Avanzi, F., Jommi, C.: Influence of installation procedures on the response of capacitance water content sensors, In: Hicks, M.A., Lloret-Cabot, M., Karstunen, M., Dijkstra, J. (eds.) Proceeding of the International Conferences on Installation effects in geotechnical engineering, ICIEGE 2013, pp. 92–98. CRC Press, Leiden (2013)

Offshore Wind Foundation Monitoring and Inspection

Jaime A. Santos⊙, Luís Berenguer Todo-Bom⊙, and Simon Siedler⊙

Abstract Deep foundations in offshore wind are challenging in terms of design, construction, and in-service inspections. This type of foundation demands new and updated methods for testing and monitoring. Today's energy transition has changed our view on deep foundations. Wind energy offers many advantages, which explains why it is one of the fastest growing renewable energy sources in the world. The drive to lower the cost has resulted in upscaling the turbine size fivefold in the last 2 decades. This increase in size, in combination with increasing water and installation depths requires special foundations bringing new engineering challenges. The design of pile elements has improved significantly in the last few years and is gradually replacing the legacy oil and gas standard methods. As monopile diameters increase to ranges beyond the databases of the common design standards of the American Petroleum Institute (API) and Det Norske Veritas (DNV), designers increasingly resort to advanced design methodologies to improve the accuracy of their models and reduce conservativism in the industry. Improvements on both the axial and lateral geotechnical models of the foundation structures are being driven by a combination of improved calculation capabilities, increased laboratory and in situ geotechnical investigations and in situ monitoring campaigns of the constructed structures during their design life. This chapter aims to present the trends and advances on monitoring and inspection of offshore wind structures with focus on the design considerations for monopile foundations.

Keywords Monopile · Offshore foundation · Wind energy · Monitoring · Pile testing

J. A. Santos (✉)
CERIS, Department of Civil Engineering, Architecture and Georesources, Instituto Superior Técnico, Universidade de Lisboa, Lisboa, Portugal
e-mail: jaime.santos@tecnico.ulisboa.pt

L. B. Todo-Bom · S. Siedler
Ramboll, Hamburg, Germany
e-mail: luis.todobom@ramboll.com

S. Siedler
e-mail: simon.siedler@ramboll.com

© The Author(s), under exclusive license to Springer Nature Switzerland AG 2023   113
C. Chastre et al. (eds.), *Advances on Testing and Experimentation in Civil Engineering*, Springer Tracts in Civil Engineering,
https://doi.org/10.1007/978-3-031-05875-2_5

1 Introduction

Offshore Wind is a key contributor to Europe's renewable energy charter and added an installed capacity of 14.7 GW in 2020 to reach a total installed 220 GW of wind capacity [1]. The Wind Turbine Generators (WTG) are mostly installed on monopile foundations, but jackets have increased their importance in recent years as wind farm developments moved into deeper waters. In recent years, offshore wind energy is starting to get exploited globally. The huge potential of wind energy has been identified particularly by countries in Asia such as China, Taiwan and Japan and these countries are strongly encouraging and promoting offshore wind developments. Jacket type foundations are playing a central role in Asia as well, not only due to deeper waters but also due to seismic activity in those regions. The requirement to increase renewable energy production has also resulted in much greater WTG structures. The amount of energy that can be harvested from wind depends on the size of the wind turbine and the length of its blades. As such, the total energy output is proportional to the dimensions of the rotor and to the cube of the wind speed, up to the wind speed of the WTG.

These developments have a significant impact on the design-life of offshore foundation structures for which limited data is available to inform the design methodologies currently in use. The result from a data gap analysis of this issue should inevitably lead to the incorporation of monitoring and inspection procedures for offshore foundation structures installed in novel sites and of untested dimensions.

In Europe, inspection and maintenance of subsea items are carried out on a scheduled basis: 5-yearly on each structure. The activites during so-called in-service inspections, consist of three levels of inspections, 'general visual inspection', 'close visual inspection' and 'non-destructive testing (NDT)'. Close visual inspection (Fig. 1) and non destructive testing is typically not required as the structures have been designed with the premise/assumption to be "inspection free".

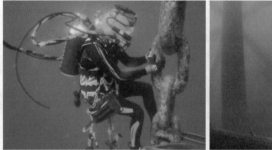

Fig. 1 Visual inspection with left: diver [2] and right: remotely operated vehicle (ROV) [3]

2 Overview

Monitoring is defined as an automated inspection and thus being a subset of inspections. A monitoring system collects and stores data automatically and continuously throughout the entire lifecycle of a WTG, from start of operation to decommissioning. Sampling rates of the sensors can be increased if a predefined threshold value is reached to increase the data quality. It continuously measures conditions without the need of offshore human operation.

Inspection, on the other hand, is when human action (offshore or onshore) is required and executed to obtain condition data from site.

The current requirements in the technical standards, e.g. DNVGL-ST-0126 [4], to carry out recurrent inspections during the operating phase serves to ensure the structural and technical safety of the structure but can also serve to improve the accuracy of future design methodologies.

2.1 Monitoring

Monitoring of the structural integrity of the foundation structure allows to derive conclusions from the measurements for the benefits of improved operation. The need is driven by business objectives and as such varies greatly depending on the operator and the wind farm site.

Monitoring is the continuous surveillance of a structure regarding structural loading, structural health and the determination of the resulting service lifetime. Monitoring is also mentioned in the rules and regulations as a procedure within the framework of in-service inspections, if at all (e.g. DNVGL-ST-0126 [4], DNV GL: RP-C210 [5], ISO 19902 [6]).

The main advantages of monitoring [7] can be summarized in the following points:

(a) Monitoring procedures can partially replace inspections (condition-based survey) and thus lead to direct cost savings.
(b) Documenting the actual load on the structure during its lifetime, allows possible reserves of use to be clearly identified at the end of the planned lifetime. This allows, under certain circumstances, the extension of the service life. This has a significantly higher savings potential compared to the potential through avoided inspections.
(c) Monitoring procedures can be used to detect damages and anomalies as they occur, so that countermeasures can be initiated quickly.

Once monitoring is in place, it enables informed decisions on mitigation and recovery actions. As such, providing faster damage assessment, which minimizes downtime during assessment and potentially over/under maintenance through simulation supported damage consequence analysis.

There are currently no standard products or approaches in the field of offshore monitoring nor can any relevant data be listed in the sense of a specification sheet.

Preparation for lifetime extension of the project or lifetime achievability after unexpected damage is typically considered in Europe and addressed with dedicated monitoring systems.

2.2 Economics of Monitoring and Inspections

The costs of implementing a monitoring program of offshore foundation structures are high when compared to onshore engineering constructions, whilst still remaining small in overall construction costs, and cannot easily be estimated in advance. This is mainly due to the rough environment. Waves, ice drift (e.g. in the Baltic Sea) and the corrosive properties of seawater can destroy both the measuring sensors and the cabling after a short time. To prevent this, extensive constructional measures are usually necessary to avoid high repair costs. The fact that these costs are not offset by any benefits in the medium term and that it is unclear what savings could be achieved in the long term through consistent monitoring, leads to a generally low willingness to invest in this area.

The advantages of monitoring listed in 2.1 should be kept in mind when the question arises to what extent the measurement results could influence cost-relevant decisions and whether this justifies the investment in such a system. The costs of monitoring are not matched by short-term benefits, but in the long term, the benefits are expected to be extremely high. To improve the long-term benefits of monitoring, the joint implementation of research projects is recommended, because there may be great potential for cost savings through development of new methodologies.

The cost effectiveness of various monitoring technologies can be established in a systematic benchmark. The starting point is a criticality analysis from where project specific needs can be derived consistently. Then based on technical readiness of the technology and cost effectiveness against other methods like inspection can be expressed even quantitatively (Fig. 2).

Finally, on the topic of inspections, the costs depend on too many parameters to be quantifiable in general terms and are subject to too much volatility, in particular due to current charter rates and the weather. Rather, comparable offers must be obtained in individual cases.

3 Virtual Sensing for Support Structures

Virtual sensing is understood as using one or more sensors in combination with system knowledge in the form of physical equations or numerical finite element (FE) models. For example, when monitoring the remaining useful lifetime of a specific hot-spot based on vibration measurements, where the hot-spot of interest is not equipped with sensors or interpreting changes in modal properties (derived from accelerations) in the form of scour, material degradation or marine growth.

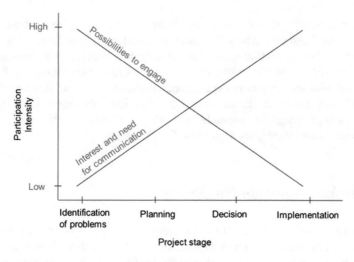

Fig. 2 Participation paradox

A prominent example of virtual sensing is the technology of ambient vibration monitoring: distributed accelerometers are applied along the structure excited by environmental loads. The vibrations measured over time are used to extract modal properties, e.g. mode shape, natural frequencies and damping on a global scale. From these modal parameters, real time stresses can be obtained using modal expansion and decomposition algorithms, and FE models as established during detailed design can be improved for residual fatigue life estimation and damage detection.

Virtual sensing has the advantage of relying on a few sensors only. An optimal sensor placement study, executed for offshore wind monopiles, revealed that accelerometers placed at two heights above water level suffice for various purposes. For offshore wind structures, virtual sensing can be established with sensors that are robust for the entire life cycle and mounted at highly accessible locations for maintenance purposes. Whenever virtual sensing can replace direct sensing, it should be the primary choice as it is a cost-effective alternative.

Typical challenges of virtual sensing lie in the strategic selection of sensor locations, and the proof of damage detection capability as normal and damaged state might produce similar monitoring results. Drawing conclusions on the monitoring data typically requires additional higher-level analytics. Thus, the setup or virtual sensing can sometimes require specifically skilled personnel with combined domain and technology knowledge, particularly when virtual sensing is to be used in a framework of digital twinning. Digital twins, once established, make use of simulation technology that is seamlessly integrated into Operation & Maintenance decision making to provide information throughout the entire life cycle, e.g. supporting operation and service with direct linkage to operation data. One important aim of feeding this data back to the structural designers is to improve the accuracy of the models used during the asset operation [8].

On the downside, virtual sensing feasibility should always be established on a conceptual level and should be validated by dedicated short term direct measurement campaigns. This requires testing in a simulation environment to identify which type of damages are detectable via modal parameter changes and which damages will remain unrevealed. Moreover, measurement uncertainties should be addressed by validating the predictions based on virtual sensing through direct sensing in the scope of a measurement campaign. For example, measuring local stresses with strain gauges to validate the stress estimations derived from virtual sensing.

3.1   Ambient Vibration Monitoring

In offshore conditions, the swell causes structural stimuli. This enables the use of the Natural Frequency Response Monitoring (NFRM) method. It is based on the measurement of the oscillations or the oscillation (decay) behaviour due to the excitation by the sea state at a single measuring point. For this purpose, the measured vibration is divided into two parts, which are orthogonally positioned to each other.

The smallest frequencies belonging to these parts are continuously recorded. If a component fails in the structure, at least one of the frequencies will decrease. In this case an alarm is given. The method has been available for a long time but is not yet widely used offshore.

With large rigid platforms, high natural frequencies occur, i.e. the oscillation periods are very short. Furthermore, the influence on the natural frequency is usually correlated with the importance of a component for strength. The method is therefore not suitable for very rigid structures with high built-in redundancy. However, it is suitable for relatively soft structures with low natural frequencies.

(a)   Technology:

- Micro-Electro-Mechanical Systems (MEMS) accelerometers.
- Plastic optical fibre-based accelerometers.
- Velocimeters.

(b)   Capabilities:

- High reliability, mature technology.
- Easy installation, robust sensors.
- Virtual sensing can be used for locations where no sensor is present using modal expansion and decomposition algorithm.
- Stable performance.

(c)   Limitations

- Experienced data analysts required to extract information for damage detection.
- Assessment of environmental and operational conditions must be fed into damage detection algorithms.

- Data must be recorded at 20 Hz at least which could cause bandwidth problems when data is transferred to on shore control centre.
- Typically, sensors are installed above water level only to allow for maintenance, subsea technologies exist.

4 Direct Sensing for Support Structures

Direct sensing is understood as a monitoring approach, where a dedicated sensor is placed at a location of interest to measure a specific physical quantity. Usually, a direct proportionality between physical quantity of interest and sensor output exists that allows to monitor the change of the physical quantity over time. It uses sensors and a simple physical equation to monitor a physical value e.g. a strain gauge application where a lookup table translates resistance to strain or an accelerometer that translates piezoelectric charges to vibration amplitudes.

An example of direct sensing is a strain gauge installed next to a circumferential weld where a fatigue crack is suspected to monitor potential increase in stresses over time. Another example is a scour sensor mounted close to the seabed that monitors via sonar if the seabed level changes over time to reach critical values.

Direct sensing has the advantage to provide direct information on a very specific physical condition, while reducing uncertainty to measurement noise or faults. Direct sensing is the preferred choice for root cause analysis. Analytics of the measurement comprise typically basic signal processing including filtering, statistical evaluation, and analysis of long-term behaviour.

Typical challenges of direct sensing lie in the interpretation of the long-term behaviour where normal and damaged behaviour might produce similar monitoring results. Drawing conclusions on the monitoring data typically requires additional higher-level analytics.

On the downside, direct sensing in the context of offshore wind support structures requires numerous distributed sensors to cover all identified critical failure modes to be monitored. Some of these sensors must be installed in unfavourable operating conditions, potentially resulting in damage of the sensor or costly maintenance activities.

5 Measuring Regions and Components

In offshore environments, the measuring point can be permanently surrounded by water or air, or alternately by one and the other in the area of the waves. The splash zone can shift due to tides and swell. The measurements as well as their inspection logistics must be carried out alternately in air, water, or both, depending on the medium in which the measuring point is located at the time of inspection.

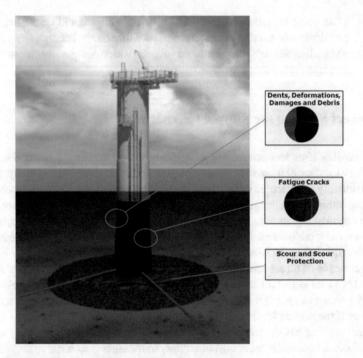

Fig. 3 Periodic inspection of offshore foundation elements

It is suggested that all sensors and cables are mounted onshore, prior to offshore installation and that all cable routings, data acquisition systems and powering of the system can be provided at any time during the phases at affordable costs. Typical inspection areas include fatigue cracks, dents, deformations, scour development as well as scour protection examination (Fig. 3).

6 Fatigue Strains and Cracks

Through the continuous measuring of the strains in the structure, the actual stresses and thus the wear of the structure or its fatigue are available. It should be noted, however, that the data collection is fundamentally more complex than the determination of the force transmission from the environmental data and the subsequent mathematical determination of the stresses. The selection of locations for the placing of strain gauges is a non-trivial exercise since the relevant areas in any structure are too numerous to have covered by instrumentation and be continuously monitored. Detailed knowledge of these locations a priori would also be required directly from the design stage in order to accurately predict the location of crack initiations. Unfortunately, however, such predictive capabilities are beyond the accuracy of current design methods.

Excessive fatigue loading can be detected via a multitude of digital twin approaches, which relies on the virtual sensing data, including FE model updating and modal expansion and decomposition. It is superior to the direct sensing approach, as even inaccessible areas like piles or subsea joints and welds can be fatigue monitored on a continuous basis in an economically feasible manner. It is recommended to validate the virtual sensing approach during a short measurement campaign using strategically placed strain gauges.

6.1   Strain Gauges

The measuring principle of strain gauges is the increase in resistance of an electrical conductor when it is stretched, as a result of elongation and reduction in cross-section. Strain gauges are usually a few square centimetres in size and are applied to the bare metal surface with a special adhesive or through spot welding. They are connected to the measuring electronics by a two-core cable, which should be as short as possible to reduce interference.

Onshore, strain gauges are the standard method in component monitoring. Offshore, waves and currents are a problem, but corrosion is the most serious threat. There are many offshore projects in which the strain gauges or the associated electronics have been completely corroded in a short time. Due to the high significance of the measured data, strain gauges are nevertheless used on many offshore structures despite needing frequent replacement.

6.2   Crack Detection

The alternation of tensile and compressive stresses in the material due to shaft loading can lead to fatigue cracks due to the high number of load cycles. These are mainly found near the welds because the material in this area is weakened due to heating and cooling during the welding process. However, for the removal of the tensile forces occurring in a structure, a conclusive connection between all components is necessary, so that cracks represent a high risk for the foundation structures.

It can be assumed that the welds of offshore foundation structures have been fully inspected during construction (e.g. by radiometry) and therefore no cracks or inclusions are present. Since weld seams only tear from the inside in the case of pre-existing defects, only the surface must be checked for incipient cracks within the scope of the periodic inspections. Since the surface can show crack-like damages, e.g. by flotsam, but also by cleaning work, it must be determined in advance from which dimension (length, depth) a surface defect is considered a crack.

For underwater inspections, the technical inspection regulations of the certification bodies are applied. As a rule, these are very openly formulated so that no procedure

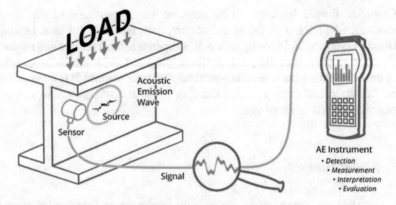

Fig. 4 Scheme of the MISTRAS acoustic emission sensors and systems [9]

exists which is approved, but that each individual case requires coordination between the operator and the certification body.

6.2.1 Time-of-Flight Diffraction (TOFD)

In the Time-of-Flight Diffraction (TOFD) method, the transit time of an ultrasonic pulse through a sample is evaluated. In an error-free section, the receiver would detect only two pulses: the directly transmitted one and the reflection from the back wall. However, if there is interference, the pulse is diffracted at its edges, which then results in additional pulses.

In principle, defects can be characterised very precisely with this procedure (length and depth), but this requires extremely experienced inspectors. TOFD (Fig. 4) has therefore been largely superseded with the market introduction of various eddy current based methods, which are much more user-friendly.

6.2.2 Magnetic Particle Inspection (MPI)

In magnetic particle inspection (MPI), the area to be examined is magnetised and then a suspension of ferromagnetic paint particles is sprayed onto the area. The nature of the pattern of the deposits provides an indication of a possible crack, i.e. false positive results. The process was state of the art until about the year 2000.

6.2.3 Alternating Current Field Measurement (ACFM)

Alternating Current Field Measurement (ACFM) is a further development of the Alternating Current Potential Difference (ACPD), which allows coatings up to approximately 5 mm thick and generally thin marine growth to remain in the area to

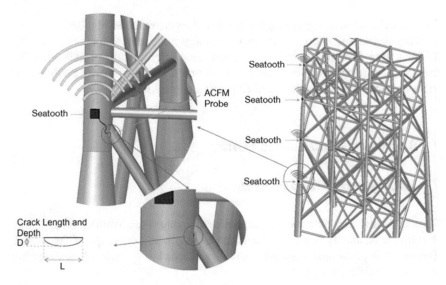

Fig. 5 Seatooth alternating current field measurement, a wireless smart NDT monitoring solution by WFS Technologies and TSC [10]

be tested. Furthermore, ACFM can be used to measure not only the crack length but also the depth, even for small defects (from 5 mm length and 0.5 mm depth). The test is performed by moving a probe over the area to be tested. By using different probes, different seam geometries can be examined (Fig. 5).

6.2.4 Eddy Current Based Processes

With all eddy current based methods, eddy currents are inductively generated in the material to be tested. These generate a magnetic field, which is measured. Defects in the material interfere with the eddy currents, which in turn causes a change in the associated magnetic field, which is detected. The inspection is carried out by passing a probe, in which magnetic coils for generating the eddy currents and sensors are combined, over the bare metal surface.

The devices are usually divided into an underwater and an above-water unit, so that the analysis of the measured values above water can be carried out by an inspector without diving skills and, on the other hand, the diver does not need any inspection knowledge.

6.2.5 Fibre Bragg Grating

A relatively new method for strain measurement, which has already proven its operational maturity in offshore applications and in rotor blades of wind turbines, is glass fibres with burnt-in Bragg gratings. Light impulses are reflected at a Bragg grating, whereby the wavelength is shifted proportionally to the elongation of the grating. Within a fibre, several gratings can be interrogated independently of each other. For example, it is possible to lead a single fibre from the topside over several diagonals to the sea floor and to monitor up to 12 points along the way.

7 Bathymetry and Seabed Scour

7.1 Bathymetry Measurements

The term bathymetry describes the submarine topography (also called hydrography). A bathymetric survey is used to determine the depth and the shape of the seabed. The survey is performed using echo sounders attached to the hulls of measurement vessels. The different types are:

- Multibeam echo sounder: The time from transmission and reception is measured and based on the velocity of sound in water the depth is determined.
- Side scan sonar: In contrast to the multibeam echo sounder, the backscatter amplitudes are evaluated. These can be interpreted together with soil samples regarding the sediment distribution on the seabed.
- Sub-bottom profiler: The sub-bottom profiler is used to gain knowledge about the uppermost meters of the seabed. The acoustic image provides information of the vertical arrangement of the seabed.

Knowledge of the properties of the seabed permits preliminary planning regarding suitable foundations as well as their possible locations, additionally changes of the submarine topography during the operation phase can be detected.

7.2 Scouring

Scour is the erosion of the seabed due to currents around obstacles on the seabed. They usually form a circle or ellipse around a structure, the diameter of which corresponds to approximately 5 times the characteristic diameter of the structure. Depending on the current, the depth increases with time and can grow up to 2.5 times the characteristic diameter.

For a jacket structure, this means that on the one hand a local scour directly at the legs, but on the other hand also a global scour, so that the sea floor near the platform sinks entirely. In the case of neighbouring objects (e.g. mother-daughter platforms), the problem increases accordingly.

Scouring represents an immediate threat to the stability of a platform and must therefore be detected and stopped at an early stage. During recurrent inspections, only large-scale events can be detected, e.g. the movement of a sand wave on the seabed. To detect this, the seabed should be surveyed before installation. Discontinuous monitoring of the scour during inspections does not cover cases in which unsustainable material is deposited again after scouring. This applies to scouring in the immediate vicinity of the foundation structures, e.g. due to storm events. For this reason, monitoring of the scour is recommended.

Only equipment that is explicitly approved for 24/7 offshore application should be used. The following devices, for example, which are permanently attached to the structure under water and use ultrasound to determine the distance to the seabed, are suitable for continuous monitoring. For these devices, marine growth does not pose a problem, since both only measure the transit time of a sound pulse and the thickness of the marine growth is small in relation to the entire measuring distance.

The more the scour depth approaches the design base, the more detailed the monitoring should be. One product that can be considered for this is the Kongsberg Dual Axis Scanning Sonar (DAS). This is a dual axis sonar in a waterproof housing, which allows a resolution of the seabed of $1°$. It is already successfully used for long term monitoring of bridges, but not yet for offshore structures.

Multibeam Echo Sounder. The Multi Beam Echo Sounder (MBES) emits an ultrasonic pulse and registers the time until reflection from objects located within a fan below the sonar. The travel time in one direction gives the distance to the object located there. When the fan-shaped echo sounder is moved through the water at a defined speed, special software can be used to calculate the distance between the objects.

This will enable a three-dimensional model of the seabed and the structures on it to be generated in real time as a point cloud from 2D sections.

Depending on the required resolution and water depth different units are used:

- mounted on the hull (forward, downward and side scanning): if a multi-beam sonar is mounted on a crew vessel, for example, the required data is determined practically "in passing".
- towed by ship: the unit is towed behind the vessel at a certain depth. In this way, disturbances caused by the sea are eliminated. A higher resolution is also possible due to the smaller distance to the seabed.

3D Sonar. The 3D sonar is a further development of the multibeam sonar. In this case not only the echoes from a fan (plane angle) are evaluated, but also in a fixed angle. In this way a 3D representation of the objects in front of the sonar can be generated immediately. With the appropriate software, these models can also be saved and later examined in the form of virtual tours.

The sonars are so compact that they can also be mounted on ROVs. This is used, for example, to work in poor visibility conditions or to monitor dangerous operations such as the installation of scour protection measures.

Light Detection and Ranging (LIDAR). The measuring principle of LIDAR is similar to the echo sounder, except that instead of a sound pulse a laser pulse is emitted, and the travel time of the light is measured in different directions in space.

The resolution of this method is approximately in the order of magnitude of the wavelength of the light, i.e. about 10^{-6} m.

While LIDAR on land (or in air) is already a common measurement method, the method under water is still in the experimental stage. This is not least due to the accuracy of LIDAR, since this method also detects a large number of suspended particles, which interfere with the measurement and must be removed by numerical filters. As of today, these methods are not yet developed to allow productive use in offshore conditions.

8 Determination of Corrosion/Wall Thickness

The material thickness of the components available for load transfer is the central parameter regarding structural integrity. It can be assumed that the surface will corrode and the cross-section available for load transfer will therefore be reduced. This effect is already considered in the design of the structures and compensated by material additions (corrosion surcharges). However, since these are based on empirical assumptions, it must be checked during the service life of a structure whether the material thicknesses assumed in the structural design are present in the installed structure.

In the offshore area, reflection sound methods are preferably used for wall thickness measurements, which differ only in the evaluation of the sound signal. The main application here is the testing of the thickness of pipelines. There are devices available on the market which can be used both above and below water.

The measurement is carried out by emitting an ultrasonic pulse from the measuring head into the material. The time of the echo reflection is measured and the thickness is calculated with the known speed of sound in the material. For this reason, ultrasonic measuring instruments must be adjusted to the material to be measured. The actual measurement takes about 1 s. Corrosion on the inside usually has no effect on the measurement result because the damping in the corrosion layer is so great that any echo is much weaker than that of the intact metal.

9 Excessive Tilting—Monopiles

Permanent tilting or deformation might result from loads occurring during extreme events. It must be noted that this failure mechanism might well occur on a very fast time scale such that preventive action is not feasible. For example, if during a storm event extreme loads are exceeding the level of expected loads, tilting occurs as direct response [11]. Thus, monitoring is to be seen mainly on the side of recovery and mitigation measures.

Excessive loading as mainly stemming from extreme wind and waves can be directly monitored via strain gauges at the base of the tower. It allows to measure the global loads and compare against design assumptions. Due to the good accessibility of the sensors, monitoring is regarded feasible although extra maintenance costs need to be considered for strain gauge repairs.

Unexpected (permanent) movements can be detected via an inclinometer mounted in the transition piece or the tower base. Inclinometers are robust and endurable sensors placed at highly accessible locations. Thus, they are regarded economically and technically feasible.

Excessive ultimate loading can be detected via the digital twin approach using modal expansion and decomposition based on accelerometers and validated with strain gauges at tower base. It is superior to the direct sensing approach, as even inaccessible areas like piles or subsea joints and welds can be load monitored (maximum loads) on a continuous basis in an economically feasible manner. It is recommended to validate the virtual sensing approach during a short measurement campaign using strategically placed strain gauges.

In addition, inclinations can be estimated from 3D accelerometers through tracking gravity. It is recommended to validate this virtual sensing approach through inclinometers.

10 Transportation

Significant amount of transportation fatigue might be accumulated depending on the transportation routes and ships. Using a temporary accelerometer on the monopile, the actual fatigue life consumed during transportation can be monitored. The findings are compared to the estimated transportation fatigue and are used for reassessment of the fatigue life extension potential.

11 Monopiles Design Issues

The monopile is the most widely used offshore wind turbine support structure type, especially in shallow water. It is basically a tubular, thick walled, steel pipe, which is easy to manufacture.

Most current applications consist of a steel pile with a diameter of between 4 and 6 m driven some 10–20 m into the seabed depending on the type of underground [12]. But today's energy transition is pushing the industry to lower the cost of renewable energies, which has resulted in upscaling the size. Monopile with a diameter between 10 and 12 m shall be common in the very near future for water depth up to 50–60 m.

The installation requires heavy duty piling equipment (i.e. hydraulic underwater hammers up to 5000 kJ energy range), and the foundation type is not suitable for locations with large boulders or hard materials in the seabed. If a large boulder is

found during piling, it is possible to drill down to the boulder and blast it with explo-
sives or heavy equipment. The installation effects have a very important influence
on the axial stiffness and bearing capacity of the monopile. In a driven, displacement pile, soil is moved radially as the pile shaft penetrate the ground with benefits
of compaction and increase of lateral stresses. In a bored, non-displacement pile,
lateral stresses in the ground are reduced during excavation and the base resistance
needs larger displacements to be mobilized.

The load-settlement curve consists of three components: load-settlement of the
shaft resistance, pile shortening and load-settlement of the pile base. The ultimate
shaft resistance occurs at very small displacement between the pile shaft and the
soil, usually a few millimetres while the ultimate base resistance only occurs for
much larger displacement depending on the installation method and pile diameter.
To illustrate the basics of pile transfer mechanisms, a simple bi-linear behaviour is
considered to represent both the lateral (R_s) and the base resistance (R_b).

Figure 6 represents a typical load-settlement curve in dimensionless scale for a
non-displacement pile in soil: (R) is the mobilized resistance, (R_u) is the ultimate
resistance, (s) is the pile head displacement and (D) is the pile diameter. The pile
exhibits a soft response after reaching the ultimate shaft resistance and so it is particularly notorious when the contribution of the base resistance is dominant (Fig. 6a).
In a bored non-displacement pile, a softened zone below the pile tip can occur and
large displacement has to occur before the soil resistance can be fully engaged.

Figure 7 represents a typical load-settlement curve for a displacement pile, driven
into the soil. The pile maintains a stiff response for load ratios even near the ultimate
resistance because both lateral and base resistance are already activated during pile
installation.

The displacement of soil during pile installation is therefore a fundamental issue
for the axial loading response of the pile [13]. Moreover, the design shall incorporate
the effects of cyclic loading from wind and waves in a safe and easy way and it is
essential to understand the soil-pile interaction with the focus on accumulation of
displacements under cyclic loading [14].

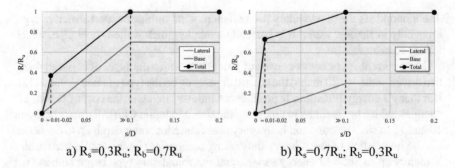

a) $R_s=0{,}3R_u$; $R_b=0{,}7R_u$ b) $R_s=0{,}7R_u$; $R_b=0{,}3R_u$

Fig. 6 Typical load-settlement curve for a non-displacement pile

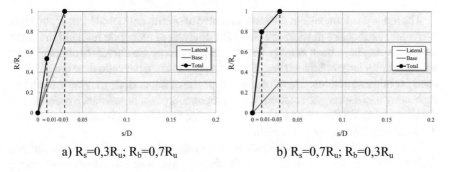

a) $R_s=0,3R_u$; $R_b=0,7R_u$ b) $R_s=0,7R_u$; $R_b=0,3R_u$

Fig. 7 Typical load-settlement curve for a displacement pile

Another issue is related to the excessive damage to the piles with regards to the flange surface as well as fatigue damage caused by excessive piling energy during the installation process. Damages can lead to malfunctioning of the sealing function and reduce the flange fatigue life. Piles driven with an impact hammer are generally subjected to high dynamic loads close to the steel yield stress, which are usually more severe than the stresses in service.

Recording of the hammer energy per blow present the easiest method of monitoring pile driving fatigue. Pile top accelerations recorded with accelerometers or strain measurements with strain gauges at the pile present additional sensors that can be used to assess the pile driving fatigue. All methods are technically feasible, however multiple sensors are needed (at least for each 90° of pile diameter) and along the pile to ensure redundancy as sensors are prone to malfunction if subject to high dynamic loads. This kind of instrumentation is common and the application of stress-wave theory and dynamic methods for designing and controlling pile installation have been used very successfully for many years.

The initial fatigue damage can be captured through a temporary monitoring campaign using accelerometers, strain gauges or recordings of hammer energy. The remaining fatigue life for operation can thus be determined based on measurements.

In addition, it is well known from previous projects and literature, that soil models, as applied in a detailed design, contribute greatly to remaining uncertainties in the fatigue life. Due to the present uncertainties, conservatism is applied in the design to ensure structural safety.

In previous projects, applied measurement campaigns have provided strong arguments for an increased soil stiffness through extracting dynamic properties of the structure, resulting in extended fatigue life. The extraction of modal parameters after pile installation is especially beneficial since the structure is less complex compared to the final setup, which includes tower, nacelle and rotor. Soil model updating is best addressed during this phase, as the contributors of uncertainty can be effectively isolated, i.e. the influence of tower and rotor-nacelle-assembly is not present. The improved models can then be further used in the next step (FE update), where other remaining uncertainties stemming from WTG installation can be addressed additionally.

The monopile foundations are also subjected to significant lateral loads from wind, wave, and earthquake in seismic zones. The analysis of monopile foundations under lateral loading have been heavily based on the p-y response curves. Monopiles have a significantly larger diameter and smaller length to diameter ratio than typical piles used for other structures. Therefore, the response of a monopile may be more like a rigid body translation and rotation, with components of shear resistance mobilized at the base and along the shaft as it rotates, affecting the lateral resistance and therefore the axial response of the foundation.

The current design of monopiles is still based on traditional methods for pile foundations with few adaptations. In the future, improved procedures may enable the use of monopiles in deeper waters, while securing a robust and cost-beneficial foundation design. The foundation represents approximately one third of the total cost of an offshore wind turbine. To develop new design procedures, it is essential to understand the soil-pile interaction under combined cyclic axial and lateral loading. Many challenges are still open for research and industry.

Acknowledgements The invaluable support from the technical staff (Dr. Ursula Smolka and Dr. Jannis Tautz-Weinert) at Ramboll is very gratefully acknowledged.

References

1. Wind Europe. Wind energy in Europe: 2020 Statistics and the outlook for 2021–2025 (2020)
2. https://www.fugro.com/images/default-source/Expertise/our-services/diving_services.jpg. Last Accessed 03 Sept 2019
3. https://www.rsdiving.de/assets/images/8/Falcon-9113a6e8.jpg. Last Accessed 03 Sept 2019
4. DNV GL: DNVGL-ST-0126 Support structures for wind turbines. DNV GL Standard, Oslo, Norway (2018)
5. DNV GL: DNVGL-RP-C210 Probabilistic methods for planning of inspection for fatigue cracks in offshore structures, DNL GL Standard, Oslo, Norway (2015)
6. ISO: ISO19902 Petroleum and natural gas industries—fixed steel offshore structures, ISO Standards (2020)
7. Cranfield University on behalf of EU Project ROMEO: D4.1 Monitoring technology and specification of the support structure monitoring problem for offshore wind farms. Doc. No.: 745625 (2008)
8. Bom, L., Siedler, S., Tautz-Weinert, J.: Validation of CPT-based initial soil stiffness in sand for offshore wind jacket piles (2020). https://doi.org/10.24355/dbbs.084-201912181435-0
9. https://mistrasgroup.co.uk/wp-content/uploads/2017/04/MP4-of-gif.gif. Last Accessed 03 Sept 2019
10. https://www.oceannews.com/images/webnews/2016/october/week10-31-16/4WFS-Crack-detection2_revised.png. Last Accessed 03 Sept 2019
11. Gourvenec, S., Randolph, M: Offshore Geotechnical Engineering, 1st edn. CRC Press, Oxon (2011)
12. Dean, E.T.R.: Offshore geotechnical engineering, 1st edn. Thomas Telford Ltd., London (2010)

13. Santos, J.A., Leal Duarte, R., Viana da Fonseca, A., Costa Esteves, E.: ISC'2 experimental site—prediction and performance of instrumented axially loaded piles. In: 16th International Conference on Soil Mechanics and Geotechnical Engineering, Osaka, Japan, pp. 2171–2174 (2005)
14. D´Aguiar, S.C., Modaressi, A., Santos, J.A., Lopez-Caballero, F.: Piles under cyclic axial loading: study of the friction fatigue and its importance in piles behavior. Can. Geotech. J. **48**(10), 1537–1550 (2011)

Transportation

Tests and Surveillance on Pavement Surface Characteristics

Elisabete Freitas , **Ablenya Barros** , **Cedric Vuye** , **Jorge Sousa** , **Leif Sjögren** , **Luc Goubert, and Véronique Cerezo**

Abstract Pavement surface characteristics have been gaining relevance with the changing mobility paradigm as they greatly contribute to the degree of safety, comfort, environmental quality and long lasting pavements. The surveillance methods of pavement surface characteristics have evolved from manual to semi- and automated methods and also to more cost-effective methods. Automated condition surveys provide objective high-quality data. At network level, pavement distresses, friction, evenness and texture are currently surveilled. More rarely surveyed and related to sustainability demands are tyre-road noise and rolling resistance. They rely on dedicated pavement data collection vehicles or trailers coupled to vehicles, equipped with high-speed digital cameras, laser systems, ultrasonic sensors, accelerometers, microphones, etc. As a complementary alternative to these methods, unmanned aerial vehicles equipped with cameras as well as smartphone-based data

E. Freitas (✉) · J. Sousa
University of Minho, ISISE, Campus Azurém, 4800-058 Guimarães, Portugal
e-mail: efreitas@civil.uminho.pt

J. Sousa
e-mail: id9829@alunos.uminho.pt

A. Barros · C. Vuye
Faculty of Applied Engineering, EMIB Research Group, Groenenborgerlaan 171, 2020 Antwerpen, Belgium
e-mail: ablenya.barros@uantwerpen.be

C. Vuye
e-mail: cedric.vuye@uantwerpen.be

L. Sjögren
Swedish National Road and Transport Research Institute, Linköping, Sweden
e-mail: leif.sjogren@vti.se

L. Goubert
Belgian Road Research Centre, Brussels, Belgium
e-mail: l.goubert@brrc.be

V. Cerezo
AME-EASE, Univ Gustave Eiffel, IFSTTAR, Univ Lyon, 69675 Lyon, France
e-mail: veronique.cerezo@univ-eiffel.fr

© The Author(s), under exclusive license to Springer Nature Switzerland AG 2023
C. Chastre et al. (eds.), *Advances on Testing and Experimentation in Civil Engineering*, Springer Tracts in Civil Engineering,
https://doi.org/10.1007/978-3-031-05875-2_6

135

collection methods are arising. In a design framework, the laboratory methods to survey pavement surface characteristics often use the same technology as in the field tests adapted to a much smaller scale and lower testing speed. In this chapter, the latest advances in laboratory and field tests, and surveillance methods of pavement surface characteristics are described and their performance is analyzed.

Keywords Surveillance methods · Road · Monitoring · Surface characteristics · Functional condition · Pavement management

1 Introduction

A road is a three-dimensional construction with several compacted layers. The road surface, also called the "top layer", "wearing course" or "pavement" is built with a flexible material, such as bituminous mixtures or with a more rigid material, such as cement concrete. Ideally, a new-laid pavement is considered adequately smooth. When it has been in use, the surface can get worn and deformed due to various factors, including the traffic load, weather influences, geological conditions and age. The road surface condition has been gaining relevance. When maintained in a good manner, it greatly contributes to safety, comfort, environmental quality and long-lasting roads. Consequently, the surveillance methods of road surface characteristics followed the same trend. They evolved from manual to semi- and automated methods and became more cost-effective.

At network level, pavement distresses, friction, unevenness and texture are the surface characteristics more often monitored. Although tyre-road noise and rolling resistance are rarely surveyed, these surface characteristics strongly related to environmental issues are essential to meet today's demands on sustainability.

In a design stage, the laboratory methods to survey pavement surface characteristics often use the same technology as field tests, adapted to a much smaller scale and lower testing speed.

In this chapter, the latest advances in laboratory and field tests, and surveillance methods of road surface characteristics, are described and their performances are discussed.

2 Pavement Distresses

2.1 *Field Surveillance Methods Using Dedicated Survey Vehicles*

Pavement imaging, profile and evenness measurements are techniques used to identify pavement distresses. To acquire this data, up-to-date tools and applications used in the industry rely mainly on dedicated survey vehicles equipped with digital cameras, laser systems, and accelerometers [1].

The acquisition of 2D images in the visible spectrum of light with Charge-Coupled Devices (CCDs) and metal–oxide–semiconductor (CMOS) sensor cameras is mature in pavement imaging, although cameras are used in commercial surveillance vehicles commonly to aid other measurement devices. Line-scanning, for example, is a distress detection imaging technique that consists of high-resolution line images combined to form single area images with greater accuracy. The Laser Road Imaging System (LRIS) is a commercially available solution based on this technique, allowing to fully capture transverse road sections up to 4 m wide with 0.5 mm resolution [2].

Laser-based instruments are a well-established technology used at network level to detect distresses that manifest in terms of height differences, as it allows accurate 3D reconstructions of the road surface. Laser triangulation is a 3D measuring technique based on the change in the shape of a laser line projected over the road surface, captured by CCD cameras. This principle is applied in the Laser Crack Measurement System (LCMS), which enables 1 mm resolution automated surveys, being adopted in many commercial tools including the Automated Road Analyzer® (ARAN), ROad Measurement Data Acquisition System® (ROMDAS), Dynatest®, PaveTesting®, Pavision®, among others [3]. 3D laser scanning based on Light Detection And Ranging (LiDAR) technology provides high accuracy by dense point cloud data collection via time-of-flight sensors. This device is used, for example, by Pathrunner® and ARRB Hawkeye® [1, 2]. Dedicated survey vehicles are commonly equipped with accelerometers to measure pavement irregularities, typically expressed as the International Roughness Index (IRI). However, these sensors generally also support the validation of other data collection systems, such as laser profilers.

2.2 Field Surveillance Methods Using Low-Cost and Opportunistic Data

In the last decades, the shift from traditional visual foot-on-ground surveys to automated methods has led to sophisticated and reliable solutions to detect pavement surface distresses, thus supporting road authorities with maintenance and repair decisions. On the other hand, the high costs of these commercial solutions, including the time-consuming data gathering and analysis, imply that the road network is not being frequently and extensively monitored. In this manner, new tools and methods to inspect surface distresses have emerged, employing less expensive devices to reduce the effort of human experts and the need for delicate hardware [4].

Unmanned Aerial Vehicles (UAV) systems are an evolving alternative platform for data acquisition to replace or complement dedicated pavement data collection vehicles for autonomous surveillance. Inzerillo et al. [5] demonstrated the reliability of using UAV image-derived 3D models of the road surface obtained from low-cost image-acquisition devices to identify distressed regions with minimum height errors

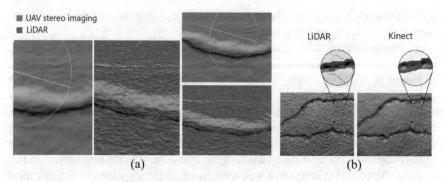

Fig. 1 Side-by-side comparison between LiDAR crack imaging and resulting images obtained from: **a** UAV stereo imaging (adapted from [5]); **b** Microsoft Kinect [7]

of around 1 cm. Li et al. [6] reported an 89% overall accuracy of pothole identification from point cloud data obtained using a low-altitude UAV LiDAR and a supervised learning algorithm.

In terms of accessible sensors able to gather 3D geometric features, Zhang et al. [7] employed a Microsoft Kinect (an infrared-based motion sensor) to identify cracking according to type and level of severity, obtaining true positive rates ranging from 53 to 88% (Fig. 1b). However, further research is needed to increase the limited field-of-view (e.g., using an array of Kinect sensors) and measurement speed.

Both vehicle on-board sensors and built-in smartphone sensors (accelerometers, gyroscopes, GPS, and cameras) enable extensive data collection that can be translated into pavement surface distresses. Using these devices for a large volume of data acquisition can be performed by a crowdsensing-based approach. The developed crowdsensing system serves as a base for Internet of Things (IoT) applications, allowing sensors with networking capacities to communicate within their environment via the Internet [8]. Most smartphone-based research and commercial apps focus mainly on estimating IRI and evenness by using the GPS and accelerometer data, as shown by Peraka and Biligiri [1]. Nevertheless, some works have been reported on the use of IoT sensors to identify surface distresses, especially potholes, such as the early Pothole Patrol [9]. More recently, El-Wakeel et al. [8] further processed the data obtained from crowdsourcing with supervised machine learning algorithms, achieving an accuracy in 86.8% of pothole detection. Van Geem et al. [10] aimed at opportunistic sensing via the on-board sensors of ordinary vehicles, communicated over the Controller Area Network (CAN). The CAN data were analyzed via an algorithm developed to report the possible presence of a road defect. This sensing method proved to allow the collection of a large amount of data that, over a long period, can indicate the development of road surface distresses.

2.3 Data Processing Using Machine and Deep Learning

In terms of data processing, developments in image processing and machine learning have significantly increased the automation in distress detection and quantification procedures. In this context, the value of the information exceeds the costs of data acquisition, and efforts are shifted towards the construction of large databases rather than precise and high-quality but limited information obtained from traditional surveillance technologies.

Many studies reported in the literature have focused on the development of vision-based automated pavement distress detection by using machine learning (ML) to process 2D images [7]. When using a traditional ML approach, a feature extraction stage is first conducted to reduce the complex and unwanted information (e.g. illumination differences, noise, objects) within these road images, thus increasing algorithm accuracy.

Deep learning, specifically deep convolutional neural networks (CNNs), is a subset of the ML techniques that applies multilayer neural networks to achieve robust learning through training. CNNs have become a popular technique for developing enhanced pavement surface distress detection algorithms, resulting in an increasing number of published articles that report a combination of low-cost sensors and promising precision levels [11].

Some examples of recent studies employing high-precision CNNs include the construction of large-scale road damage data sets to train the networks, obtained via simple methods and devices, such as a smartphone installed on a car [12], Google Street-View images [13], and black-box cameras [14].

Much effort has been put into crack and potholes detection, yet a lack of research on pavement distress detection exists related to microtexture analysis, such as raveling [3]. Some of the literature that assesses this distress include the collection of 1000 image samples to train and verify a machine learning approach to detect raveling, leading to a classification accuracy rate of 88% [15]. Chatterjee et al. [16] used data obtained from LiDAR as the ground truth labels to train a CNN for pavement condition estimation using the vibration response of a smartphone. The distresses were sorted into five categories: rutting, raveling, cracking, potholes, and IRI. While rutting and IRI showed promising results for practical use, the performance on cracking, raveling, and potholes was considered unsatisfactory.

2.4 Outlook and Conclusions

Advances in the field of computer vision, artificial intelligence (AI), low-cost sensors, and IoT offer new possibilities to provide information to road agencies and cities about their complete road network. However, most cost-effective data collection methods are limited to detecting specific distress types, such as potholes and cracks, and evenness, e.g. IRI estimation. Table 1 contains a summary of the most widely-

Table 1 The accuracy level of techniques and devices employed to collect pavement distress data regarding distress type (based on Coenen and Golroo [3])

	Cracking	Potholes	Rutting	Raveling
Line-scan imaging	+++	+++		
Stereovision imaging	+	+++	+	+
3D laser profiling sensors	+++	+++	+++	++
LiDAR	+++	+++	+++	++
Kinect	+	++	+	
Smartphone and IoT-based imaging sensors	++	++		+
Smartphone and IoT-based vibration sensors		++	+	
CAN bus data		++	+	

Notes +++ = very good,++ = satisfactory,+ = moderate

used techniques described in this section and how accurately the different types of pavement distress can be detected. This table is partly based on Coenen and Golroo [3] and completed based on more recent research studies. The low number of studies and reported results related to raveling makes it more challenging to provide a conclusive summary of this type of distress.

3 Evenness on Roads

For many years, the degree of evenness of the pavement has been used as an indicator of the wear and deformation for maintenance strategies and performance control of contractors. Evenness can be defined as unwanted irregularities in the pavement and it exists in all directions and areas of the road pavement. To simplify quantifying the degree of irregularities, indicators have been developed separately for the transverse and longitudinal directions of the road. Of course, this simplification is also based on the need to separate relevant information, such as deformation from heavy traffic and wear from cars, expressed as rut depth calculated from the profiles in the transverse direction and unevenness related to bumpiness and comfortable ride in the longitudinal direction. Different levels of e.g. wavelengths of evenness influence several functional characteristics of roads, such as safety, environmental effects such as internal and external noise, particulates and other emission, ride quality, vehicle operating costs and road construction durability [17]. This really emphasizes the importance of monitoring evenness!

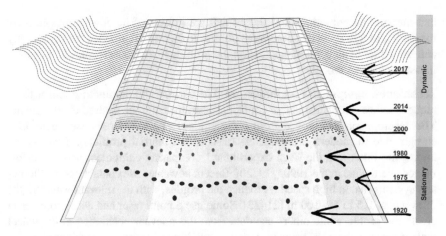

Fig. 2 Historical development of the coverage and degree of detail when measuring road profiles (figure courtesy of Leif Sjögren, VTI)

3.1 Measuring Evenness

Methods and indicators have continuously been improved to measure the road surface quality by means of evenness. The measuring methods can, depending on purpose, be divided into stationary and dynamic methods (Fig. 2). Stationary methods are used at surfaces not suitable for the dynamic methods, such as parking lots and closed road work areas while dynamic methods are a must when monitoring road networks.

The main differences with the two principles are that the dynamic method implies measuring at traffic speed while stationary is standing still. In Fig. 2 one can see the development from stationary measurements using only one point to indicate unevenness to today's possibility to calculate many indicators from an almost true 3D description of the actual road surface and the surrounding of the road.

Originally, simple physical rulers, straightedge methods were used. This is a stationary method that only gives data about a single irregularity from a specific section. The traditional straightedge has still its practical use in many cases [18]. The straightedge indicator that is defined as the vertical distance from the bottom of the ruler to the surface (more specifically the maximum distance of all possible distances) is the unevenness indicator. Nowadays, there are new ways that do the calculation digitally, based on profiles from contactless laser scanners. The straightedge method can be used in any direction of the road.

3.2 Longitudinal Evenness

Dynamic monitoring and assessment of the longitudinal unevenness of roads use two principles. One is detecting the response of the unevenness with accelerometers that measure the vibrations and shocks that are induced. The other principle is by measuring longitudinal and transverse profiles to digitally reproduce the topography and use effect models/filters applied on the profiles to assess outcomes of a desired measure or effect model.

Techniques, equipment and indicators. The current and dominating sensor technique is to use laser sensors that contactless can measure the desired characteristic. The development has moved from static equipment as the previously mentioned straightedge to modern profilometers (also called inertial profilers). Two experiments, the Filter and the Even experiment,[1] comparing equipment and its performance, were done in the 1990s [19, 20]. Recent advances use contactless profilometers systems that in high speed reproduce the profiles, with the relevant wavelengths of evenness 0.5 to 50–100 m [21–23]. Some use a point laser and accelerometer to create the profile while others use data from the relevant position, e.g. the wheel path of the transverse scanning laser data in combination with an accelerometer. The accelerometer is used to compensate for the vehicle body movement to make the profile independent of the measurement speed.

In Europe, three types of indicators are used, IRI, waveband analysis and the Weighted Longitudinal Profile (WLP). IRI is a standardised filter that simulates a quarter of a car traveling at 80 km/h on the measured longitudinal profile [24]. The car body's vertical movement and the corresponding wheel movement are summarised per travelled distance and presented as mm/m [25]. Waveband analysis is pure signal frequency processing of the measured profile dividing it into three bands: short, medium and long wavelength unevenness. The limits defining the bands varies with what country it is used in. WLP is also a frequency process giving two values, one for general and another for local unevenness [24].

3.3   Transversal Evenness

Response type methods are not useful in the case of transverse unevenness where the focus is more on the degree of durability. For transverse unevenness, the focus is on wear and deformations caused by loads from the traffic, expressed as rut depth [26].

Techniques, equipment and indicators. Profilometers able to measure the transverse profile have evolved from using several point measuring lasers to one or two scanning lasers covering almost 4 m road width with a continuous profile of high resolution (almost below a millimeter longitudinally and transversely). Currently, the LiDAR technique is used to digitally reproduce the road surroundings [21, 23], but is more and more used to measure profiles of the road surface as well. It is expected that soon this technique will be reliable and have enough resolution to replace traditional profilometer techniques.

---

[1] Arranged by FEHRL and PIARC in 1999.

The main indicator is rut depth but there are a lot of variants of possible definitions, e.g. ridge height and edge drop [26]. With the availability of continuous profiles, indicators such as profile shape, water depth and deformation areas are possible to calculate.

3.4 Recent Developments, Connected Cars Big Data and Sensor Fusion

With the increasing availability of connected cars, a new possibility has arisen to complement traditional monitoring. Modern cars are equipped with several sensors, among them accelerometers. This can be used as a complement to the less frequent monitoring done by profilometers today. A day to day information about acute road surface problems will be improved and can optimize the operational maintenance activities and, for example, direct them to prioritised places.

Connected car data can also be used to monitor and assess the perceived ride quality. The disadvantages are that no standard or specification exists. There are many depending variables to take into account such as car- and tyre-types and condition, speed of the car, the skill of the driver, the position of the car on the road, both transverse and section-wise. It is important to remember that cars in traffic will only give data on the smoothness of the road. Data is needed from many cars to be reliable. It requires major efforts with data management and knowledge how to deal with or harmonise the use of several cars and car types, how to manage the data gathering (crowdsourcing), business matters with car manufacturers (data owners), and the integrity issues to make this possible to implement in production.

Despite the challenges mentioned previously this development with digital, time and position marked data (Big Data) opens for new smart indicators, more directed to give objective answers to functionality rather than technical condition. The goal in sensor fusion [27] is to utilize information from spatially separated sensor networks and sensors of different kinds with the combination of ground level sources as soil and climate conditions achieved via geographical information systems (GIS). In the long run this may reduce the importance and need of traditional evenness measurement and data.

4 Texture

Road surface texture is defined as the deviations from a planar and smooth surface. These irregularities are due to the mix composition (size of aggregates, binder and fines content) and the laying process.

In the road field, three main scales are considered: megatexture, macrotexture and microtexture [28]. Megatexture corresponds to surface irregularities with wavelengths of a pavement profile lying between 50 and 500 mm and vertical amplitude ranging from 0,1 to 50 mm. Wavelengths of megatexture are in the same order of size as tyre/road interface. Megatexture is mostly related to longitudinal unevenness (see Sect. 3) and low frequency noise generated inside and outside the vehicle (see Sect. 6). Macrotexture is defined as surface irregularities whose dimensions range between 0.1 mm and 20 mm vertically, and between 0.5 mm and 50 mm horizontally, whereas microtexture is defined as surface irregularities whose dimensions range between 0.001 mm and 0.5 mm vertically, and below 0.5 mm horizontally. Macrotexture plays a great role in water surface drainage. Once the majority of the water is evacuated from the tyre/road contact area, a thin water film remains (few micrometers). This residual water film is perforated by the summit of the asperities forming the microtexture, allowing direct contact between aggregates and the rubber tyre. Several methods and indicators are available to assess microtexture and macrotexture in the laboratory and on the field. The most popular are described below.

4.1 Field Surveillance Methods

Measuring methods and devices for macrotexture. Macrotexture can be assessed directly by volumetric and profilometric methods.

The volumetric methods consist of evaluating the average height of a circular patch of diameter D obtained thanks to a given volume V of standardized glass beads. The volumetric method provides an indicator called Mean Texture Depth (MTD), ranging from 0.2 mm (which is the size of the glass beads) to more than 4 mm. Despite the fact that this assessment is punctual and highly operator dependent, it remains a reference for pavement acceptance [29].

The profilometric method consists of recording 2D profiles along a measuring line, obtained with contact-less laser-based method. The most popular indicator in the road field is Mean Profile Depth (MPD) [28]. This standard was recently revised and the new version introduced changes in the filtering process, the peak removal procedure for outliers and data sampling. It generates slight changes in the MPD values but the classification of the pavement performances remains unchanged.

Moreover, MTD and MPD exhibit a linear correlation (Fig. 3), which is not perfect but satisfying considering the fact that this means to relate an indicator based on volume and one based on two-dimensional profiles [30, 31].

Two main families of devices exist for macrotexture measurements: circular and linear profilometers. Circular profilometers (Fig. 4.) like the Circular Texture Meter®® (CTM) or Elatexture® provide a static, circular and punctual laser scan of the surface [32, 33]. Linear profilometers measure macrotexture continuously on an itinerary [34]. They can be mounted on vehicles driving in the traffic flow (RUGO®,

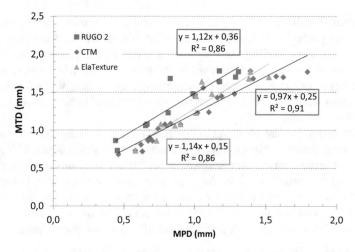

Fig. 3 Linear correlation between MPD and MTD measured on Université Gustave Eiffel test tracks with several profilometers

Fig. 4 Circular profilometers: **a** Circular texture meter and **b** elatexture. Photos by EASE laboratory, University Gustave Eiffel

Dynatest®) or pushed by hand (Protex®). All these devices are easy to implement but the measurements are sensitive to glossy coating and to wetness on the pavement.

More recently, progress is made to obtain 3D cartographies of the pavement surface. One way is to use a beam with a laser-based measurement head mounted on it. The profilometer makes 2D profile measurement along the beam and it is progressively moved perpendicularly to the main direction of the beam. The profiles are then combined to obtain a 3D representation of the surface. This characterization is static, punctual and takes several hours for sections of 1 to 5 m long [33, 35–38].

Lastly, 3D images can also be obtained with optical stationary measuring devices. They capture the surface three-dimensionally by the combination of images obtained with a high-resolution camera and various lighting conditions [33, 39, 40].

Other texture indicators. It is possible to use other texture indicators to correlate with surface properties [33, 41, 42]. Parameters like skewness of the profile (rsk), kurtosis, maximum profile peak height (Rp), mean profile peak height (Rpm) or Abbott "bearing curve" are widely used in the industrial domain [43, 44]. Despite the fact that they offer additional relevant information to understand physical phenomena in the contact area, these indicators are not extensively used in the road sector except for research purposes.

Lastly, the profiles can be analyzed by digital filtering technique in order to determine the magnitude of its spectral components at different wavelengths. This spectral analysis (or texture wavelength analysis) is performed with constant narrow bandwidth frequency analysis. Thus, the texture spectrum of the profile is calculated with a texture level, given in dB, for each path of measurement. The most common bandpass filters are one-third-octave bands. Recent studies demonstrated that texture spectral levels are well correlated with the rolling resistance coefficient [33].

Microtexture estimation. In operating conditions, microtexture is assessed indirectly by means of a friction coefficient measured on a wet surface at low speed (see Sect. 5). Several researchers aimed at measuring microtexture directly with laser profiles or 3D images [39, 40] recorded at very high resolution but they are not available yet in a commercial way.

4.2 Laboratory Tests

Macrotexture measuring methods and devices. All the static measuring methods developed for in-field measurements can be used in the laboratory (Sect. 4.2). Additionally, specific laboratory devices based on digital imagery, like computed tomography (CT) scans [30], topographical maps measured by a focus-variation sensor Infinite Focus from Bruker Alicona with various lens [45], Laser Scanning Confocal Microscope or Atomic Force Microscopy [46] are used to measure 3D surface characteristics. Thus, current laboratory characterization focuses on 3D cartography analysis considering technological developments which allow higher accuracy in measurements (vertical resolution of a few decades of μm) and more complex calculus (filters, indicators).

Microtexture measuring methods and devices. Laboratory devices can also provide a direct measurement of microtexture if the accuracy of the sensors is close to a few micrometers. Measurement principles are the same as the ones used for macrotexture and they enable to obtain micro- or even nanotexture scale (see above).

Several authors proposed microtexture indicators well correlated to pavement properties. They can be grouped as follows: (a) volume parameters based on Abbott bearing curve; (b) hybrid parameters; (c) roughness parameters [47, 48]. Thus, Nataadmadja et al. [49] demonstrated that height root-mean-square (Rq), average curvature of peaks (Spc), density of summits (Sds) and average quadratic slope of asperities (Sdq) can explain friction evolution on aggregates. Cerezo et al. [50] showed that this result can be extended to bituminous mixtures and more especially at an early age period. Additionally, analysis based on the geometry of the texture profile (height, density and shape) can be considered. This approach has been extended to the indentors method [51] with a slightly different definition of height and shape.

Nevertheless, Do et al. [52] pointed out some limits of current approaches based only on road surface texture (i.e. geometry) and highlighted the fact that wear phenomenon, debris detachment and bitumen movement must be also considered (tribological circuit).

5 Friction

Pavement friction results mainly from two mechanisms: adhesion and hysteresis. Adhesion emanates from the attractive binding force between the tire rubber and contacting surface of the pavement. The hysteresis component is usually due to energy loss resulting from deformation of the tyre [53].

Friction force developed at the contact zone between tire and pavement is called skid resistance. The skid resistance is related to many factors. It is known to be a function of pavement construction materials, pavement roughness, surface conditions [54] and tyre rubber characteristics.

5.1 Measurement of Skid Resistance

Skid resistance is generally measured through the force generated when a locked tire slides on a pavement [55]. Some methods can be applied in the laboratory, others only in the field and, in some cases, both situations. The main skid resistance measuring approaches are represented in Fig. 5 and their advantages and disadvantages in Table 2. To measure skid resistance in the longitudinal direction there are three methods. In the locked wheel method, the slip speed is equal to the vehicle speed, which means that the test wheel is locked and unable to rotate. The main disadvantages are the impossibility to measure continuously and the high cost of the equipment [53].

In the fixed and variable slip methods, the friction coefficient is a function of the slip of the test wheel while rolling over the pavement. Fixed-slip devices maintain a constant slip, typically between 10 and 20 percent, as a vertical load is applied to the test tyre. These devices are more sensitive to microtexture at low speeds. The difference between fixed and variable slip is that the second case uses a predetermined set of slip values for measuring the frictional force [55].

The sideway force is a method whose output is indicative of vehicles' ability to maintain control in curves. In this test, the test wheel must maintain a constant angle

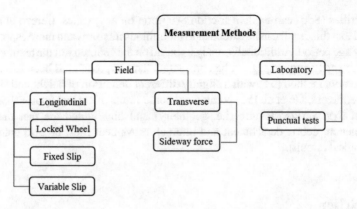

Fig. 5 Main skid resistance measuring methods

to the direction of motion. Side force testers are sensitive to changes in the pavement microtexture, but they are generally insensitive to the pavement macro-texture [56].

Laboratory tests only assess friction on a small pavement sample. The most used are the British Pendulum and the Dynamic Friction Tester.

5.2 Advances in Friction Measurement Methods

Based on the current methods, there are new proposals to measure pavement friction. Figure 6 shows the main requirements for new measurement methods.

New measurement methods are targeted to estimate friction based on other road surface characteristics extracted from pavement images or profiles. Therefore, to indirectly determine friction. One example is the use of LiDAR sensor technology [57]. The operating methodology has already been covered in Sect. 2.1. The proposed LiDAR-based method is able to assist for rapid, economical and automatic estimation. However, the results obtained do not have the same reliability that traditional/field measurement methods offer.

Other emerging methods are related with the relevance of friction control in autonomous vehicles (AV). AV are not currently designed to measure the skid resistance of roads, and traditional measurements are too inefficient for real-time AV control [58]. Currently, a dynamic method to estimate pavement friction level using computer vision is under development based on the introduction of a framework regarding the car-following behavior and turning movements. The anti-skid control framework provides new insights into enhancing the AV safety performance.

Table 2 Advantages and disadvantages of friction measurement methods. Adapted from [55]

Method	Measurement	Advantages	Disadvantages
Locked wheel	Obtain the coefficient of friction by measuring the resistive drag force and the wheel load applied	The systems are relatively simple	Cannot be used on curves. Might miss slippery sections. Continuous measurement is impossible. The equipment has high primary and operating costs
Fixed slip	It is based on measuring the resistive drag force and the wheel load. The friction is reported as a Friction Number	Continuous measurement. Can be used for network and project level friction monitoring	The system needs large amounts of water for continuous measurement. Slip speed (especially on snow) covered surfaces does not always coincide with the critical slip speed value
Variable slip	It is similar to the fixed slip method. The only change is the possibility to set different slip values	Performs continuous measurement	Equipment is large and complex. Maintenance costs of equipment are high. Data processing and analysis is complex
Sideforce	Measure the average of the side force perpendicular to the plane of rotation	Skid condition is relatively well controlled. Can be used on straight sections and curves. Performs continuous measurement	Is very sensitive to road potholes and these defects can destroy tires quickly
British pendulum	It is based on the return height of pendulum after a low speed sliding contact with the pavement surface. This method provides the British Pendulum Number (BPN)	Can be used for both field and laboratory evaluation. Is highly portable and easy to handle	Only measures a friction property of surface at low speed. Has unreliable behaviour when tested on surfaces with coarse texture. Operator procedures and wind can affect results

(continued)

Table 2 (continued)

Method	Measurement	Advantages	Disadvantages
Dynamic friction tester	It is based on horizontal spinning disk fitted with three spring-loaded rubber sliders which contact the surface as the disk rotational speed decreases due to the friction generated between the sliders and the surface	It is highly repeatable and is unaffected by operators or wind. Results of this methods are representative of high-speed values. The IFI correlates well with BPN	Needs traffic control and lane closure when used in field

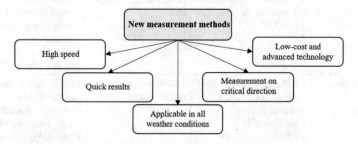

Fig. 6 Requirements for new measurement methods

6 Tyre-Road Noise

Tyre-road noise measurement methods advanced from simple sound pressure level tests measured at the roadside, on-board tests, to sound field holography in the laboratory, and more recently to beamforming. Each method is applied in different conditions, with its own advantages and disadvantages. In Fig. 7, the methods are divided

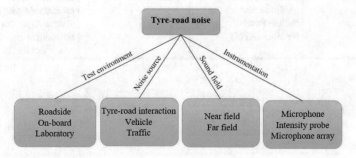

Fig. 7 Main categories of tyre-road noise measurement methods. Adapted from [59]

into four categories [59]: test environment, noise source, sound field, and instrumentation. Next, the existing standardized methods and alternative methods are briefly presented according to the test environment. The other categories will be referred to in particular.

6.1 Roadside or Pass-By Tests

With the roadside measurement test methods, also known by pass-by methods, the far-field noise of a vehicle moving at the roadside is measured. Typically, the maximum sound pressure level (SPLmax) is measured by a microphone placed on the roadside when a vehicle passes by. There are many methods, some of them standardized, dedicated to measuring the average total traffic noise of a given section, as the Statistical Pass-By (Fig. 8a), the noise of a vehicle (Controlled Pass-By), and the tyre-road interaction (Coast-By). One of the main differences among those methods is the operating condition of the vehicles.

For the first method, the traffic flow is used, the second uses a test vehicle, and the last goes further using a test vehicle with the engine switched off near the measurement area and using specific test tyres. Another difference is the number of microphones and their position or the acoustic equipment adopted for the sound measurement (Fig. 8b). For further details about variants of these main tests, see [59].

Microphone arrays produce spatial samplings of a sound field. Advanced noise source identification analysis is possible using microphone arrays and improved signal processing techniques as near field acoustic holography and beamforming, which is suitable for identifying moving sources [60]. Besides that, with the beamforming algorithm application, it is possible to map the distribution of the sources at a certain distance from the array and, therefore, locate the strongest one.

| (a) | (b) | (c) |

Fig. 8 Roadside tests: **a** statistical pass-by test; **b** pass-by test with an acoustic camera; **c** controlled pass-by test with a Head and Torso Simulator. Photos by Elisabete Freitas

Other measurement equipment as Head and Torso Simulators (HATS, see Fig. 8c) are used for analysis in the psychoacoustic domain and sound auralization. Auralization is the process to synthetize an artificial sound environment exactly as the target sound field. As perceptually it is not possible to be distinguished as virtual or real, auralization is used, e.g. to simulate traffic environments in experiments with humans in advanced simulators [61].

6.2 On-Board or on Vehicle Tests

The on-board measurement test methods, also identified by on vehicle, measure the near-field noise (from the contact of a reference tyre with the surface) with the measurement equipment assembled to a wheel (Fig. 9a) or in a trailer (Fig. 9b) with a specific geometry. There are two main standardized methods, the Close Proximity Method (CPX) and the On-board Sound Intensity method (OBSI). While the first measures the sound pressure level (SPL), the second uses intensity probes to measure sound intensity. A combination of these methods is also possible [62].

Another example of alternative methods is the A-CPX [63]. It was designed for evaluating the sound power level emitted by a rolling tyre, employing sound pressure level measurements. Other measurement methods try to measure the noise inside the tyre instead of outside. The Tyre Cavity Microphone method (TCM) consists of fitting a TCM in the wheel. On the one hand, this method screens the exterior noise, but on the other hand, tyre cavity noise is difficult to interpret. Nevertheless, it was used to classify pavements [64].

(a)   (b)

Fig. 9 On-board tests devices: **a** measurement system installed on a wheel (photo by Elisabete Freitas); **b** semi-anechoic trailer (photo by Luc Goubert)

6.3 Laboratory Tests

There are several laboratory tests to measure the acoustic properties of road surfaces. The Drum method (DR) is the most common to assess the sound generated by the tyre-road interaction. Basically, a test tyre rotates over a cylindrical bituminous sample (often composed of several curved bituminous slabs), and a microphone is placed at a certain distance between them. The DR method can be divided into three categories, according to the wheel position: outer rotating drum or inner rotating drum, when the wheel is fixed (Fig. 10 a), and inner stationary drum, when the wheel rolls against the drum (similar to Fig. 13, Sect. 7). For further details, see [59].

With the DR method, despite the possibility to control test parameters (speed, loading, temperature) and the ease to use advanced instrumentation as microphone arrays, accelerometers, lasers, etc., the tonal noise due to the joints between the pavement slabs together with the effect of their curvature distort the final tyre-road noise [59]. The testing speed, usually low, about 15 km/h, is also a disadvantage of most testing equipment. Modern drums can reach high speeds [65] (Fig. 10b). Tyre noise measurements are also possible in accelerated loading facilities as circular tracks (Fig. 10c).

One example of an emerging laboratory test is the Tyre Rolling method. It is a method in which a tyre is accelerated while rolling in a fixed rail with a pre-set angle and impacts an bituminous sample with a certain speed [66]. The sound generated is measured by two microphones near the bituminous sample. The low testing speed (around 20 km/h) and the uncertainty associated with measurements limit the use of this test.

(a)   (b)   (c)

Fig. 10 Laboratory tests: **a** outer rotating drum test facility with microphone array installed [67]; **b** high speed outer stationary drum test facility (photo courtesy of Fwa Fang); **c** CPX assembly installed in the tyre of a circular test track (photo by Elisabete Freitas)

6.4 Other Tests

Based on opportunistic sound and vibration measurements, in-vehicle tests have drawn the researcher's attention [68]. The measurement system, composed of a microphone, accelerometer, and GPS sensor, can be installed in the trunk of vehicle fleets covering broad areas. However, these measurements have many confounders and modifiers affecting sound frequency that must be identified and taken into account in the data analysis, when the objective is noise mapping.

There are other acoustic related tests and also non-acoustical tests used to assess tyre-road noise. Sound absorption is a property that can be measured either in situ or in the laboratory by different methods. One consisted of modifying the impedance tube to be mounted vertically with an open end at the road surface [69]. Another method, adaptable to vehicles to measure acoustic impedance, consists of positioning a p-u sensor at a certain distance from the road to measure sound pressure and sound particle velocity simultaneously, generated by a loudspeaker [70].

Non-acoustical tests are used as input to models to predict and assess the noise of certain bituminous mixtures, as surface texture, flow resistivity, water permeability and dynamic stiffness (see [71] for more details).

7 Rolling Resistance

The engine of a vehicle driving on a road delivers work to overcome energy losses caused by the so-called vehicle driving resistance. This vehicle driving resistance can be split into three main subcategories: propulsion resistance, aerodynamical resistance and vehicle rolling resistance [72]. These can be divided further into subcategories. Vehicle rolling resistance consists of bearing resistance, transmission resistance, tyre/road rolling resistance and suspension resistance. This chapter deals with the "tyre/road rolling resistance", which is influenced by tyre characteristics, the wheel load, inflation pressure and temperature of the tyre, but depends as well on the macro and mega texture of the pavement [73, 74].

The tyre/road rolling resistance is the force which counteracts the motion of the tyres originating from the interaction between the tyres and the pavement. It is directed in the opposite direction of the speed vector of the vehicle. The rolling resistance coefficient (Cr) is the ratio of the tyre/road rolling resistance divided by the tyre load, yielding a dimensionless figure. Typically, the rolling resistance of a car tyre is about 20 to 40 N, whereas the tyre load is about 3000 to 4000 N. The Cr is hence of the order of 1% [75]. The measurement of the Cr with the same tyre and under the same conditions on different road surfaces, makes it possible to assess and compare the energy efficiency of pavements. The Cr is indeed directly related to the energy consumption and CO_2 emission of vehicles: as a rule of thumb, an increase of x % of the rolling resistance increases the fuel consumption and CO_2 emission of the vehicles with x/5% [76]. The tyre/road rolling resistance or Cr of a given pavement with a given tyre can be measured on the road and in the laboratory.

7.1 Measuring the Rolling Resistance on the Road

On the road, it is possible to measure the rolling resistance directly or indirectly by means of a common vehicle, a dedicated vehicle (e.g. a car equipped with a fifth wheel) or a trailer. The following measuring methods can be used [72].

- *Coast down measurements*—Are carried out with common vehicles, equipped with four identical test tyres. The vehicle is accelerated to an initial speed and the wheels are uncoupled from the engine. The vehicle coasts down on the pavement section to be tested and the loss of speed is measured after driving a given distance or time. The vehicle driving resistance can be calculated from the deceleration. By carrying out this test on different pavements but with an identical measurement vehicle and under identical measuring conditions, it is possible to compare the tyre/road rolling resistance of the pavements.
- *Force measurements*—Can be subdivided, depending on how the force is measured, in different subcategories: measurement of the reaction force in the spindle of a wheel, measurement of the drag force or measurement of the reaction force of an axle equipped with two wheels with identical tyres. All force measurements are carried out at constant speed.
- *Torque measurements*—The tyre/road rolling resistance can also be assessed by measuring the torque in the spindle of a test wheel mounted in a common or dedicated vehicle or a trailer. Dividing the torque by the dynamic diameter of the wheel yields the tyre/road rolling resistance force.
- *Angle measurements*—The rolling resistance coefficient C_r can directly be measured as an angle [76]. A test wheel is integrated in a dedicated vehicle or in a trailer (Fig. 11). The assembly wheel equipped with a test tyre and an own suspension is towed at the end of a bar which is fixed at the wheel spindle. The towing bar is loaded to provide a representative load on the test wheel via the suspension (not depicted). The inclination of the bar with respect to the vertical axis, the angle θ, equals C_r provided it is expressed in radians, as θ is a small angle.

Devices for measuring the rolling resistance of pavements are very rare. The Technical University of Gdansk, Poland, and the Belgian Road Research Centre in Brussels, Belgium, both operate trailers based on the angle measurement principle (Fig. 12).

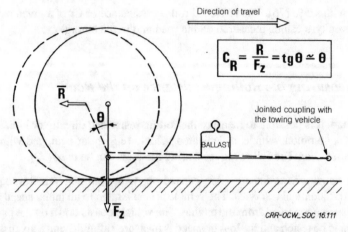

Fig. 11 Measurement principle of the rolling resistance trailer of the Belgian Road Research Centre

(a)   (b)

Fig. 12 Trailers for rolling resistance measurements, based on the angle measurement principle: **a** TU Gdansk trailer; **b** BRRC trailer. Photos by Anneleen Bergiers

7.2 Measuring the Rolling Resistance in the Laboratory

Rolling resistance can be measured in the laboratory as well [77, 78]. Generally, one uses a drum on which the pavement is applied, at the inside or at the outside of the drum. Both types are depicted in Fig. 13. Several methods can be used:

- *The force method*—Measuring the reaction force in the spindle of a testing wheel rolling on the pavement applied on a drum which is turning at a constant speed. The method is like the reaction force measurement on the road.
- *The torque method*—Measuring the torque at the spindle of the test wheel running in/on the drum at constant rotational speed.
- *The power method*—Measuring the power needed to keep the drum running at a constant rotational speed while the test wheel is running in/on it.

(a)   (b)

Fig. 13 Laboratory drums suitable for measuring rolling resistance in the laboratory: **a** 4.5 m diameter internal drum of the Federal Institute for Highway Research in Köln, Germany (photo by Luc Goubert); **b** the 2.0 m diameter external drum at TU Gdansk, Poland (photo courtesy of Jurek Ejsmont)

- *The deceleration method*—The tyre/pavement rolling resistance force of a wheel running in/on a drum with switched off power causes a rotational deceleration of the drum, from which the rolling resistance force can easily be deduced.

8 Conclusions

Advances in pavement surface characteristics surveying methods are being stimulated by the development of improved instrumentation and increased usage of laser and image-based techniques, but also by new methods that make use of low-cost instrumentation, artificial intelligence and other advanced data processing and communication protocols, e.g. using Internet of Things. While the traditional methods are becoming more sophisticated and accurate, allowing for multidimensional analysis, the new ones, e.g. for pavement distresses surveillance, provide information to road agencies about their complete road network but with many limitations. Connected vehicles and autonomous driving give way to new approaches that are being explored, especially to complement the traditional data with real time information, e.g. enabling acute maintenance to be optimized and to support the future digital twin asset management systems. Nevertheless, advances in laboratory tests are more limited than in field tests as the replication of road pavements for surface characteristics analysis remains a limitation.

References

1. Peraka, N., Biligiri, K.: Pavement asset management systems and technologies: a review. Autom. Constr. **119** (2020)
2. Pavimetrics. http://www.pavemetrics.com/. Last Accessed 19 Nov 2020
3. Coenen, T., Golroo, A.: A review on automated pavement distress detection methods. Cogent Eng. **4**(1), 1374822 (2017)
4. Souza, V., Giusti, R., Batista, A.: Asfault: A low-cost system to evaluate pavement conditions in real-time using smartphones and machine learning. Pervasive Mob. Comput. **51**, 121–137 (2018)
5. Inzerillo, L., Di Mino, G., Roberts, R.: Image-based 3D reconstruction using traditional and UAV datasets for analysis of road pavement distress. Autom. Constr. **96**, 457–469 (2018)
6. Li, Z., Cheng, C., Kwan, M., Tong, X., Tian, S.: Identifying asphalt pavement distress using UAV LiDAR point cloud data and random forest classification. Int. J. Geo-Inf. **8**(1), 39 (2019)
7. Zhang, Y., Chen, C., Wu, Q., Lu, Q., Zhang, S., Zhang, G., Yang, Y.: A kinect-based approach for 3D pavement surface reconstruction and cracking recognition. IEEE Trans. Intell. Transp. Syst. **19**(12), 3935–3946 (2018)
8. El-Wakeel, A., Li, J., Noureldin, A., Hassanein, H., Zorba, N.: Towards a practical crowdsensing system for road surface conditions monitoring. IEEE Internet Things J. **5**(6), 4672–4685 (2018)
9. Eriksson, J., Girod, L., Hull, B., Newton, R., Madden, S., Balakrishnan, H.: The pothole patrol: using a mobile sensor network for road surface monitoring. In: MobiSys'08 - Proceedings of the 6th International Conference on Mobile Systems, Applications, and Services, pp. 29–39 (2008)
10. Van Geem, C., Bellen, M., Bogaerts, B., Beusen, B., Berlémont, B., Denys, T., De Meulenaere, P., Mertens, L., Hellinckx, P.: Sensors on vehicles (SENSOVO)—proof-of-concept for road surface distress detection with wheel accelerations and ToF camera data collected by a fleet of ordinary vehicles. Trans. Res. Procedia **14**, 2966–2975 (2016)
11. Gopalakrishnan, K., Khaitan, S., Choudhary, A., Agrawal, A.: Deep convolutional neural networks with transfer learning for computer vision-based data-driven pavement distress detection. Constr. Build. Mater. **157**, 322–330 (2017)
12. Maeda, H., Sekimoto, Y., Seto, T., Kashiyama, T., Omata, H.: Road damage detection and classification using deep neural networks with smartphone images. Comput.-Aided Civil Infrastruct. Eng. **33**, 1127–1141 (2018)
13. Majidifard, H., Adu-Gyamfi, Y., Buttlar, W.: Deep machine learning approach to develop a new asphalt pavement condition index. Constr. Build. Mater. **247**, 118513 (2020)
14. Bang, S., Park, S., Kim, H., Kim, H.: Encoder–decoder network for pixel-level road crack detection in black-box images. Comput.-Aided Civil Infrastruct. Eng. **34**(8), 713–727 (2019)
15. Hoang, N.: Automatic detection of asphalt pavement raveling using image texture based feature extraction and stochastic gradient descent logistic regression. Autom. Constr. **105**, 102843 (2019)
16. Chatterjee, A., Tsai, Y.: Training and testing of smartphone-based pavement condition estimation models using 3D pavement data. J. Comput. Civil Eng. **34**(6) (2020)
17. Sayers, M.W., Karamihas, S.W.: The Little Book of Profiling: Basic Information About Measuring and Interpreting Road Profiles. University of Michigan, Ann Arbor, USA (1998)
18. EN 13036–7: Road and airfield surface characteristics-test methods-Part 7: Irregularity measurement of pavement courses: the straightedge test. CEN (2003)
19. FILTER: Investigation of longitudinal and transverse evenness of roads. FEHRL project, FEHRL, Brussels, Belgium (2002)
20. EVEN: International experiment to harmonise longitudinal and transverse profile measurement and reporting procedure. PIARC Committee on Surface Characteristics (C1), no. 01.07.B, World Road Association (PIARC), Paris, France (2002)
21. Lundberg, T., Andrén, P., Wahlman, T., Eriksson, O., Sjögren, L., Ekdahl, P.: New technology for road surface measurement, Transverse profile and rut depth. VTI report 961A, 2018, Linköping (2002)

22. PIARC: State of the art in monitoring road condition and road/vehicle interaction. 2016R17EN (2016)
23. PIARC: State of the art in monitoring road condition and road/vehicle interaction. 2019R14EN (2019)
24. EN 13036–5: Road and airfield surface characteristics: test methods: Part 5: determination of longitudinal unevenness indices. CEN (2019)
25. Sayers, M.W.: On the calculation of international roughness index from longitudinal road profile. Transp. Res. Rec. **1501**, 1–12 (1995)
26. prEN 13036–8: Road and airfield surface characteristics- test methods- Part 8: Transverse unevenness and irregularities, definitions, methods of evaluation and reporting. CEN (2008)
27. Bridgelall, R., Huang, Y., Zhang, Z., Tolliver, D.: A Sensor Fusion Approach to Assess Pavement Condition and Maintenance Effectiveness. Department of Civil Engineering North Dakota State University Fargo, Upper Great Plains Transportation Institute (2016)
28. EN ISO 13473–1: Characterization of Pavement Texture by Use of Surface Profiles – Part 1: Determination of Mean Profile Depth. CEN (2020)
29. EN 13036–1: Road and airfield surface characteristics—test methods—part 1: measurement of pavement surface macrotexture depth using a volumetric patch technique. CEN (2009)
30. Fisco, N., Sezen, H.: Comparison of surface macrotexture measurement methods. J. Civ. Eng. Manag. **19**(Supplement 1), S1531–S2160 (2013)
31. Flintsch, G., De Leon, E., McGhee, K., Al-Qadi, I.: Pavement surface macrotexture measurement and applications. Trans. Res. Rec. **1860**, 168–177 (2013)
32. ASTM E2147: Standard test method for measuring pavement macrotexture properties using the circular track meter. American Society for Testing and Materials, Volume 04.03 Road and Paving Materials, Vehicle-Pavement Systems (2004)
33. Gottaut, C., Goubert L.: Deliverable D4.2: texture-based descriptors for road surface properties and how they can be used in the appropriate standards. Project ROSANNE (ROlling resistance, skid resistance, and noise emission measurement standards for road surfaces), 143p, (2016)
34. Choubane, B., McNamara, R.L., Page, G.C.: Evaluation of high-speed profilers for measurement of asphalt pavement smoothness in Florida. Transp. Res. Rec. **1813**, 62–67 (2002)
35. Vilaça, J., Fonseca, J., Pinho, A., Freitas, E.: 3D surface profile equipment for the characterization of the pavement texture—TexScan. Mechatronics **20**(6), 674–685 (2010)
36. Wang, W., Yan, X., Huang, H., Chu, X., Abdel-Aty, M.: Design and verification of a laser based device for pavement macrotexture measurement. Transp. Res. Part C Emerg. Technol. **19**(4), 682–694 (2011)
37. Klein, P., Cesbron, J.: A 3D envelopment procedure for tyre belt radiated noise level prediction. In Proceedings of Inter-Noise 2016, 223041, Hamburg, Germany (2016)
38. Edmondson, V., Woodward, J., Lim, M., Kane, M., Martin, J., Shyha, I.: Improved non-contact 3D field and processing techniques to achieve macrotexture characterisation of pavements. Constr. Build. Mater. **227**, 116693 (2019)
39. Ben Slimane, A., Khoudeir, M., Brochard, J., Do, M.T.: Characterization of road microtexture by means of image analysis. Wear **264**(5–6), 464–468 (2008)
40. Sun, L., Wang, Y.: Three-dimensional reconstruction of macrotexture and microtexture morphology of pavement surface using six light sources–based photometric stereo with low-rank approximatio. J. Comput. Civil Eng. **31**(2) (2017)
41. Stout, K.J., Sullivan, P.J., Dong, W.P., Mainsah, E., Luo, N., Mathia, T., Zahouani, H.: The development of methods for the characterisation of roughness in three dimensions. Publication EUR 15178EN of the Commission of the European Communities. Available at www.bookshop.europa.eu (1993)
42. Goubert, L., Do, M.T., Bergiers, A., Karlsson, R., Sandberg, U.: Deliverable D4.1: State of the art concerning texture influence on skid resistance, noise emission and rolling resistance. Project ROSANNE (ROlling resistance, Skid resistance, And Noise Emission measurement standards for road surfaces), 122p. (2016)
43. EN ISO 4287: Geometrical Product Specifications (GPS)—Surface Texture: Profile Method—Terms, Definitions and Surface Texture Parameters. CEN (1998)

44. ISO 13473–2: Characterization of pavement texture by use of surface profiles—Part 2: Terminology and basic requirements related to pavement texture profile analysis. CEN (2002)
45. Do, M.T., Cerezo, V.: Road surface texture and skid resistance. Surf. Topogr. Metrol. Prop. **3**, 1–16 (2015)
46. Blom, J., Soenen, H., Katsiki, A., Van den Brande, N., Rahier, H., Van den Bergh, W.: Investigation of the bulk and surface microstructure of bitumen by atomic force microscopy. Constr. Build. Mater. **177**, 158–169 (2018)
47. ISO 25178–2: Geometrical product specifications (GPS)—Surface texture: Areal—Part 2: Terms, definitions and surface texture parameters. CEN (2012)
48. ISO EN 4287: Geometrical Product Specifications (GPS)—Surface Texture: Profile Method—Terms, Definitions and Surface Texture Parameters. CEN (1998)
49. Nataadmadja, A., Do, M.-T., Wilson, D., Costello, S.: Quantifying aggregate microtexture with respect to wear—case of New Zealand aggregates. Wear **332–333**, 907–917 (2015)
50. Cerezo, V., Ropert, C., Hichri, Y., Do, M.-T.: Evolution of the road bitumen/aggregate interface under traffic-induced polishing. Proc. IMechE Part J. J. Eng. Tribol. (2019)
51. Do, M.-T. Cerezo, V., Ropert, C.: Questioning the approach to predict the evolution of tire/road friction with traffic from road surface texture. Surf. Topogr. Metrol. Prop. (STMP) **8**(2) (2020)
52. Do, M.-T., Zahouani, H., Vargiolu, R.: Angular parameter for characterizing road surface microtexture. Transp. Res. Rec. **1723**, 66–72 (2000)
53. Yu, M., You, Z., Wu, G., Kong, L., Liu, C., Gao, J.: Measurement and modeling of skid resistance of asphalt pavement: a review. Construct. Build. Mater. **260** (2020)
54. Kogbara, R.B., Masad, E.A., Kassem, E., Scarpas, A., Anupam, K.: A state-of-the-art review of parameters influencing measurement and modeling of skid resistance of asphalt pavements. Constr. Build. Mater. **114**, 602–617 (2016)
55. Mataei, B., Zakeri, H., Zahedi, M., Nejad, F.M.: Pavement friction and skid resistance measurement methods: a literature review. Open J. Civil Eng. **6**, 537–565 (2016)
56. Wasilewska, M., Gardziejczyk, W., Gierasimiuk, P.: Comparison of measurement methods used for evaluation the skid resistance of road pavements in Poland–case study. Int. J. Pavement Eng. **21**, 1662–1668 (2020)
57. Du, Y., Li, Y., Jiang, S., Shen, Y.: Mobile light detection and ranging for automated pavement friction estimation. Transp. Res. Rec. **2673**, 663–672 (2019)
58. Du, Y., Liu, C., Song, Y., Li, Y., Shen, Y.: Rapid estimation of road friction for anti-skid autonomous driving. IEEE Trans. Intell. Transp. Syst. **21**, 2461–2470 (2020)
59. Li, T.: A state-of-the-art review of measurement techniques on tyre–pavement interaction noise. Measurement **128**, 325–351 (2018)
60. Chiariotti, P., Martarelli, M., Castellini, P.: Acoustic beamforming for noise source localization–reviews, methodology and applications. Mech. Syst. Sign. Process. **120**, 422–448 (2019)
61. Soares, F., Silva, E., Pereira, F., Silva, C., Sousa, E., Freitas, E.: The influence of noise emitted by vehicles on pedestrian crossing decision-making: a study in a virtual environment. Appl. Sci. **10**, 2913 (2020)
62. Kozak, P., Matuszkova, P., Radimsky, M., Kudrna, J.: Measuring tyre rolling noise at the contact patch. IOP Conf. Ser. Mater. Sci. Eng. **216**(1) (2017)
63. Campillo-Davo, N., Peral-Orts, R., Campello-Vicente, H., Velasco-Sanchez, E.: An alternative close-proximity test to evaluate sound power level emitted by a rolling tyre. Appl. Acoust. **143**, 7–18 (2019)
64. Masino, J., Pinay, J., Reischl, M., Gauterin, F.: Road surface prediction from acoustical measurements in the tyre cavity using support vector machine. Appl. Acoust. **125**, 41–48 (2017)
65. Han, S., Peng, B., Chu, L., Fwa, T.F.: In-door laboratory high-speed testing of tyre-pavement noise. Int. J. Pavement Eng. 1–11 (2020)
66. Ren, W., Han, S., Fwa, T.F., Zhang, J., He, Z.: A new laboratory test method for tyre-pavement noise. Measurement **145**, 137–143 (2019)
67. Clar-Garcia, D., Velasco-Sanchez, E., Campillo-Davo, N., Campello-Vicente, H., Sanchez-Lozano, M.: A new methodology to assess sound power level of tyre/road noise under laboratory controlled conditions in drum test facilities. Appl. Acoust. **110**, 23–32 (2016)

68. Van Hauwermeiren, W., David, J., Dekoninck, L., De Pessemier, T., Joseph, W., Botteldooren, D., Martens, L., Filipan, K., De Coensel, B.: Assessing road pavement quality based on opportunistic in-car sound and vibration monitoring. In: 26th International Congress on Sound and Vibration (ICSV 2019), Canadian Acoustical Association., pp. 1–8 (2019).

69. Freitas, E.; Raimundo, I.; Inácio, O; Pereira, P: In situ assessment of the normal incidence sound absorption coefficient of asphalt mixtures with a new impedance tube. In: 39th International Congress and Exposition on Noise Control Engineering—INTER-NOISE 2010. The Institute of Noise Control Engineering of the USA, Inc., Lisbon (2010)

70. Bianco, F., Fredianelli, L., Lo Castro, F., Gagliardi, P., Fidecaro, F., Licitra, G.: Stabilization of a pu sensor mounted on a vehicle for measuring the acoustic impedance of road surfaces. Sensors **20**(5), 1239 (2020)

71. Mikhailenko, P., Piao, Z., Kakar, M.R., Bueno, M., Athari, S., Pieren, R., Heutschi, K., Poulikakos, L.: Low-noise pavement technologies and evaluation techniques: a literature review. International Journal of Pavement Engineering 1–24 (2020).

72. Zöller, M.: State of the art on rolling resistance measurement devices. FP7 project ROSANNE deliverable D.3.1 (2014)

73. Goubert, L., Do, M.-T., Bergiers, A., Karlsson, R., Sandberg, U., Maeck, J.: State-of-the-art concerning texture influence on skid resistance, noise emission and rolling resistance. FP7 project ROSANNE deliverable D.4.1 (2014)

74. Ejsmont, J., Taryma, S., Ronowski, G.; Świeczko-Żurek, B., Dujardin, N., Sjögren, L.: Parameters influencing rolling resistance and possible correction procedures. FP7 project ROSANNE deliverable D.4.1 (2015)

75. Bergiers, A., Goubert, L., Anfosso-Lédée, F., Dujardin, N., Ejsmont, J., Sandberg, U., Zöller, M.: Comparison of rolling resistance measuring equipment—pilot study. Report MIRIAM SP1 Deliv. #3 (2011)

76. Descornet, G.: Road-surface influence on tire rolling resistance. Surface characteristics of roadways: international research and technologies, ASTM STP 1031. In: Meyer, W.E., Reichert, J. (eds.) American Society for Testing and Materials, pp. 401–415. Philadelphia, USA (1990)

77. Regulation No 117 of the Economic Commission for Europe of the United Nations (UNECE)—Uniform provisions concerning the approval of tyres with regard to rolling sound emissions and/or to adhesion on surfaces and/or to rolling resistance (2016)

78. Ejsmont, J., Świeczko-Żurek, B., Ronowski, G., Taryma, S.: Results of rolling resistance laboratory drum tests. FP7 project ROSANNE deliverable D.3.2 (2014)

Full-Scale Accelerated Pavement Testing and Instrumentation

José Neves⍟, Ana Cristina Freire⍟, Issam Qamhia⍟, Imad L. Al-Qadi⍟, and Erol Tutumluer⍟

Abstract Road and airfield pavements play an important structural and functional role of ensuring safe, economic, and environmentally sustainable mobility of people and freight. Promoting circular economy in transportation infrastructure construction requires using innovative and non-traditional materials and technologies. However, validation is needed prior to their implementation. Hence, construction and monitoring of instrumented trial sections exposed to vehicular and environmental accelerated loading have been usually considered to assess their performance. Pavement embedded instruments allow monitoring the pavement health through measuring layer deformations, strains, stresses, and environmental conditions. Advantages and limitations of various instrument types are presented. In addition, two case studies are discussed.

Keywords Accelerated pavement testing · Instrumentation · Pavement materials · Pavements sensors

J. Neves (✉)
CERIS, Department of Civil Engineering, Architecture and Georesources, Instituto Superior Técnico, Universidade de Lisboa, Lisboa, Portugal
e-mail: jose.manuel.neves@tecnico.ulisboa.pt

A. C. Freire
LNEC, Nacional Laboratory for Civil Engineering, Lisboa, Portugal
e-mail: acfreire@lnec.pt

I. Qamhia · I. L. Al-Qadi · E. Tutumluer
Illinois Center for Transportation, Department of Civil and Environmental Engineering, University of Illinois at Urbana, Champaign, USA
e-mail: qamhia2@illinois.edu

I. L. Al-Qadi
e-mail: alqadi@illinois.edu

E. Tutumluer
e-mail: tutumlue@illinois.edu

© The Author(s), under exclusive license to Springer Nature Switzerland AG 2023   163
C. Chastre et al. (eds.), *Advances on Testing and Experimentation in Civil Engineering*, Springer Tracts in Civil Engineering,
https://doi.org/10.1007/978-3-031-05875-2_7

1   Introduction

Pavement design involves material selection for various layers and determining the required layer thicknesses. The current depletion of natural resources and recent innovation in pavement technologies led to the utilization of new materials in pavement systems, which require assessment.

A pavement structure exhibits a complex mechanical response to loading because of variation in layer materials properties, traffic characteristics, and environmental conditions. This necessitates pavement monitoring. Conventional methods for structural evaluation (e.g., core drilling and deflection) are invasive, inaccurate, discrete, and have low coverage.

In general, pavement monitoring currently consists of using wired sensors or instruments. Recent advances in information and sensing technologies have led to significant development and improvement in pavement condition evaluation. The implementation of pavement smart sensors allows for intelligent and continuous pavement monitoring systems. This allows instrument measurements to be integrated into pavement management systems allowing for adopting optimum pavement maintenance and rehabilitation strategies.

Pavement instrumentation and full-scale Accelerated Pavement Testing (APT) are useful techniques to promote the implementation of emerging and non-traditional pavement technologies. They allow capturing pavement potential performance, in a relatively short period of time, in response to complex traffic loading, use of new materials and construction techniques, innovative pavement design concepts, and effect of asphalt material ageing and healing. There are several APT machines used by agencies and universities, having different designs and capabilities. The APT load application may be fixed or mobile; linear or circular; and uni- or bi-directional.

This chapter presents advanced and innovative techniques of pavement instrumentation related to stress, strain, displacement, moisture content, and temperature measurements. It includes a brief summary of worldwide full-scale APT. Further, two case studies are presented to illustrate the main advantages and limitations of various instruments and full-scale APT: (1) description of the test sections instrumented during the SUPREMA project in Portugal focused on using recycled aggregates from construction and demolition materials in unbound granular layers; and (2) description of the Accelerated Transportation Loading Assembly (ATLAS) at the Illinois Centre for Transportation (ICT), including some examples of layouts of constructed and instrumented test sections on utilizing quarry fines in pavement subsurface layers.

2 Advancements in Pavements Instrumentation

2.1 General Principles

Pavement stiffness is affected by pavement material and geometry. Hence, it is affected by vehicular loading and environmental conditions. Amplitude and frequency of traffic loads impact the asphalt concrete (AC) layers stiffness, while moisture and temperature affect the various layers of pavements. The pavement structural capacity may be evaluated by embedded instruments measuring layer strains, stresses, and deflections. Temperature, frost depth, and moisture content are measured using environmental sensors.

Pavement instruments must be compatible with the heterogeneous nature and thickness of the pavement layers. The instrument should cause the least disturbance to the pavement layer while able to capture the constituent effect of the material. In addition, sensors should withstand potential aggressive construction and in-service conditions (e.g., high temperature and excessive compaction). The sensors should also be resistant to corrosion and humidity conditions, while having a relatively long fatigue life. Proper pavement instrumentation should be performed during pavement construction. Thus, the disturbance of the tested pavement material while the sensors are being embedded should be minimized.

Figure 1 presents a general scheme of the most common parameters to be measured in the pavement structure: deformation, pressure, strain, temperature, frost depth, and moisture content. Measured parameters depend on pavement type, layer composition, and location. In general, the critical responses of a pavement structure are due to the

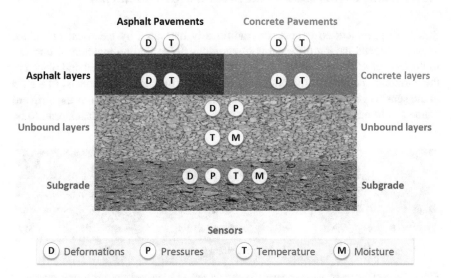

Fig. 1 Common instrumentation types for pavement structures

passage of traffic, including transverse horizontal and longitudinal strains at the bottom of bound (AC and concrete) layers; vertical and horizontal strains at the top, bottom, and within unbound layers, including subgrade; and pressure at the bottom of all layers and occasionally within the subgrade. In addition, environmental sensors are installed at various depths within each layer to measure temperature and moisture content. In the case of rigid pavements, additional measurements related to joints and cracks associated with curing and curling may be essential to assess the pavement performance.

In bound layers, strain gauges, multi-depth deflectometers, and temperature measuring sensors are often used. In the case of unbound layers, including the subgrade soil, strain gauges, pressure cells, moisture sensors, and multi-depth deflectometers (MDD) are the most commonly used instruments. The choice of the most suitable sensors should be mainly based on the measurement needed. Sensors should meet the corresponding requirements, including capacity, measuring range, accuracy, repeatability, simplicity of installation, survivability, and durability [1, 2].

Traditional sensors mainly consist of wired sensors, which are costly and time consuming to install. Recent advances in pavement sensors focus mainly on attributes related to continuous, autonomous and accurate sensing; size and robustness of the sensors; use of wireless communication to transmit data; power supply; data acquisition system; and suitability for network integration in conjunction with other monitoring systems (e.g., environmental conditions, traffic characteristics, surface inspection).

2.2 Strain, Deformation, and Pressure Measurements

Strains in pavement structures are traditionally measured by electrical resistance gauges designed having different configurations depending on location of interest. In the case of AC bounded layers, H-type strain gauges are often used and placed in the horizontal position at the bottom of the layers (Fig. 2a). They consist of a strip in the sensitive part where the strain gauge is fixed. The extremities of the strip are connected to metal bars functioning as anchors inside the pavement material to be

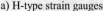

a) H-type strain gauges   b) Vertical strain gauge   c) Pressure Cell

Fig. 2 Sensors for pavement instrumentation: strain, deformation, and pressure measurements

monitored, while allowing relatively free movement at the bottom of the gauge to allow proper measurements.

Similarly, in the case of unbound granular materials, the gauges, which are usually linear variable differential transformer (LVDT) based could be placed in any orientation. Figure 2b shows a vertical strain gauge composed of the sensitive part with two circular plates in the extremities.

Pressure cells can measure the compressive loads within the pavement structure (unbound granular layers and subgrade). Two types of pressure cells are commonly used: hydraulic cells and diaphragm-based cells. Figure 2c shows a hydraulic cell composed of two plates forming a cavity to be filled with viscous liquid and received inside a strain sensor related to the compression load through calibration.

Optical fibre has been successfully applied in pavement sensors to replace conventional strain sensors. The sensor is generally composed of an optical fibre connected to a sensing element responsible for the measurements. These types of sensor present a suited performance to adverse conditions of temperature, moisture, compaction, and loading.

Various other sensors are being developed for pavement instrumentation [3]. In a perspective of the continuous and long-term monitoring of pavement structures, advanced micro and wireless sensor systems are being developed. Recent achievements in micro- and nano-electromechanical technologies made it possible to manufacture sensors using micro-fabrication techniques. This advanced/smart-sensing technology, including wireless sensors, shows vast potential for pavement monitoring, mainly in concrete structures.

Piezoelectric transducers are sensors used to evaluate materials' modulus through the measurement of shear waves propagation [4]. Bender Elements (BE) are piezoelectric transducers suitable for small strain modulus evaluation of granular materials in geotechnical applications [5]. Recently, the University of Illinois has developed a new field sensor based on BE and suitable for the direct measurement of the modulus of unbound granular layers [6]. Figure 3 shows the general frame where the BE sensors are installed, including the protection mechanism.

Power supply for wireless sensors is a concerning challenge. Piezoelectric transducers are being used to self-power wireless sensors, and significant research is being carried out [7, 8]. The Michigan State University recently developed a small battery-less strain sensor based on the integration of piezoelectric transducers with an array of ultra-low-power floating computational gate circuits (Fig. 4a) [7, 9]. Another example of a modern sensor is the system for measuring plastic and elastic strains in unbound granular layers that was developed by the Canterbury Accelerated Pavement Testing Indoor Facility (CAPTIF) of the New Zealand Transport Agency. It is based on a wired inductive coil sensor that consists of a circular plastic disk with a groove cut at the perimeter where the copper wire is wound to form an inductive circuit (Fig. 4b) [10].

Further, SmartRocks sensors are 3D-printed devices, similar to real stones, equipped internally with a triaxial accelerometer, gyroscope and magnetometer capable of recording the translation, rotation and orientation of particles in nine degrees of freedom (Fig. 4c). The newest edition also contains sensors for temperature

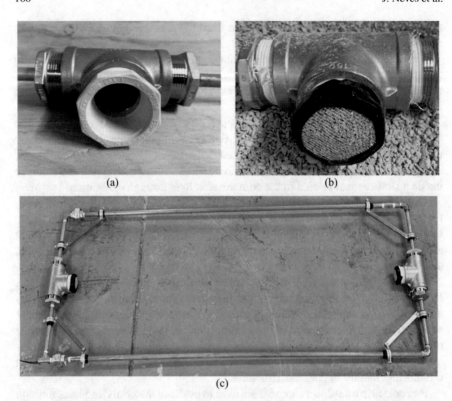

(a)   (b)

(c)

Fig. 3 **a** BE protection module; **b** BE protection module with coupling material and protection cover; **c** BE sensor frame [6]

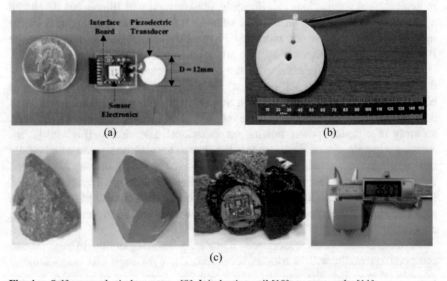

(a)   (b)

(c)

Fig. 4 **a** Self-powered wireless sensor [9]; **b** induction coil [10]; **c** smart rocks [11]

and normal stress measurements. Successfully applied in granular layers, SmartRock sensors were also successfully adapted for AC layers [11].

Overall, the most used sensors for deflection measurements are LVDTs, geophones, and accelerometers. The MDDs are the most common device used to measure pavement layer deflections using LVDTs. It is installed inside the pavement structure through a previous core and adequately anchored on the base and top. Accelerometers and geophones are simple and robust sensors used to measure accelerations and velocities, respectively, within the pavement. These sensors do not need fixed references and can be installed at any point of the pavement structure. They are more frequently used to measure pavement deflections when subjected to dynamic loads by integrating their signal. However, research is being developed to use this type of sensor for continuous measurements of displacements inside the pavement structure [12]. Deflection measurements on the pavement surface can also be performed with laser profilometers, often used in pavement monitoring at the network level. Rutting of pavement, as a long-term evaluation, can be easily and accurately assessed using this equipment.

2.3 Assessment of Environmental Conditions

The most important environmental conditions that significantly affect the behaviour of the pavement structure are moisture, frost depth, and temperature.

Time Domain Reflectometry (TDR) could be used to predict moisture content in unbound layers of pavements [13]. The working principle of TDR involves the determination of electromagnetic pulse travel time along the rods of the TDR probes inserted into the tested material where moisture content is being investigated. The reflected signal carries information on the dielectric characteristics of the tested material. The contact between the rods and the tested material should be precise and permanent to allow reliable readings. Prior to installation, TDR probes should be calibrated in controlled conditions using the same material at various moisture contents. Different TDR probes have been developed over the years [14]. Figure 5a shows a TDR sensor composed of three-rod probes.

Thermocouples are the most used instruments for temperature measurements in pavement layers. They are made of two different metal wires (commonly copper and constantan) connected together (generally by welding). The voltage at each junction is different and influenced by temperature. The sensitivity of a thermocouple depends on the type of metal. Thermocouples are often used in pavement instrumentation because they do not require an energy supply, are inexpensive and easy to install, and provide a quick response over a wide range of temperatures. Resistance Temperature Detectors (RTD) are types of temperature resistors, consisting of a wire wrapped around a ceramic or glass core, capable of measuring temperatures in a range compatible with pavement construction and service conditions. Thermistors are another type of resistance thermometers also suitable for temperature measurements. These sensors are made from a semiconductor material sensitive

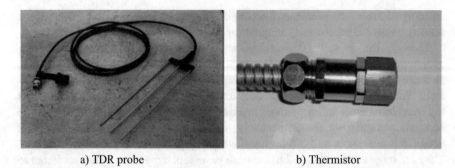

a) TDR probe   b) Thermistor

Fig. 5 Sensors for pavement instrumentation: moisture content and temperature measurements

to temperature fluctuation. Thermistors are inexpensive, easy to install, and accurate in a wide operating range of temperatures. Figure 5b shows a type of thermistors.

Wireless micro-electromechanical sensors and other smart sensing devices are also used for real-time remote monitoring of moisture content and temperature [15].

3 Advances on Full-Scale Accelerated Pavement Testing

Full-scale APT facilities allow, using specific wheel load conditions, for the evaluation of the pavement response and performance under a controlled and accelerated accumulation of damage in a reduced time period. The APT, usually designed to study the performance of real pavements under accelerated heavy traffic, presents major advantages over simulated laboratory small-scale testing. Pavement construction and structure, boundaries, loading and other conditions that correspond to real conditions, are directly applied in APT sections. On the other hand, temperature and different degrees of moisture application can be controlled to meet test requirements. On the other hand, The Long-Term Pavement Performance (LTPP) focuses on evaluation of the in-service performance of pavements, combining typical traffic loads and environmental conditions affecting the pavement in a normal mode [16, 17].

The first APT experiment was the Detroit Circular Test Track built in 1909 and used to test different construction materials. A loading device was built to simulate "horse and wagon traffic of the day". Since the early 1900s, other APT experiments were developed around the world allowing the evaluation and selection of proper pavement materials and designs, studying the effect of axle loading on flexible and rigid pavements performance, and developing of serviceability concepts, among other applications [18].

Nowadays, there are more than 40 APT facilities around the world. Some are located in Europe, namely in Spain, France, Germany, Switzerland, the Netherlands, United Kingdom, Romania, Slovakia, Finland, Sweden, and Norway. Other APT facilities exist in Canada, United States (U.S.), Brazil, South Africa, People's

Republic of China, South Korea, Japan, Australia, India, Saudi Arabia, and New Zealand [18, 19]. A selection of various types of APT facilities is presented in Table 1. The APT facilities varied based on layout (circular or linear), indoors vs. outdoors, testing length, axle load, speed, and environmental control.

Table 1 Examples of APT facilities around the world (adapted from [18, 19])

Country	Brief description
Australia	The Australian ALF facility applies dual loads from 40 to 80 kN over a dual-tire/ single-wheel assembly to a 12 m test length, under a constant speed of 20 km/h. A specific aspect concerns the application of loads in one direction only and wander is 1.4 m. ALF facility is a transportable machine designed to be applied on in-service highways or on specially built test pavements
Canada	The IRRF test road is in northeast Edmonton, Alberta, and serves as the new access road to the Edmonton Waste Management Centre. The test road has two 20-m-long monitoring sections approximately 100 m apart. Each of the sections are instrumented allowing to study the pavement response to traffic and environmental loading. Full-scale instrumentation of the IRRF with AC strain gauges, pressure cells, and soil compression gauges allows evaluating the pavement response submitted to heavy traffic
Denmark	The Danish Road-Testing Machine is a linear track facility which can test full-scale pavements under loads up to 65 kN at speeds up to 30 km/h. The pavement is built in a pit with temperature control between -10 °C and + 30 °C
France	The French largest circular test track in operation is near Nantes and has a four-arm rotating loading system, running two-wheel assemblies on an inner track, 30 m in diameter, and an outer track, 40 m in diameter. Loads from 40 to 75 kN on a dual wheel can be applied at speeds of up to 105 km/h. Tandem axles of 280 kN at lower speeds and single wheel-loading can also be simulated
Finland and Sweden	Finland and Sweden have a joint APT program operating a Heavy Vehicle Simulator (HVS) Mark IV. The HVS–Nordic is a linear full-scale accelerated pavement testing machine that allows dual or single wheels with standard or wide-base tires. The lateral movement is ± 750 mm and the wheel load vary from 20 to 110 kN, with speeds up to 15 km/h. This APT has a unique characteristic by being mobile with full temperature control and loading
Spain	Spain has a major APT facility near Madrid. The APT presents two straight sections of 75 m each joined by two circular arcs with a 25 m radius. The test sections are constructed with regular building equipment in a U-shaped concrete box, 8 m wide by 2.6 m deep, allowing the control of moisture content in the subgrade soil. Loads are applied by two guided bogies, which can be equipped with one-, two-, or three-wheel, single- or dual-tire half axles. The load is applied at speeds up to 60 km/h, by gravity, and applied axle loads are between 110 and 150 kN. The operation of the two bogies and the monitoring of applied instrumentation is done in a control centre
South Korea	The HAPT is a linear test track, having a pavement testing length of 12.5 m, and a testing length of 9 m, at constant speed. The range of axle loading is 70–250 kN, with a lateral wander of 350 mm at maximum speed of 12 km/h. The environmental characteristics, air and pavement temperature, water table level and rainfall can be under control

(continued)

Table 1 (continued)

Country	Brief description
USA	The Advanced Transportation Research and Engineering Laboratory (ATREL) at the University of Illinois at Urbana-Champaign developed the Accelerated Transportation Loading System (ATLAS) to evaluate multiple transportation support systems. The wheel carriage of ATLAS can be fitted with single or dual wheels used for highway trucks, an aircraft wheel, or a single-axle rail bogie. The structural gantry is mounted on four crawler tracks to facilitate positioning of the device. The 40-m-long APT can test pavement sections up to 28-m-long. Loading can be unidirectional or bidirectional. ATLAS was recently upgraded to provide braking/acceleration, shear angle, and tandem loading capabilities

Several studies have been and continue to be developed with different APT facilities to evaluate pavements' performance related to new and innovative materials and techniques. APT also contributes to the development and calibration of design methods, maintenance and rehabilitation techniques, effects of vehicle variables on pavement behaviour, and pavement performance [20–24].

Numerous studies have been conducted using APT. The Federal Highway Research Institute of Germany used APT to compare the structural performance of flexible pavements [21]. Strain gauges were used as pavement instrumentation and surface deflections were measured. Extensive APT programs were developed in the Western Cape province of South Africa. The APT program provided an accelerated assessment of major pavement structural behaviour. The results were linked to real pavement behaviour to enable outputs to be calibrated in pavement design [22]. The French APT facility used to evaluate the performance of different weight-in-motion sensors under a wide variety of loading conditions [23]. The Florida Department of Transportation used its HVS to evaluate the long-term rutting potential of Superpave AC mixtures and Styrene–Butadiene–Styrene (SBS)-modified AC mixtures, simulating 20 years of traffic within a short period of time [24]. Recently, APT was used to investigate the impact of autonomous vehicles on transportation infrastructures [25]. The following are two case studies on utilizing APT.

4 Case Studies

4.1 Use of Recycled Aggregates for Pavement Unbound Granular Layers

General Description. The SUPREMA research project (Sustainable Application of Construction and Demolition Recycled Materials in Road Infrastructures) was developed in Portugal between 2010 and 2014 by the National Laboratory for Civil Engineering (LNEC), in collaboration with the University of Lisbon (IST). The main goals of the project were to promote sustainable use of recycled materials and identify optimal corresponding technologies [26, 27].

The project focussed of the following: (1) assessment of the geomechanical and geoenvironmental characteristics of various recycled materials, considering their origin, sorting methodology and final composition; (2) comparison of unbound layers with and without recycled aggregate; (3) determination of pavement design parameters when a recycled material is utilized; and (4) evaluation of the construction techniques when considering the use of recycled materials in unbound granular layers, including base, subbase and subgrade [28].

Materials and Testing. Four recycled and virgin aggregates were selected for laboratory characterization and for the construction of the experimental pavement sections, namely (1) crushed mixed concrete, composed of a mixture of concrete and clay masonry units; (2) crushed reclaimed asphalt pavement (RAP); (3) milled RAP; and (4) crushed natural limestone aggregates (as control material). In the case of milled RAP, a mixture of RAP (30%) and crushed limestone (70%) was used [26]. The constituents of the recycled aggregates were evaluated according to EN 933–11 [29]. The grain size distribution of the recycled materials was determined in accordance with EN 933–1 [30]. Additional laboratory tests were performed to characterize the chemical, geometrical, physical, and mechanical properties of the materials. Leaching tests were also performed on the materials to verify their compliance with the legal limits of acceptance of waste for disposal in landfills [26, 28, 31]. In situ tests were performed on instrumented experimental pavement test sections built with selected recycled and virgin aggregates, respectively [27, 32].

Construction and Instrumentation. After laboratory characterization of the materials, full-scale pavement sections were constructed: 100-m-long and 6-m-wide. The sections were constructed at a new industrial park dedicated to recycling activities, involving waste from diverse origins, and including construction and demolition materials [31].

The pavement structures of the sections comprised of unbound granular material (UGM) base layers, composed of recycled and virgin aggregates, at 0.30-m-thick average, and hot-mix asphalt (HMA) wearing course at 0.07-m-thick average (has asphalt binder AC 20 at a pen 50/70). Four sections – T1 to T4—were constructed with the geometry presented in Fig. 6 and composed of the following materials: crushed RAP (T1), crushed mixed concrete (T2), crushed virgin aggregates (T3), and milled RAP (T4). The UGM base layer was constructed in two lifts: 0.18 and 0.12 m. Each test section had length of 25 m [32]. The same construction procedure was applied to all sections. Density and moisture content measurements were conducted during the preparation of the subgrade and the construction of UGM. The quality control of the compaction was based on modified Proctor test.

The instrumentation of the four full-scale test sections (T1 to T4) was carried out during the construction. Instruments were installed in the UGM layer and the HMA layer. Each experimental test section was instrumented in two zones 1 and 2 as illustrated in Fig. 7 [28]. The position of all the sensors was referenced using GPS coordinates.

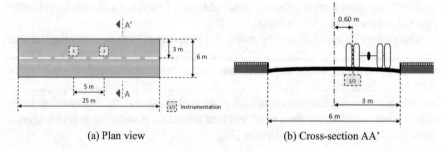

(a) Plan view   (b) Cross-section AA'

Fig. 6 a Plan view and **b** cross-section AA' of the experimental test sections [27, 28]

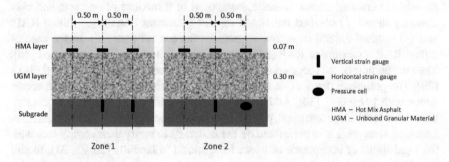

Fig. 7 Instrumentation layout [26, 32]

The following instruments were used: strain gauges in the HMA layer and subgrade soil, pressure cells in the subgrade, and thermistors in the HMA layer for temperature measurements. Figure 7 displays the details of the instrumentation of the two subsections: vertical strain gauges (LVDTs) and pressure cells on top of the subgrade, and horizontal strain gauges at the bottom of the HMA layer. The horizontal strain gauges were placed in the wheel paths of the traffic at the bottom of the HMA wearing course (Fig. 6b). The central position was considered as a reference. The strain gauges were manufactured at LNEC and they are composed of a slender element of araldite, and the strain gauge fixed inlaid (sensitive element) [32].

Figure 8 illustrates some of the construction activities. Figure 8a shows the spreading of the UGM over the sensors in the subgrade. Figure 8b illustrates the installation of thermistors inside the HMA layer. Figure 8c shows the final constructed test sections after the construction and compaction of the wearing course HMA layer.

Pavement Monitoring. The responses of the test sections' instruments due to Falling Weight Deflectometer (FWD) loading were monitored. A load impact of 30 kN was applied on a circular plate with a 0.45-m-diameter. The deflections were measured through geophones located at the following distances from the centre of the circular plate: 0, 0.30, 0.45, 0.60, 0.90, 1.20, 1.50, 1.80 and 2.10 m. During loading, the HMA layer's temperature was measured by the thermistors.

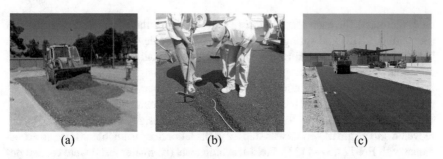

Fig. 8 Instrumentation and construction of the full-scale experimental sections [32]

Three sets of FWD measurements were performed during the construction phase of the full-scale test sections: on subgrade, UGM base layer, and HMA layer. The first set of FWD tests allowed the evaluation of the subgrade stiffness. This approach allows better estimation of the layer moduli.

Performance Evaluation. FWD tests were carried out during and after the construction of the pavement sections. The backcalculation analyses of FWD results in conjugation with instrumentation responses allows validation of the bearing capacity of the recycled materials used in the test sections [27, 32]. Figure 9 presents the resilient modulus of the subgrade, and UGM and HMA layers. The resilient moduli were

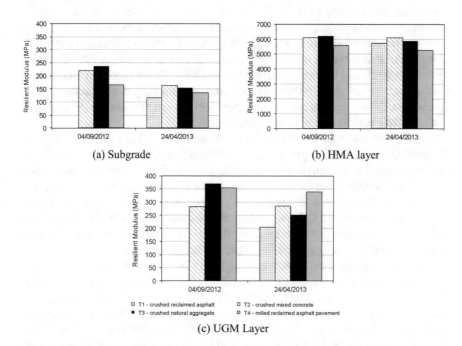

Fig. 9 Resilient moduli backcalculated from FWD tests

backcalculated for each test section (T1 to T4) from the FWD results in the fall (September 2012) and spring (April 2013). It was evident that recycled aggregates behaved different than virgin aggregates. Nevertheless, all test sections with recycled materials demonstrated acceptable performance.

In addition to the structural assessment of the pavement sections, the environmental impact was evaluated to control potential contamination of surface water, groundwater, and soil due to the use of recycled materials. The leachability of the recycled and virgin materials was studied in the laboratory using batch tests, in accordance with EN 12,457–4 [33]. Field-leaching tests (lysimeter tests) were carried out over a period of seven years; and the results were satisfactory [26–28].

4.2 Utilization of Quarry By-Products in Pavement Foundation Layers

General Description. Quarry by-products (QB), or quarry fines, are by-product materials produced at rock production sites during blasting, crushing, screening, and washing operations. QB materials typically comprise particles smaller than 6 mm in size, and are mostly a mixture of coarse-, medium-, and fine-grained sand particles with a small fraction of silts and clays. QB can exist in three distinct types: screenings, pond fines, and baghouse fines [34]. During the crushing stages, QB are generally carried out in three stages, i.e. primary, secondary, and tertiary crushing.

Over the years, billions of tons of QB were generated worldwide. In the U.S. alone, more than 4 billion tons of QB accumulated from more than 3000 operating quarries in 2013. The importance of utilizing aggregate QB in pavement applications thus stems from the vast quantities that are produced annually and are underutilized. In addition, QB stockpiling and disposal represent is a serious issue for the aggregate industry; they interfere with quarry operations [35, 36].

Research efforts at the Illinois Center for Transportation (ICT) in the U.S. focused on evaluating sustainable applications of QB or blends of QB with marginal and recycled aggregate materials in pavements. Applications include unbound or chemically-stabilized pavement subgrade and subbase and base layers. A laboratory study was conducted to characterize the engineering properties of QB fines produced from different crushing stages. QB were collected from four quarries in Illinois [36], and index tests were conducted to determine grain size distribution, morphological shape properties, and mineralogy for each sample and crushing stage.

Further, unconfined compressive strength (UCS) tests were conducted on the virgin QB fines and on QB stabilized with 2% Type I Portland cement or 10% Class C fly ash [36]. The study concluded that QB collected from different sources and crushing stages had different properties such as aggregate gradation, shape, texture, angularity, and mineralogy. Virgin QB samples had a low compressive strength (less than 76 kPa), while the UCS of the stabilized QB samples was 10–30 times greater. Lastly, a packing study to aid the construction of subgrade improvement layers with large primary crushed rocks (up to 152 mm top size) and QB materials filling the inherent void structure was conducted. A packing box with 0.61 by 0.61 by 0.53 m was used to understand packing, optimize QB content and mixing ratios, investigate proper compaction techniques, and study QB percolation by gravity [37].

Ultimately, the promising QB applications studied in the laboratory were used to construct full-scale pavement test sections for further performance evaluation under heavy loading using APT. Sixteen sections—four construction platforms and 12 flexible pavements—were evaluated with the Accelerated Transportation Loading Assembly (ATLAS) at ICT for rutting and fatigue performance.

University of Illinois ATLAS. ATLAS (Fig. 10a) is a linear APT system, 37.8 m long, with a 25.9 m loading length and a 19.8 m span of constant wheel velocity. Concerning ATLAS' features, it has a wheel carriage that can accommodate a single or dual-wheel tire, an aircraft tire, or a single-axle rail bogey (Fig. 10b). Heavy loads up to 356 kN can be applied through a hydraulic ram attached to the wheel carriage. The carriage can travel at speeds up to 16 km/h in the constant loading length. Further, the wheel carriage can wander up to 0.9 m laterally to simulate real traffic patterns. Pavements under ATLAS can be loaded either in unidirectional or bidirectional modes. When temperature control is critical, aluminium panels are attached around ATLAS (Fig. 10c) and heater elements are used to control and

(a) Accelerated Transportation Loading Assembly (ATLAS)

(b) ATLAS wheel carriage   (c) Insulation panels   (d) Heater elements

Fig. 10 Some features of ATLAS accelerated pavement tester

maintain constant temperatures (Fig. 10d). Recently, the system went through major modification to include braking/acceleration, torque force, turning force, and tandem loading capabilities.

Applications for Quarry By-Products. The QB applications studied included both unbound and chemically-stabilized pavement subsurface/foundation layers. Details for the 16 test sections are presented in Table 2. The QB pavement applications can be divided into five main categories:

- Using QB for filling voids between large stones as aggregate subgrade on soft subgrade soils.
- Increased fines content (i.e., 15% QB fines passing No. 200 sieve) in dense-graded aggregate subbase over soft subgrade soils.
- Using QB as a cement- or fly ash-treated subbase (e.g., in inverted pavements).
- Using QB as a cement- or fly ash-treated base material.
- For base course application, blending QB with coarse aggregate fractions of recycled materials and stabilizing the blends with 3% cement or 10% class C fly ash.

Instrumentation and Testing Program. The following is a summary of the testing program to assess the performance of the constructed QB applications [38–41]:

- Quality assurance and quality control (QA/QC) during and after construction. Tests included Dynamic Cone Penetration (DCP) testing to check subgrade strength and uniformity, layer modulus estimates using Geogauge® (Fig. 11a), Lightweight Deflectometer (Fig. 11b), and FWD, as well as layer density measurements using a nuclear gauge (Fig. 11c).
- Instrumentation: Soil pressure cells (Fig. 11d) were installed on top of the subgrade in four of the test sections; specifically, in the wheel path in sections C2S1, C2S4, C3S2, and C3S4. Eight Geokon® model 3500 soil pressure cells, with a 400 kPa range and a 230 mm diameter were installed. Thermocouple wires were also installed by drilling holes adjacent to the wheel path and installing them mounted on sticks at specified depths in the subgrade, mid-subbase height, mid-base height and various depths in the HMA layer. The holes were filled with fine silica sand and sealed with a crack sealant material on the surface.
- Accelerated pavement testing: Each 'Cell', comprising four test sections, marked one location for ATLAS. Accordingly, Cells 1S, 1 N, 2 and Cell 3 were loaded separately. The testing of Cell 1S, having four construction platform sections, was completed first, followed by HMA-paved Cells 1 N, 2, then 3. A wide-base tire (455/55R22.5) was used to load the test sections. A constant unidirectional wheel load of 44.5 kN, a tire pressure of 760 kPa, and a constant speed of 8 km/h were assigned. Channelized wheel loading was applied with no wander considerations. Once Cells 2 and 3 received 100,000 passes, the wheel load and pressure were increased to 62.3 kN and 862 kPa, respectively, and additional 35,000 passes were applied to better differentiate rutting progression trends of the various sections.
- APT data collection: The performance trends of the test sections was monitored through periodic measurements of surface rutting. For construction platforms, the

surface profile was measured using a customized rut measurement device, shown in Fig. 11e, consisting of slide callipers and a perforated channel with holes every 50 mm. For HMA test sections, surface profiles were measured with the automated laser profiler shown in Fig. 11f. Every time ATLAS was stopped between runs, the wheel load deviator stress on top of the subgrade was measured with static wheel loading on top of the pressure cell locations. Sections were also inspected and mapped regularly for the appearance and/or propagation of cracks. Data for air and pavement layer temperatures, and precipitation were logged on an hourly basis.

- Post-APT activities: Following APT, several forensic analysis tests were conducted to better understand and assess the performance trends of the studied applications. These include: (1) FWD testing (Fig. 11g) to track changes in deflections before and after ATLAS testing; (2) HMA coring, to obtain accurate wheel path HMA thicknesses; (3) DCP testing of the subgrade/subbase/base (Fig. 11h) to assess the strength profiles of the substructure layers; (4) partially flooded pavement tests for the ASI test sections, to assess the effect of flooding on FWD deflections; (5) trenching of the test sections (Fig. 11i) to determine the as-constructed layer thicknesses and assess construction uniformity; and (6) extraction of intact chunks of stabilized QB bases/subbases (Fig. 11j) to saw-cut samples for UCS, and wet-dry and freeze- thaw durability testing.

Selected Results from APT and Post-APT Stages. A few results from the APT study were selected and presented in this section [38–41]. Figure 12a presents a comparison of the average wheel path rut progression for the stabilized QB pavement section. Generally, a satisfactory performance was observed for all the chemically stabilized QB test sections, with the two sections stabilized with 10% class C fly ash (i.e., C2S3 and C3S3) consistently accumulating higher surface rutting and rutting progression rates. Higher stress levels were applied in the QB base and subbase layers when the load levels were increased, thus accelerating the rut accumulation. A comparison of the stress levels on top of the subgrade is presented in Fig. 12b.

The stiffer chemically-stabilized test sections consistently recorded lower pressure values on top of the subgrade compared to the control section; both at the original and increased ATLAS load levels. Clearly, the stiffer stabilized base materials are changing the mechanism of stress distribution in the pavement structure, resulting in reducing subgrade pressures and rutting potential.

A unique feature of APT is the opportunity to collect data through core samples and trenching after the completion of the test. For example, DCP tests were conducted, at core locations, to compare the strength profile of the bound base and subbase layers and relate strength to surface rut accumulation. Figure 13a shows the surface rut accumulation after 40,000 passes plotted with the normalized DCP penetration rates. The strength profiles were found to correlate inversely with rutting progression. Stabilized samples were also extracted from the test sections and saw-cut to evaluate their wet-dry and freeze–thaw durability. Figure 13b summarizes results of wet-dry durability testing. Samples with cement-stabilized QB2 had significantly low soil–cement loss and slow rates of deterioration (2% or less average soil–cement loss

Table 2 Descriptions of the constructed QB test sections

Sample Name	QB Field Application	Description [T]
Cell 1[a]		
C1S1	Aggregate Subgrade Improvement (ASI)	Primary Crushed Rocks (PCR) blended with 25% QB1 by weight, constructed in two lifts using a vibratory roller compactor
C1S2	ASI	PCR with 16.7% QB1 by weight, constructed in one lift in a similar manner as C1S1
C1S3	ASI	Dense-graded CA 6 aggregates with 15% plastic fines content (passing No. 200 sieve) with a Plasticity Index (PI) of 8
C1S4	ASI	Dense-graded CA 6 aggregates with 15% nonplastic fines content (passing No. 200 sieve)
Cell 2[b]		
C2S1	Base Material	A blend of 70% QB2 and 30% FRAP[+], stabilized with 3% cement
C2S2	Base Material	A blend of 70% QB2 and 30% FRCA[++], stabilized with 3% cement
C2S3	Base Material	A blend of 70% QB2 and 30% FRAP, stabilized with 10% Class C fly ash
C2S4	Base Material	QB2 stabilized with 3% cement
Cell 3[b]		
C3S1	Base Material	QB3 stabilized with 3% cement
C3S2	Subbase Material	QB2 stabilized with 3% cement
C3S3	Subbase Material	QB2 stabilized with 10% Class C fly ash
C3S4	NA	Control section—Conventional aggregate base

[T] QB1, QB2, and QB3 refer to QB from three different sources. QB1 and QB3 are limestone fines. QB2 comprise dolomite fines.

[a] Cell 1 studies aggregate subgrade improvement (ASI) applications of QB. In situ subgrade CBR is 1%. Construction platform sections have 533 mm of ASI material topped with 76 mm of capping. Flexible pavement sections are further topped with 102 mm of HMA.

[b] Cells 2 and 3 study base and subbase applications of QB. In situ subgrade CBR is 6%. Sections have 305 mm of base; or 152 mm of base and subbase each; topped with 102 mm of HMA.

[+] FRAP = Fractionated Reclaimed Asphalt Pavements

[++] FRCA = Fractionated Recycled Concrete Aggregates

Fig. 11 Pavement Testing Process: **a** Geogauge® testing, **b** lightweight deflectometer, **c** nuclear density gauge, **d** pressure cell installation, **e** rutting measurement in construction platforms, **f** rutting measurement in paved sections with a laser profiler, **g** FWD, **h** DCP testing through cores, **i** a cross section of an open trench in C2S1, and **j** saw-cut durability samples extracted from the field

after 12 cycles); indicating durable applications. On the other hand, fly ash–stabilized samples had significantly higher soil–cement losses and rates of deterioration. Shown also in Fig. 13b are the UCS results of field-extracted samples. The average UCS and wet-dry durability results are well correlated (coefficient of determination, $R^2 = 0.852$).

Fig. 12 **a** Surface rut
accumulations with traffic
for test sections in Cells 2
and 3, and **b** pressure cell
readings on top of the
subgrade for the
instrumented test sections

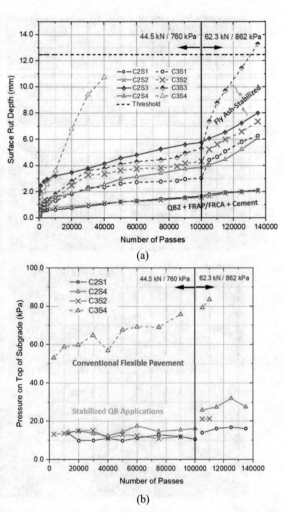

(a)

(b)

5 Final Remarks and Future Challenges

This chapter provides a general overview of instrumentation and full-scale accelerated pavement testing (APT) that provides basic understanding of pavement structural
response, especially when new and innovative materials and techniques are applied.

The main sensors used in pavement instrumentation were used to measure
displacements, stresses, strains, and environmental conditions (moisture content and
temperature). Optical fibre sensors and piezoelectric transducers were presented as
promising technologies that could be evolved for long-term pavement sensing.

The chapter presents the main characteristics of selected APT devices and their
capabilities for research potential to assess effectively and efficiently the potential
long-term performance of highway and airfield pavement systems under control

Fig. 13 a DCP penetration rates into stabilized base and subbase layers, and **b** Wet-dry durability and unconfined compressive strength results for field samples (1 psi = 6.89 kPa)

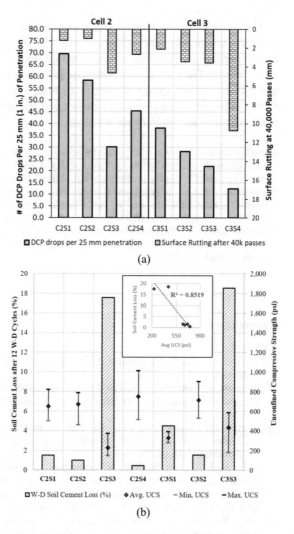

(a)

(b)

conditions within a short period of time. Two case studies are presented to illustrate the application of APT. The SUPREMA project (Portugal) was used to assess the geomechanical and geoenvironmental characteristics of recycled aggregates when used as unbound granular layers in pavements; and the Illinois Center for Transportation (U.S.) was used to demonstrate the utilization of quarry fines in pavement subsurface/foundation layers.

Smart sensing of pavement is expected to be adapted for continuous and long-term monitoring; especially when considering autonomy, robustness, long-term health monitoring, wireless communication, data acquisition, and cost. Recent research in the use of smart aggregate to quantify the layer stiffness characteristics or moduli of constructed aggregate layer is promising.

The development of wireless sensor networks would significantly modify current pavement instrumentation and monitoring. The integration of low-cost and miniaturized sensors in pavement network would support network pavement management systems. The ease of data collection [e.g., by Radio Frequency (RF) and Bluetooth (BLE) technologies] could improve the rapid prediction of potential pavement distresses before they appear on the surface and provide the means for better planning of maintenance and rehabilitation actions.

Acknowledgements The authors gratefully acknowledge the support of the Center of Civil Engineering Research and Innovation for Sustainability (CERIS), the National Laboratory for Civil Engineering (LNEC), and the financial support of the R&D project "PTDC/ECM/100931/2008—SUPREMA—Sustainable Application of Construction and Demolition Recycled Materials (C&DRM) in Road Infrastructures" by FCT (Foundation for Science and Technology). The Illinois Department of Transportation (IDOT) is acknowledged for funding the quarry by-products APT research as part of the Illinois Centre for Transportation R27-168 project. José Neves is grateful for the Foundation for Science and Technology's support through funding FCT-UIDB/04625/2020 from the research unit CERIS.

References

1. Barriera, M., Pouget, S., Lebental, B., Rompu, J.V.: In situ pavement monitoring: a review. Infrastructures **5**, 18 (2020)
2. Graziano, A., Marchetta, V., Cafiso, S.: Structural health monitoring of asphalt pavement using smart sensor networks: a comprehensive review. J. Traffic Transp. Eng. (English Edition) **7**(6), 639–651 (2020)
3. Maeijer, P.K., Luyckx, G., Vuye, C., Voet, E., Bergh, W., Vanlanduit, S., Braspenninckx, J., Stevens, N., Wolf, J.: Fiber optics sensors in asphalt pavement: state-of-the-art review. Infrastructures **4**, 36 (2019)
4. Ji, X., Hou, Y., Chen, Y., Zhen, Y.: Fabrication and performance of a self-powered damage-detection aggregate for asphalt pavement. Mater. Design **179** (2019)
5. Lee, J.S., Santamarina, J.C.: Bender Elements: Performance and signal interpretation. J. Geotech. Geoenviron. Eng. **131**(9), 1063–1070 (2005)
6. Kang, M., Qamhia, I., Tutumluer, E., Hong., W.T., Tingle, J.: Bender element field sensor for the measurement of pavement base and subbase stiffness characteristics. Transp. Res. Record 1–14 (2021)
7. Lajnef, N., Chatti, K., Chakrabartty, S., Rhimi, M., Sarkar, P.: Smart pavement monitoring system. Report No. FHWA-HRT-12–072. Federal Highway Administration (2013)
8. Ji, X., Hou, Y., Chen, Y., Zhen, Y.: Fabrication and performance of a self-powered damage-detection aggregate for asphalt pavement. Mater. Des. **179**, 107890 (2019)
9. Alavi, A.H., Hasni, H., Lajnef, N., Chatti, K.: Continuous health monitoring of pavement systems using smart sensing technology. Constr. Build. Mater. **114**, 719–736 (2016)
10. Greenslade, F.R.: Development of a new pavement strain coil measuring system at CAPTIF. In: J.P. Aguiar-Moya et al. (eds.) The Roles of Accelerated Pavement Testing in Pavement Sustainability, pp. 633–643. Springer (2016)
11. Wang, X., Shen, S., Huang, H., Zhang, Z.: Towards smart compaction: particle movement characteristics from laboratory to the field. Constr. Build. Mater. **218**, 323–332 (2019)
12. Duong, N.S., Blanc, J., Hornych, P., Menant, F., Lefeuvre, Y., Bouveret, B.: Monitoring of pavement deflections using geophones. Int. J. Pavement Eng. **21**(9), 1103–1113 (2020)

13. Bhuyan, H., Scheuermann, A., Bodin, D., Becker, R.: Soil moisture and density monitoring methodology using TDR measurements. Int. J. Pavement Eng. **21**(10), 1263–1274 (2020)
14. Suchorab, Z., Widomski, M.K., Lagód, G., Barnat-Hunek., Majerek, D.: A noninvasive TDR sensor to measure the moisture content of rigid porous materials. Sensors **18**, 3935 (2018)
15. Godoy, J., Haber, R., Muñoz, J.J., Matía, F., García, A.: Smart sensing of pavement temperature based on low-cost sensors and V2I communications. Sensors **18**, 2092 (2018)
16. Gilchrist, M.G., Hertman, A.M., Owende, P.M.O., Ward, S.M.: Full scale accelerated testing of bituminous road pavement mixtures. Key Eng. Mater. **204–205**, Trans Tech Publications, Ltd., pp. 443–45 (2001)
17. Steyn, W., Anochie-Boateng, J., Fisher, C., Jones, D., Truter, L: Calibration of full-scale accelerated pavement testing data using long-term pavement performance data. In: 4th International Conference on Accelerated Pavement Testing. Davis, California, USA (2012)
18. Greene, J.: FWD User's Group. Accelerated Pavement testing to Assess Performance of Pavement Systems. APT, Florida's Experience (2010)
19. TRB Committee. Full-scale accelerated pavement testing (AFD40) https://sites.google.com/site/afd40web/apt-facility-links. Last Accessed 12 Dec 2020
20. Kozel, M., Kyselica, M., Mikolaj, J., Herda, M.: Accelerated pavement testing in Slovakia. Procedia Eng. **153**, 310–316 (2016)
21. Ritter, J., Rabe, R., Wolf, A.: Analysis of the long-term structural performance of flexible pavements using full-scale accelerated pavement tests. Procedia. Soc. Behav. Sci. **48**, 1244–1253 (2012)
22. Anochie-Boateng, J.K., Steyen, W., Fisher, C. Truter, L.: A link of full-scale accelerated pavement testing to long term pavement performance study in the Western Cape province of South Africa. In: J.P. Aguiar-Moya et al. (eds.) The Roles of Accelerated Pavement Testing in Pavement Sustainability, pp. 67–79. Springer (2016)
23. Hornych, P., Simonin, JM., Piau, J.M., Cottineau, L.M., Gueguen, I., Jacob, B.: Evaluation of weight in motion sensors on the IFSTTAR accelerated testing facility. In: Aguiar-Moya J., Vargas-Nordcbeck A., Leiva-Villacorta F., Loría-Salazar L. (eds.) The Roles of Accelerated Pavement Testing in Pavement Sustainability. Springer, Cham (2016)
24. Sirin, O., Kim, H., Tia, M., Choubane, B.: Comparison of rutting resistance of unmodified and SBS-modified Superpave mixtures by accelerated pavement testing. Constr. Build. Mater. **22**, 286–294 (2008)
25. Steyn, W., Maina, J.: Guidelines for the use of accelerated pavement testing data in autonomous vehicle infrastructure research. J. Traffic Transp. Eng. (English Edition) **6**(3), 273–281 (2019)
26. Freire, A.C., Neves, J., Roque, A., Martins, I.M., Antunes, M.L.: Feasibility study of milled and crushed reclaimed asphalt pavement for application in unbound granular layers. Road Mater. Pavement Design **22**(7), 1500–1520 (2021)
27. Neves, J., Freire, A.C., Roque, A.J., Martins, I.M., Antunes, M.L.: Performance of C&DW materials for road application validated by field monitoring. In: International HISER Conference on Advances in Recycling and Management of Construction and Demolition Waste, Delft University of Technology. Delft, The Netherlands (2017)
28. Freire, A.C., Neves, J., Roque, A.J., Martins, I.M., Antunes, M.L., Faria, G.: Use of construction and demolition recycled materials (C&DRM) in road pavements validated on experimental test sections. In: 2nd international Conference WASTES 2013: Solutions, Treatments and Opportunities, pp. 91–96 (2013)
29. EN 933–11. Tests for geometrical properties of aggregates—Part 11: Classification test for the constituents of coarse recycled aggregates. Brussels (Belgium): Comité Européen de Normalisation (CEN) (2009)
30. EN 933–1. Tests for geometrical properties of aggregates—part 1: determination of particle size distribution—sieving method. Brussels (Belgium): Comité Européen de Normalisation (CEN) (2012)
31. Roque, A.J., Martins, I.M., Freire, A.C., Neves, J., Antunes, M.L.: Assessment of environmental hazardous of construction and demolition recycled materials (C&DRM) from laboratory and field leaching tests application in road pavement layers. Procedia Eng. **143**, 204–211 (2016)

32. Neves, J., Freire, A.C, Roque, A.J., Martins, I., Antunes, M.L., Faria, G.: Utilization of recycled materials in unbound granular layers validated by experimental test sections. In: 9th International Conference on the Bearing Capacity of Roads, Railways and Airfields, Norway (2013)

33. EN 12457–4. Characterisation of waste—leaching—compliance test for leaching of granular waste materials and sludges—part 4: one stage batch test at a liquid to soil ratio of 10 l/kg for materials with particle size below 10 mm (Without or with size reduction). Brussels (Belgium): Comité Européen de Normalisation (CEN) (2002)

34. Chesner, W.H., Collins, R.J., MacKay, M: User guidelines for waste and by-product materials in pavement construction. Publication FHWA-RD-97–148. FHWA, U.S. Department of Transportation, Washington D.C. (1998)

35. Hudson, W.R., Little, D.N., Razmi, A. M., Anderson, V., Weissmann, A. J.: An investigation of the status of by-product fines in the United States. University of Texas at Austin. International Center for Aggregates Research, Austin (1997)

36. Stroup-Gardiner, M., Wattenberg-Komas, T.: Recycled materials and by-products in highway applications. Volume 4: mineral and quarry by-products. NCHRP Synthesis of Highway Practice (Project 20–05, Topic 40–01), Transportation Research Board, Washington D.C. (2013)

37. Tutumluer, E., Ozer, H., Hou, W., Mwumvaneza, V.: Sustainable aggregates production: Green applications for aggregate by-Products. Final Report. Illinois Center for Transportation and Illinois Department of Transportation, Rantoul (2015)

38. Qamhia, I., Tutumluer, E., Ozer, H.: Field performance evaluation of sustainable aggregate by-product applications. Final Report FHWA-ICT-18–016. Illinois Center for Transportation/Illinois Department of Transportation, Rantoul (2018)

39. Qamhia, I., Tutumluer, E., Ozer, H., Shoup, H., Beshears, S., Trepanier, J.: Evaluation of chemically stabilized quarry byproduct applications in base and subbase layers through accelerated pavement testing. Transp. Res. Record J. Transp. Res. Board **2673**(3), 259–270 (2019)

40. Qamhia, I.: Sustainable pavement applications utilizing quarry by-products and recycled/nontraditional aggregate materials. Doctoral Dissertation, University of Illinois at Urbana-Champaign (2019)

41. Qamhia, I., Tutumluer, E., Ozer, H., Boler, H.: Durability aspects of stabilized quarry by-product pavement base and subbase applications. Final Report FHWA-ICT-19–012. Illinois Center for Transportation/Illinois Department of Transportation, Rantoul (2019)

Monitoring of Pavement Structural Characteristics

Simona Fontul⊙**, José Neves**⊙**, and Sandra Vieira Gomes**⊙

Abstract Testing and monitoring of transport infrastructures are challenging tasks due to the increase in traffic intensity and load. Non-destructive tests are used for pavement diagnosis, most of them at traffic speed, in order to avoid interaction with users. With the development of new technologies and enhanced quality of data collected, a combination of different monitoring methods can be used to assess the pavement condition along its entire length. Equipment, able to collect significant amounts of data are used, such as traffic speed deflectometers, ground penetrating radar, laser 3D, remote sensing, digital cameras, etc. One of the challenges is the data processing, due to the significant amount collected, sensitivity to test condition and correlations between different tests. This chapter addresses the main equipment and testing methodologies available nowadays for structural evaluation and integrity detection. Traffic considerations are additionally covered, as they are responsible for the pavement loading. The knowledge about its magnitude and the way that they are transmitted from different vehicles configurations are also presented. Additionally, future trends for data integration, such as Building Information Modelling are referred herein, with some examples. In addition, approaches that are still at prototype phase, such as remote sensing detection of settlements are mentioned. The main challenges and future perspectives of continuous structural monitoring are referred herein.

Keywords Ground penetrating radar · Non-destructive tests · Pavement · Structural evaluation · Traffic characterisation · Traffic speed deflectometer

S. Fontul (✉) · S. V. Gomes
LNEC, National Laboratory for Civil Engineering, Lisbon, Portugal
e-mail: simona@lnec.pt

S. V. Gomes
e-mail: sandravieira@lnec.pt

J. Neves
CERIS, Department of Civil Engineering, Architecture and Georesources, Instituto Superior Técnico, Universidade de Lisboa, Lisboa, Portugal
e-mail: jose.manuel.neves@tecnico.ulisboa.pt

© The Author(s), under exclusive license to Springer Nature Switzerland AG 2023   187
C. Chastre et al. (eds.), *Advances on Testing and Experimentation in Civil Engineering*, Springer Tracts in Civil Engineering,
https://doi.org/10.1007/978-3-031-05875-2_8

1 Introduction

The evaluation of pavements structural conditions plays an essential role in designing adequate maintenance and rehabilitation strategies at the project level and identifying sections with structural deficiencies at the network level. In addition to functional indicators (e.g., roughness, rutting, skid resistance, and cracking), structural monitoring can also be essential to Pavement Management Systems (PMS) to obtain a more accurate pavement assessment condition.

The primary indicator of the structural condition is the bearing capacity of the pavement and subgrade. Since the first experiments on pavement deflection tests in the 1950s, significant efforts were made to develop non-destructive tests (NDT) adapted to structural monitoring. This chapter presents a literature review on deflectometer tests, focusing on the latest advancements on high-speed deflectometers suitable for the continuous monitoring of structural pavement behaviour. These continuous deflectometers are more representative of the traffic influence on the pavement response and more adapted to network-level evaluations.

Another essential information for structural assessment is the pavement geometry and its characteristics, and their variation along the length. In the last decade, the Ground Penetrating Radar (GPR) has been increasingly used for pavement evaluation. The main advantage of this test consists in performing measurements at traffic speed, continuously, without contact with the pavements and with the possibility of visualising the data during the measurement. Layer thickness, for both bonded and unbounded materials, is the main information obtained. The location of changes in pavement geometric structure are identified with precision. More recently, other pavement characteristics such as bituminous mixture density variation, existing cracking, layer interface debonding, water trapped in granular layers and settlements of the subgrade can be located and diagnosed with confidence. The information obtained with GPR, together with loading tests performed with TSD are crucial information for the structural evaluation of pavements.

Traffic considerations are not made nowadays during structural monitoring campaigns, only in case of design of new pavements and reinforcement. With the development of new technologies, it is possible to detect and track the traffic, mainly heavy load vehicles per pavement section. The influence of traffic into the structural behaviour of the pavement and the main factors that affect it, from overloading to tyre configuration and pressure, are presented herein. In addition, possible tools for traffic counting and identification are referred as opportunities for data integration with structural monitoring tests. In this way, a more realistic evaluation of pavement condition along its life cycle becomes possible.

As future trends in structural monitoring, this chapter addresses two aspects that are promising for data processing and manipulation at network level: the integration between TST and GPR and the implementation of the monitoring results on Building Information Modelling (BIM). Some examples of pavement monitoring data implemented into BIM are also presented. Along with this chapter, the main challenges and possible future perspectives on structural monitoring of pavements are addressed. The previous chapter regarding the pavements' instrumentation has described the most recent advances on sensors for deflection and load measurements that are very important in pavement structural monitoring.

2 Continuous Load Testing

The pavements' structural monitoring has been currently assessed by traditional deflectometers, e.g., Benkelman beam and the Falling Weight Deflectometer (FWD).

The FWD is the most widely NDT to assess the road and airfield structural condition in terms of the bearing capacity (Fig. 1a) [1]. It consists of measuring the pavement surface deflections generated by the impact of a falling mass by sensors (e.g., geophones) mounted along the circular loading plate's centreline at various distances. Figure 2a shows a schematic representation of the FWD. The deflection basin corresponds to the pavement response's vertical deformations due to the load test and can be related to pavement and subgrade stiffness through back-calculations procedures.

The FWD can be performed according to standard test methods of ASTM D4694 and ASTM D4695 [2, 3]. Since the first prototype created in France in the early 1960s, the FWD has been extensively studied and applied for decades. Significant experience in its use and back-calculations is available worldwide. Manufacturers have

Fig. 1 Deflectometers devices: **a** FWD [7]; **b** LWD; **c** RWD [4]; **d** TSD [8]

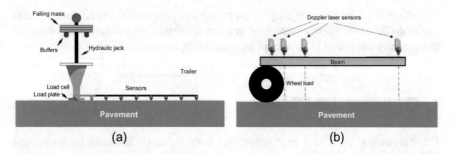

Fig. 2 Comparison of measuring principles of non-continuous and continuous deflectometers (adapted from [7, 9]): **a** FWD; **b** TSD

provided different FWD versions with similar operating principles. The Light Weight Deflectometer (LWD) is a portable model producing a limited load test (Fig. 1b). It is commonly used for the in-situ quality control of unbound granular layers and subgrade soil. The Heavy Weight Deflectometer (HWD) can apply higher loads, and is more suitable for the monitoring of thicker runway pavements in airfield infrastructures.

The FWD provides at the test site the deflection basin generated by a dynamic load pulse and characterized by a haversine shape, which is essential to understand the pavement structural behaviour. However, the main disadvantage of FWD is related to being a non-continuous loading test unable to be performed at traffic speed. Efforts have been made to develop alternative techniques to assess the continuous monitoring of the structural pavement behaviour, more representative of the traffic conditions influence on the structural response of the pavement, and more adapted to network-level evaluations. Also, continuous methods can detect singularities of the pavement, such as zones affected by section joints, cracks, and potholes.

Advanced deflectometers were developed, focusing on continuous deflection profiles during the testing vehicle movement at traffic speed. These circumstances are more realistic both in terms of the traffic loads and the pavement and subgrade soil response. The continuous monitoring without disturbing the traffic flow and in a short period of time is also a significant advantage to enhance the testing productivity.

Since the preliminary prototypes emerged in the 1990s in Europe and the USA, different continuous high-speed deflectometers have been in constant development. Currently, the Rolling Wheel Deflectometer (RWD) (Fig. 1c) and the Traffic Speed Deflectometer (TSD) (Fig. 1d) are the most worldwide investigated and applied devices, recognized as the most promising for pavement structural evaluation [4–6]. In general, these devices consist of large trucks or semi-trailers able to apply the load directly and measure the pavement's deflections.

Table 1 summarizes the literature review on RWD and TSD devices' most essential characteristics, mostly related to the load and deflection measurements (type, location and accuracy) systems and the operational speed [4–6].

The Applied Research Associates, Inc. (ARA) has developed the Rolling Wheel Deflectometer for the Federal Highway Administration in the USA. The RWD uses

Table 1 Characteristics of the continuous deflection devices [5, 11]

Parameter	Device	
	RWD rolling weight deflectometer	TSD traffic speed deflectometer
Manufacturer	Applies research associates, Inc. (ARA)	Greenwood Engineering A/S (GE)
Country	USA	Denmark
Load system	Single-axle; dual-tire 80.068 kN	Single-axle; Dual-tire 48.930 kN
Deflection system	Distance laser sensors (deflection)	Doppler laser sensors (deflection velocity)
Deflection location	Behind the load wheel	Behind and in front of the load wheel
Defection accuracy	0.0635 mm	0.1016 mm
Operation speed	16–104 km/h	60–80 km/h

distance laser sensors mounted behind the longitudinal centreline's load wheel to directly measure the deflections produced by the dual-tire (80.068 kN) of a single axle. This device can operate at speeds ranging from 16 to 104 km/h [4].

In Denmark, the Greenwood Engineering A/S (GE) developed the Traffic Speed Deflectometer in collaboration with the Transport Research Laboratory and the Danish Road Institute. The TSD is equipped with Doppler laser sensors, mounted on the support beam located beneath the test vehicle and on the load midline, to measure pavement motion based on the Doppler effect principle. Load test is applied by the dual-tire (48.930 kN) of the rear right single axle. The operation speed of TSD is in the range of 60 km/h to 80 km/h. Figure 2b presents schematically the arrangement of the sensors in the rigid beam of the TSD. Since the first device appears in 2004, more TSD are actually operating in Australia, China, Denmark, Italy, Poland, South Africa, the USA, and United Kingdom [8].

The RWD and TSD devices use laser sensors to measure pavement surface deflections but according to distinct principles [1, 8, 10]. The RWD uses a spatial methodology based on distance laser sensors to evaluate the pavement surface profile. The deflection corresponds to the difference between the loaded and unloaded pavement surface. The TSD uses velocity-sensing lasers based on the Doppler principle to measure the instantaneous pavement deflection velocity. The Doppler lasers, installed at an angle from the vertical, measure the sensor's horizontal speed (equal to the driving speed of the test vehicle) and the vertical deflection velocity. The deflection slope—the ratio of the vertical to the horizontal speeds—allows the mathematical calculation of the deflection basin of the pavement surface.

The RWD and TSD collect and process data (load and deflections) in real-time by the acquisition system installed inside the test vehicle's cabin. At traffic speed, the two devices can test large road network extensions per working day. Additional equipment can be installed simultaneously, like imaging systems and Ground Penetrating Radar. Both devices can synchronize the data collected by these complementary systems with structural behaviour analysis.

Other continuous deflection devices were developed over time [6], although some prototypes need further validation to be implemented in current practice. The Rolling Dynamic Deflectometer (RDD), developed at the Center for Transportation Research of the University of Texas at Austin (USA), applies a vibrator load and uses accelerometers for pavement deflections measurements. The Road Deflection Tester (RDT), developed at the Swedish National Road and Transport Research Institute (VTI), consists of a moving truck equipped with laser sensors to measure pavement deflections. The Image-Based Deflection (IMD) was developed by LCPC (France), and this device uses a structured light pattern on the pavement surface. In the case of airfield infrastructures, the Airfield Rolling Weight Deflectometer (ARWD) was created especially for the US Air Force for the structural monitoring of runway pavements. The Laser Dynamic Deflectometer (LDD), developed at the Transportation Research Center of Wuhan University, China, uses Doppler laser sensors but, unlike TSD, it is equipped with a gyroscope for measuring and compensating the vibration of the support beam and reducing the calibration time consuming [9].

The great current challenge on continuous structural monitoring is the balance between the high costs of the investments in high-speed deflection devices and the benefits they can bring to PMS implementation. Intensive research has been carried out on demonstrating the feasibility of the high-speed deflection devices for the continuous structural monitoring of road pavements at network level. In general, these studies use the FWD as a reference device to assess the accuracy and repeatability of the data and to understand the real capabilities of the continuous deflectometers [11, 12]. Advanced studies are also being performed on more adapted back-calculation of high-speed deflectometer tests data to evaluate pavement bearing capacity [13].

3   Continuous Pavement Geometry and Characteristics Assessment

Ground Penetrating Radar (GPR) represent an essential tool for pavement structural assessment as allows for continuous detection of the pavement layer interfaces through electro-magnetic waves (see Fig. 3) [1]. A detailed description of the methodology applied to pavement assessment and a comprehensive theoretical background is presented in [14, 15]. The Standard and technical guides that address the GPR application in pavements are: ASTM D6432 [16], ASTM D4748 [17], DMRB 7.3.2 [18] and GS1601 [19].

There are several applications of GPR to road and airfield pavements evaluation, with different type of antennas. An extensive review of the main GPR application

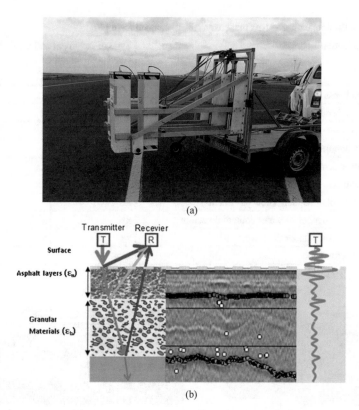

(a)

(b)

Fig. 3 Ground penetration radar: **a** Air coupled antennas; **b** measuring principle on pavements

to flexible and rigid pavements evaluation is presented by Solla et al. [20] and are summarized in Table 2.

The main characteristics of GPR equipment used for structural monitoring of pavements are briefly presented herein [1, 20]. Antennas, specifically developed for pavement monitoring, are functioning suspended at 40–50 cm above the pavement, allowing for measurements at traffic speed. Generally, they are horn antennas with frequencies are ranging between 1 and 2.5 GHz. These frequencies provide a high vertical resolution, of about 5–2 cm, while the depth penetration is between 1.2 and 0.4 cm [1, 21, 22]. The measurement is continuous (usually a reading is taken every 25 cm) and in this way, information regarding layer thickness is gathered and an accurate identification of changes in the pavement structure can be made.

On one hand, the lower frequencies antennas enable the assessment of base and subbase layers, and subgrade [21–23] (see Fig. 4), and rebar detection in concrete pavements [24] (see Fig. 5).

On the other hand, the high frequencies antennas enable detecting distresses in asphalt layers, such as cracking [25], layer debonding [26], density [27, 28] and concrete permittivity [29]. Generally, specific signal processing is required for

Table 2 GPR application to pavement assessment, adapted from [20]

GPR application	Type of pavement and level of concept development[a]	
	Flexible	Rigid
Layer thickness	xxx	xx
Detection of cracks and voids in pavement	xx	x
Quality control of asphalt (new pavements)	xx	
Layers debonding	x	
Rebar cover depth		xx
Rebar location		xx
Quality control of asphalt, (new pavements)	xx	
Moisture and changes of water content detection	xx	x
Detection of voids under the pavement	xx	x
Subgrade settlements	xxx	

[a] x—low …xxx—high

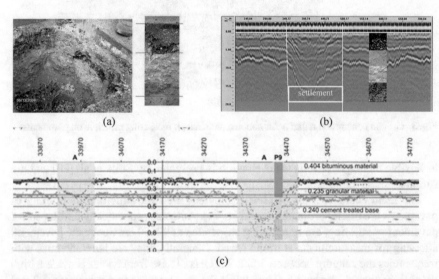

(a)   (b)

(c)

Fig. 4 Subgrade settlement detection of flexible pavement with GPR, adapted from [21]: **a** test pit, **b** radargram, **c** interpretation

detecting distresses and materials characteristics and condition, such as time and frequency domain processing [20, 27, 29, 30].

For structural monitoring, the main information needed from GPR is the layer thickness and changes in pavement structure along the road. This information is used together with load testing results for bearing capacity evaluation [1, 21, 31–33]. Its use enables a continuous view of pavement structure and a confident structural

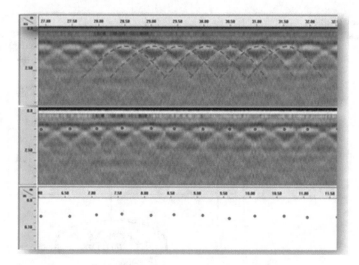

Fig. 5 Rebar detection on concrete reinforced runway with GPR [24]

assessment. Bezina et al. [33] presents a spatial representation of GPR that enable a better visualization of the layer thickness.

4 Integrated Approach of Continuous Monitoring

The standard procedure for bearing capacity evaluation is to measure the deflections in a considerable number of points located along several lines parallel to the pavement axis and spaced in 50–200 m. The GPR tests are performed along the same test lines as the FWD. In this way, confident information of the pavement structure is obtained for each FWD test point. The combination of results from Falling Weight Deflectometer and Ground Penetrating Radar tests provide a major improvement to the quality of the results of pavement structural evaluation. Furthermore, the use of artificial intelligence, e.g., Artificial Neural Networks (ANN) is an interesting option for the back-calculation of layer moduli. In this way, it became possible to perform the interpretation of all FWD test results and consequently to evaluate the pavement structure in all FWD test points. For this purpose, the GPR results are processed in order to get an average value of layer thickness over a 5 m interval around each FWD test point [1, 21]. Due to the large amount of data involved in the process, an efficient tool is needed for the back calculation based on Artificial Intelligence (AI). The flowchart of the structural interpretation approach is presented in Fig. 6. The same approach can be used in case of TSD results combined with GPR.

Despite being a recent technology, as already presented in Sect. 2, due to its continuous measurement, there are studies of TSD application together with GPR, in order to estimate the remaining life of the pavements. [34]. This represent an

Fig. 6 Flowchart for
pavement structural
evaluation

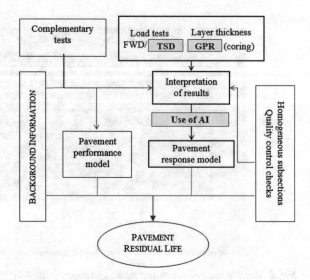

important step forward on the Pavement Management Systems, mainly at network
level, as they are nowadays mainly based on surface distress. The combination of
both monitoring techniques provides information on pavement layer mechanical
properties, namely the elastic modulus.

An example of this approach is presented herein [34]. The main inputs and
processing steps used during remaining life estimation are: (1) Deflections (from
TSD slopes); (2) Layer Thickness (GPR); (3) Layer Moduli calculation (steps $1 + 2$
using linear elastic program); (4) Effective Structural Number $(2 + 3)$; (5) Required
Structural Number for 20-year life $(3 + \text{ESALs } 80 \text{ kN})$; (6) Required Overlay Thickness (from 5); 7. Remaining Service Life (from 3, 4, and ESALs 80kN), where ESALs
80kN is the predicted loading. An example of the results obtained on the Pilot Project
on Idaho District, available via statewide geodatabase is presented in Fig. 7.

Another integrated approach of continuous monitoring is performed by
combining, loading tests (FWD), thickness measurement (GPR) and road surface
profile, roughness (IRI), assessed continuously with laser profilometer [35, 36]. It
is shown that this approach enables a better identification of subgrade condition,
based in IRI data and therefore proper maintenance actions. In other words, as IRI
indicates heterogeneous condition along the pavement, due to settlements caused by
the lack of bearing capacity of subgrade, GPR can confirm the real cause of high
unevenness (IRI), detecting these subgrade settlements. Based on GPR diagnostic,
proper maintenance measures can be adopted addressing subgrade improvement and
not only milling and overlaying processes.

Fig. 7 Remaining life map (years) based on TDS and GPR monitoring tests. Adapted from [34]

5 Traffic Considerations

Traffic is responsible for the main forces exerted on the pavement, and therefore it is of high importance the knowledge its magnitude. A significant diversity of technologies is available to perform traffic volumes data collection, as described in another chapter, which focuses on intelligent traffic monitoring systems. However, the availability of this information is scarce, only being collected when needed and not generally collected in a systematic way for all the networks. Urban areas usually present a higher coverage, as these systems are commonly used for traffic management, for the purpose of reducing congestion.

To overcome the lack of data, surrogate indicators are frequently used, as it is the vehicle fleet characterization, which allows to keep up to date the variation of the number and characteristics of heavy vehicles. Figure 8 presents an example the heavy vehicle percentage evolution in Portugal from 1975 to 2019. A clear reduction over time can be identified.

An increased level of detail is also possible in terms, for instance, of the type of vehicle and number of axles, as presented in Fig. 9. This type of information may be used to improve the estimates about the loads that are induced in the road pavements.

Nevertheless, the forces exerted in the pavement may be significantly increased if the vehicles run overloaded [37]. It is well known that overweight traffic can be one of the main sources of damage in road pavements. It reduces the pavement's life expectancy and also increases the maintenance costs [38]. Other negative effects are related to road safety due to increased stopping distances and lower vehicle

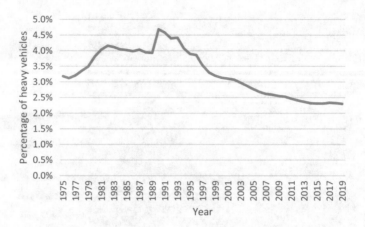

Fig. 8 Heavy vehicles percentage evolution in Portugal from 1975 to 2019

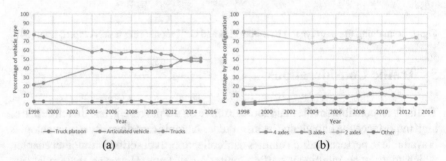

Fig. 9 Evolution of heavy vehicles percentage in Portugal from 1998 to 2016 adapted from [INE—www.ine.pt, 37]: **a** by type of vehicle; **b** by number of axles

maneuverability. Capacity may also be compromised due to the generally slower speeds of the overweight vehicles, particularly in hilly roads.

Nevertheless, the forces exerted in the pavement may be significantly increased if the vehicles run overloaded [37]. It is well known that overweight traffic can be one of the main sources of damage in road pavements. It reduces the pavement's life expectancy and also increases the maintenance costs [38]. Other negative effects are related to road safety due to increased stopping distances and lower vehicle maneuverability. Capacity may also be compromised due to the generally slower speeds of the overweight vehicles, particularly in hilly roads.

Studies developed in South Africa concluded that legally loaded heavy vehicles causes a small amount of pavement damage (40%) when compared to overloaded heavy vehicles that cause approximately 60%, as shown in Fig. 10 [39].

A common way to collect information about overloading is through Weigh-in-Motion (WIM) systems [40], which is an efficient system for evaluation the traffic composition and weight without interfering with the users and an important tool for

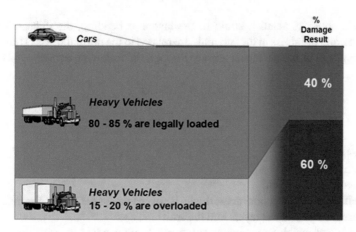

Fig. 10 Damage Caused by Overloaded Vehicles [39]

overweight control and potentially for enforcement as well. It is recommended that, in order to obtain useful data, a WIM network with a good coverage is required to provide effective results in the necessary enforcement [41].

To adequately design a pavement, adequate loading characteristics must be considered about the expected traffic it will encounter. In the majority of the approaches, this traffic is characterized by using [37]:

(1) an estimate of the annual average daily truck traffic (AADTT);
(2) an estimate of the annual truck traffic growth rate;
(3) a truck factor (TF) which converts the mixed heavy-traffic stream into a number of equivalent single-axle loads (ESALs). This conversion requires equivalent single-axle load factors (EALFs) which represent the damage caused by the passage of the actual axle load relative to the damage caused by each pass of the ESAL on the pavement.

These estimations are not constant either in time or in space: they vary by country and in time, and should be systematically reevaluated in order to get the best fitted estimates.

Pavement design should be made for expected lifetime traffic loading, that is less biased with more fitted estimates. Heavy-traffic data is of course crucial information for pavement design, as they are responsible for the higher impacts on the pavement.

Nevertheless, several factors affect the way that these loads are applied in the pavement: axle or wheel load, interaxial distance between wheels, the number of tyres in the wheel, tyre pressure, type of tyre and suspension and the vehicle speed.

To evaluate the impact of the load transmitted by the wheel on the durability of the pavement, it is first necessary to determine the dependence of the deflection of the pavement, the relative strain or stress at bending of the layer, as well as the stress in the subgrade from this load. Then, the comparative impact of loads of different magnitudes on the durability of road pavement is determined based on the test results of road-building materials on fatigue at different values of stress or strain.

Tensile stress or relative strain in the layers at bending, the deflection of the pavement, and the stress in the subgrade are related to the normal stress Q transmitted from the wheel to the pavement surface through the following equation [42]:

$$\sigma, \varepsilon, \omega \approx c \times Q^{\alpha} \tag{1}$$

where

σ stress;

ε relative strain;

ω pavement deflection;

c   coefficient that depends on the thickness of the pavement layers, the mechanical characteristics of the layers and the soil;

α coefficient which varies between 0.5 and 1, since when the load on the wheel increases, the area through which it is transmitted usually also increases.

The capacity to resist to fatigue damage is determined by testing the durability of material samples under laboratory conditions at controlled amplitude of stresses σ or strains ε.

As previously mentioned, the tyre air pressure is also a factor that affects the way that loads are applied in the pavement. As the tyre pressure increases, the wheel treadprint area reduces, i.e. the load on the tyre surface becomes more concentrated—see Fig. 11. As a result, the stress in the road surface increases.

As increasing air pressure in the tyres reduces their life-time costs [43], truck owners tend to do so, in order to reduce costs. However, this action may reduce until 90% of rolling resistance of pneumatic tyres [44].

However, an increase in pressure has an adverse effect on the durability of the road surface. Tests performed in the University in Texas [45] revealed that an increase of air pressure in tyres from 0.517 MPa (normative pressure in the calculations of road pavements) to 0.862 MPa (actual air pressure in tyres of 85% trucks in Texas) leads to an increase in relative deformation at bending of asphalt concrete surface with thickness of 2.5 cm by 20–30%, and for surface thickness of 10 cm—by 10%. In addition, the paper notes that the generally accepted assumption of the uniform

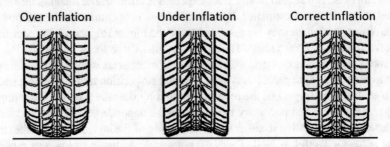

Fig. 11 Treadprint area for different tyre pressure. Adapted from https://www.bridgestone. sg/en/tyre-clinic/tyre-talk/tyre-tread-wear-causes

distribution of normal pressure over the tread print area leads to an underestimation of the relative deformation of the surface by about 2 times compared with its relative deformation in its actual non-equal distribution.

Sousa et al. [46] studied the influence of the suspension type of twin and triple axes on the road pavement durability, using the horizontal relative elongation in the pavement to obtain the influence of dynamic load on the service life. The authors found that the most unfavorable suspension type was the free-beam, while the most favorable was torsion bar suspension and cloverleaf suspension. It was also stressed that if modal frequencies caused by irregularities on the surface are close to the suspension's natural frequency, the dynamic part of the load transmitted by the wheel to the road surface increases.

Interaxial distance has also shown to have an impact on road surface in several studies performed by Radovsky [42]. His experiments showed that the impact curves of closely placed axes on road surface overlap interaxial distance between 2 and 2.5 m. This overlapping is not symmetrical, which is explained by the viscoelastic properties of road construction materials and soil.

This author also observed that the vertical stresses and displacements of the moving single axis wheel are greater than those in front of it. Therefore, when two equally loaded wheels follow in quick succession, the maximum vertical stress and movement under the rear wheel should be greater than that under the front wheel.

Vertical movement and stress caused by the second and third (twin) axes overlap. However, the impact of the second axis wheels on stress and movement under the third axis wheels is greater than the impact of the third axis wheels on stress and movement under the second axis wheels, although the load on these wheels is equal.

Axle loads are not the only factor that impacts road pavement: the number of axles and their position, and the number of wheels in the axle also have an influence.

In case of dual tires being equally loaded, and the wheel tread print centers are at a normal distance of 32–38 cm from each other, the vertical stresses on the surface of foundation and subgrade are on average 20–25% less than in the case of a single tire (referred in [47]).

The use of paired cylinders can distribute the wheel load on the surface of the pavement. This means that the permissible load on the axle with twin wheels can be increased by 40–50% compared to the permissible load on the axle with single wheels without impairing the road surface.

It is not an easy task to consider an exact characterization of traffic into pavement design. This chapter briefly covered several aspects that should be considered to do so, namely:

- an adequate characterization of the heavy vehicles within a country, in terms of the characteristics related to their weight and load distribution to the pavement;
- quantification of the heavy vehicles traffic volumes that circulate within a road network, or at least in a specific road;
- characterization of the magnitude of overloading vehicles problem in order to improve the knowledge about the effective loads that are being applied to the pavements of a specific road network;

- adequate analysis of the way that loads are transmitted to the pavement, in order to improve the knowledge of the impact in their structure.

The adequate consideration of all this information within pavement design should be made in a systematic way, to allow and updated consideration of all the factors and their evolution in time.

6 Building Information Modelling Tool for Pavement Monitoring. Examples of Application

Building Information Modelling (BIM) is a methodology that allows the representation of the structural and functional characteristics of a construction, including activities and other information. The main feature of BIM is the three-dimensional modelling system that includes the management, sharing and exchanging data across the entire life cycle of a structure, where each element or object has information of its physical data. [48, 49].

Due to its characteristics BIM seems to be a promising tool to integrate the continuous structural monitoring, and to enable the immediate updating of pavement structure after each monitoring campaign. Nevertheless, studies are needed to enable the direct import of monitoring data into BIM.

Several steps were done during the last years on applying BIM to transport infrastructures both in pavements and railways [50–53]. Some examples of these applications for roads and airfields are presented in Fig. 12.

Studies were performed using monitoring data and construction information of in service pavements and airfields. Two examples of the studies developed are briefly exemplify: one on airfields and the other one on under rehabilitation pavement. The evaluation of the practical applicability of the BIM concept was studies based on its implementation to these case studies, in order analyse the mechanisms to frame the data for the establishment of a model of recording the information of the construction in a database structured by objects.

The airfield study addressed an in service runway and several monitoring campaigns of structural and functional characteristics were implemented in a BIM model to test the methodology and to study its efficiency [52]. After constructing the pavement structural model in BIM, FWD data and the results of pavement structural characteristics, obtained through back calculation were implemented regarding three different monitoring campaigns performed in different years, before and after reinforcement. Some BIM outputs, represented in range of colours along the runway [53] are presented in Fig. 11, namely: (a) the Pavement Classification Number, a quantification of the bearing capacity, before (1998 and 2000) and after reinforcement (2002); (b) presents the back calculated asphalt layer moduli before (1998) and 8–10 years after rehabilitation (2010 and 2012); (c) Fiction coefficient assessed by Grip Tester (surface characteristics) and (d) Texture depth assessed through sand patch method.

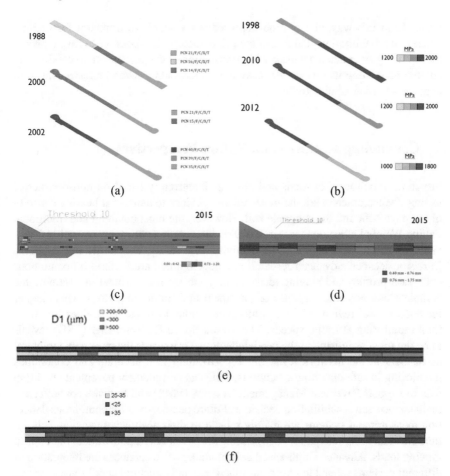

Fig. 12 Examples of BIM application to pavement monitoring data of: a runway pavement [53] **a** PCN classification of runway pavements; **b** asphalt layers elasticity moduli; **c** friction coefficient; **d** texture depth and of a road loading tests [54] **e** overall pavement condition along the section, reflected by central deflection (D1) and **f** subgrade condition along the pavement, reflected by defection measured at 1.8 m from load centre (D7)

Another study aimed to model a road pavement during rehabilitation. One of the tasks was to represent the structural characteristics of the pavement measured through load tests with Falling Weight Deflectometer [54]. Some BIM outputs are presented in Fig. 11, namely: (e) the central deflection D1, that reflects the general structural condition of the pavement and (f) the farther deflection D7, located at 1.8 m from the centre of the load plate, which reflects the stiffness of the subgrade.

If updated regularly, the modelling of pavement structural characteristics in BIM provides a visual indicator of the pavement condition evolution in time, allowing for a better management and maintenance planning. For example, a precise condition of pavement structure and a thorough historic information of maintenance is becoming available in this way, at any time, representing valuable information for maintenance and rehabilitation planning of the infrastructure. It is possible to save time and money using an updated database with information of the pavement real condition at any moment along its service life, easy to be accessed by several intervenient from engineers to road administration.

7 Concluding Remarks and Future Perspectives

Structural monitoring of roads and airfields is currently based on non-destructive testing. Deflectometers are the most suitable devices to assess the bearing capacity of the pavement and the subgrade soil. However, the most common and traditional Falling Weight Deflectometer test (FWD) is discrete and with some operating disadvantages in practice, e.g., low productivity, low safety, and traffic interruptions. In recent decades, advanced research and development are focused on continuous deflection profiles and moving loads, as they are the most critical in obtaining the complete and accurate response of pavement and traffic conditions. This chapter included a brief review of the common continuous load testing devices for structural monitoring at traffic speeds. The Traffic Speed Deflectometer (TSD) reveals to be the most promising high-speed deflectometer towards the pavement structural assessment at the network level. Since performing fast, accurate and continuous monitoring in safe operating conditions, TSD can complement pavement condition data to support Pavement Management Systems (PMS) and to make cost-effective maintenance and rehabilitation decisions. Future perspectives for continuous deflection measurement systems are mainly related to their accuracy, precision, calibration, and data back-calculation for assessing the pavement response under high-speed moving loads. Advanced high-speed deflectometers also enhances the integration of different systems related to pavement functional and structural conditions.

Ground Penetrating Radar (GPR) is nowadays a consensual tool for continuous layer thickness assessment, at traffic speed. Additionally, several other important pavement characteristics for structural monitoring can be detected, such as settlements, layers debonding and water presence. A brief description of GPR advantages is presented in this chapter.

The main challenge for continuous structural monitoring is the data processing and integration, due to the significant volume of information gathered during tests. The main approach for loading and layer thickness joint interpretation is presented herein. Also, an example of an advanced methodology to calculate the pavement residual life based on TSD and GPR data is described. Other possible integrations for structural monitoring of FWD, GPR and IRI are also referred.

The quantification of heavy vehicles traffic volumes, as responsible for the main forces exerted on the pavement, are also analysed. A significant diversity of technologies is available to perform data collection, but due to their high cost, they are not usually implemented in all the network, or being collected in a systematic way. Alternatives are presented that consider vehicle fleet characteristics. The specificity of the way the loads are transmitted from different vehicles configurations is also presented, as differences in the tyre pressure, suspension type or interaxial distance induce different impacts into the pavement.

All these aspects are suitable to be integrated in an overall system that is constantly feed, creating a dynamic database suitable to be used by any infrastructure manager.

The developments of Building Information Modelling (BIM) as a tool that can integrates the evaluation results in an updated pavement model is also addressed. The BIM model, automatically integrating the tests results can reflect the evolution of pavement condition along life cycle and can represent a base for performance indexes development and validation. Examples of structural monitoring data implemented in BIM are also given.

Another future trend is the use of remote sensing to monitor settlements of pavements and is useful as it is available free of costs by European satellites (Copernicus European Program). Several studies performed on pavement evaluation area have proven the efficiency of this approach, mainly on subsidence and road profile [21, 54–57].

The importance of advanced monitoring technology, integrated interpretation and modelling are highlighted as tools for pavement management systems, by providing a confident diagnosis of structural condition of existing pavements and enabling an economically efficient maintenance, that guarantee a proper behaviour of the pavement and an extended life cycle.

References

1. Fontul, S.: Structural evaluation of flexible pavements using non-destructive tests. Ph.D. Thesis. University of Coimbra, Portugal (2004)
2. ASTM D4694–09 2015: Standard test method for deflections with a Falling-Weight-Type impulse load device (2015)
3. ASTM D4695–03 2015: Standard guide for general pavement deflection measurements (2015)
4. Steele, D., Lee, H., Beckemeyer, C.: Development of the rolling wheel deflectometer (RWD). Report No. FHWA-DTFH-61–14-H00019, Federal Highway Administration, U.S. Department of Transportation (2020)
5. Flintsch, G., Ferne, B., Diefenderfer, B., Katicha, S., Bryce, J., Nell, S.: Evaluation of traffic-speed deflectometers. transportation research record. J. Transp. Res. Board **2304**; Transportation Research Board of the National Academies, Washington, D.C., pp. 37–46 (2012)
6. National Academies of Sciences: Engineering, and Medicine: Assessment of Continuous Pavement Deflection Measuring Technologies. The National Academies Press, Washington, DC (2013)
7. Neves, J., Cardoso, E.: Uncertainty evaluation of deflection measurements from FWD tests on road pavements. In: Kopacik, A., Kyrinovic, P., Henriques, M.J. (eds.) INGEO 2017, 7th International Conference on Engineering Surveying, pp. 59–68. LNEC, Portugal (2017)
8. Greenwood Engineering Homepage. https://greenwood.dk. Last Accessed 08 Apr 2021
9. Li, Q., Zou, Q.: Efficient calibration of a laser dynamic deflectometer. IEEE Trans. Instrum. Meas. **62**(4) (2013)
10. Xiao, F., Xiang, Q., Hou, X., Amirkhanian, S.: Utilization of traffic speed deflectometer for pavement structural evaluations. Measurement 178 (2021)

11. Katicha, S., Shrestha, S., Flintsch, G., Diefenderfer, B.: Network level pavement structural testing with the traffic speed deflectometer. Report No. VTRC 21-R4, Virginia Transportation Research Council (2020)
12. Rada, G., Nazarian, S., Visintine, B., Siddharthan, R., Thyagarajan, S.: Pavement structural evaluation at the network level: Final report. Report No. FHWA-HRT-15–074, Federal Highway Administration, U.S. Department of Transportation (2016)
13. Deng, Y., Luo, X., Gu, F., Zhang, Y., Lytton, R.: 3D simulation of deflection basin of pavements under high-speed moving loads. Constr. Build. Mater. **226**, 868–878 (2019)
14. Maser, K.R.: Condition assessment of transportation infrastructure using ground-penetrating radar. J. Infrastruct. Syst. 94–101 (1996)
15. Scullion, T., Saarenketo, T.: Implementation of ground penetrating radar technology in asphalt pavement testing. In: Ninth international conference on asphalt pavements, pp.874–890. Copenhagen, Denmark (2002)
16. ASTM D6432–19: Standard Guide for Using the Surface Ground Penetrating Radar Method for Subsurface Investigation, ASTM International, West Conshohocken, PA, (2019). https://doi.org/10.1520/D6432-19
17. ASTM D4748: Standard test method for determining the thickness of bound pavement layers using short-pulse radar. Non-Destr. Test. Pavement Struct. (2004)
18. DMRB 7.3.2.: Design manual for roads and bridges, data for pavement assessment—annex 6 HD 29/2008: Ground-Penetrating Radar (GPR), UK, Highway Agency (2008)
19. GS1601: The European GPR Association Guidelines for Pavement Structural Surveys (2016)
20. Solla, M., Pérez-Gracia, V., Fontul, S.: A Review of GPR application on transport infrastructures: troubleshooting and best practices. Remote Sens. **13**, 672 (2021). https://doi.org/10.3390/rs13040672
21. Fontul, S.: Future trends in transport infrastructure monitoring. In: Scientific Symposium Future Trends in Civil Engineering, Zagreb, Croatia, 17 October 2019, pp.262–275 (2019). https://doi.org/10.5592/CO/FTCE.2019.12
22. Pajewski, L.; Fontul, S.; Solla, M.: Ground-penetrating radar for the evaluation and monitoring of transport infrastructures. In: Innovation in Near-Surface Geophysics. Instrumentation, Application, and Data Processing Methods. Elsevier (2019). https://doi.org/10.1016/B978-0-12-812429-1.00010-6
23. Marecos, V., Solla, M., Fontul, S., Antunes, V.: Assessing the pavement subgrade by combining different non-destructive methods. Constr. Build. Mater. **135**, 76–85 (2017). https://doi.org/10.1016/j.conbuildmat.2017.01.003
24. Marecos, V.; Fontul, S.; Antunes, M.L.; Solla, M.: Assessment of a concrete pre-stressed runway pavement with ground penetrating radar, In: Proceedings of 8th International Workshop on Advanced Ground Penetrating Radar (IWAGPR), 7–10 July Florence, Italy, pp. 1–4 (2015). https://doi.org/10.1109/IWAGPR.2015.7292635
25. Fernandes, F.M., Pais, J.C.: Laboratory observation of cracks in road pavements with GPR. Construct. Build. Mater. **154**, 1130–1138 (2017). https://doi.org/10.1016/j.conbuildmat.2017.08.022
26. Dérobert, X., Baltazart, V., Simonin, J.-M., Todkar, S.S., Norgeot, C., Hui, H.-Y.: GPR monitoring of artificial debonded pavement structures throughout its life cycle during accelerated pavement testing. Remote Sens. **13**, 1474 (2021). https://doi.org/10.3390/rs13081474
27. Pedret Rodés, J.; Martínez Reguero, A.; Pérez-Gracia, V.: GPR spectra for monitoring asphalt pavements. Remote Sens. **12**, 1749 (2020). https://doi.org/10.3390/rs12111749
28. Wang, S., Zhao, S., Al-Qadi, I.L.: Real-time monitoring of asphalt concrete pavement density during construction using ground penetrating radar: theory to practice. Transp. Res. Rec. **2673**(5), 329–338 (2019). https://doi.org/10.1177/0361198119841038
29. Zadhoush, H., Giannopoulos, A., Giannakis, I.: Optimising the complex refractive index model for estimating the permittivity of heterogeneous concrete models. Remote Sens. **13**, 723 (2021). https://doi.org/10.3390/rs13040723
30. Fontul, S., Paixão, A., Solla, M., Pajewski, L.: Railway track condition assessment at network level by frequency domain analysis of GPR data. Remote Sens. **10**(4), 559 (2018). https://doi.org/10.3390/rs10040559

31. Marecos, V., Fontul, S., Antunes, M.L., Solla, M.: Evaluation of a highway pavement using non-destructive tests: falling weight deflectometer and ground penetrating radar. Constr. Build. Mater. **154**, 1164–1172 (2017). https://doi.org/10.1016/j.conbuildmat.2017.07.034
32. Marecos, V.: Optimisation of ground penetrating radar testing at traffic speed for structural monitoring of pavements—Tese de Doutoramento em Engenharia Civil, Universidade de Vigo (2018)
33. Bezina, Š., Stančerić, I., Domitrović, J., Rukavina, T.: Spatial representation of GPR data—accuracy of asphalt layers thickness mapping. Remote Sens. **13**, 864 (2021). https://doi.org/10.3390/rs13050864
34. Maser, K., Schmalzer, P., Gerber, A., Poorbaugh, J.: Implementation of the Traffic Speed Deflectometer (TSD) for Network Level Pavement Management, Roanoke, VA (2019). https://www.vtti.vt.edu/PDFs/PE-2019/Maser2.pdf
35. Gkyrtis, K., Loizos, A., Plati, C.: Integrating pavement sensing data for pavement condition evaluation. Sensors **21**, 3104 (2021). https://doi.org/10.3390/s21093104
36. Uzarowski, L., Henderson, V., Rizvi, R., Mohammad, K., Lakkavalli, V.: Use of FWD, GPR and IP in combination on complex pavement projects—including case studies. In: Conference and Exhibition of the Transportation Association of Canada (2016). http://tac-atc.ca/sites/tac-atc.ca/files/conf_papers/uzarowski_0.pdf
37. Almeida, A., Moreira, J., Silva, J., Viteri, C.: Impact of traffic loads on flexible pavements considering Ecuador's traffic and pavement condition. Int. J. Pavement Eng. **22**(6), 2021 (2019)
38. Pais, J.C., Hélder, F., Pereira, P., Kaloush, K.: The pavements cost due to traffic overloads. Int. J. Pavement Eng. **20**(12), 2019 (2018)
39. Yassenn, O., Endut, I., Hafez, M., Ishak, S., Yaseen, H.: Overloading of heavy vehicles around the world. Int. J. Eng. Sci. Res. 34–45 (2015)
40. Jacob, B., O'Brien, E., Jehaes, S.: Weigh-in-motion of road vehicles—final report of COST 323 Action, LCPC, Paris (2002)
41. NCHRP: Use of Weigh-in-Motion Data for Pavement, Bridge, Weight Enforcement, and Freight Logistics Applications. The national Academies of Sciences Engineering Medicine, Austin (2020)
42. Radovsky, B.: The effect of vehicle loads on the durability of pavements. Highways **10**, 6–8 (1984)
43. Kim, O., Bell, C., Wilson, J.: Effect of increased truck tire pressure on asphalt concrete pavement. J. Transp. Eng. ASCE **115**(4), 329–350 (1989)
44. Clark, S.: Mechanics of rubber, designing and testing of rubber goods. In: Kyiv, International Rubber Conference (1978)
45. Chen H., Marsheck, K., Saraf, C.: Effects of truck tire contact pressure distribution on the design of flexible pavements: a three-dimentional finite element approach. Transp. Res. Rec. 72–78 (1986)
46. Sousa, J., Lysmer, O., Chen, S., Monismith, C.: Effect of dynamic loads on performance of asphalt concrete pavements. Transp. Res. Rec. **1207**, 145–168 (1988)
47. Breemersch, T., Boschmans, S., Vyrozhemskyi, V., Pechonchyk, T., Raikovskyi, V., Koval, P., Gameliak, I., Fontul, S., Cardoso, J. L., Vieira Gomes, S., Knight, I.: Overweight vehicles—impact on road infrastructure and safety. Intermediate report of the PIARC project Overweight vehicles (2020)
48. Salvado, F., Silva, M., Couto, P., Azevedo, Á.: Standardization of BIM objects: development of a proposal for Portugal. Open J. Civil Eng. **6**, 469–474 (2016). https://doi.org/10.4236/ojce.2016.63039
49. Silva, M., Salvado, F., Couto, P., Azevedo, Á.: Roadmap proposal for implementing building information modelling (BIM) in Portugal. Open J. Civil Eng. **6**, 475–481 (2016). https://doi.org/10.4236/ojce.2016.63040
50. Fontul S., Couto P., Silva M.J.F.: BIM Applications to Pavements and Railways. Integration of Numerical Parameters. In: Rodrigues H., Gaspar F., Fernandes P., Mateus A. (eds.) Sustainability and Automation in Smart Constructions. Advances in Science, Technology & Innovation (IEREK Interdisciplinary Series for Sustainable Development). Springer, Cham (2021). https://doi.org/10.1007/978-3-030-35533-3_10

51. Neves, J., Sampaio, Z., Vilela, M.: A case study of BIM implementation in rail track rehabilitation. Infrastructures **4**(1), 8 (2019)
52. Lopes, J., Fontul, S., Silva, M.J., Couto, P: Avaliação de infraestruturas aeroportuárias. Proposta para integração de dados em BIM. 2º Congresso Português de Building Information Modelling 17 e 18 de maio de 2018, Instituto Superior Técnico, Universidade de Lisboa (in Portuguese, 2018), pp.311–320. https://ptbim.org/wp-content/uploads/2021/02/LivroDeAt asDoPTBIM-2018.pdf
53. Manico, H.: Flexible road pavements in Angola. Characterization and application of BIM methodologies. MsC dissertation. Nova University of Lisbon, Portugal (in Portuguese, 2018). http://hdl.handle.net/10362/57616
54. Biancardo, S.A., Viscione, N., Oreto, C., Veropalumbo, R., Abbondati, F.: BIM Approach for modeling airports terminal expansion. Infrastructures **5**, 41 (2020). https://doi.org/10.3390/inf rastructures5050041
55. Fiorentini, N., Maboudi, M., Leandri, P., Losa, M., Gerke, M.: Surface motion prediction and mapping for road infrastructures management by PS-InSAR measurements and machine learning algorithms. Remote Sens. **12**, 3976 (2020). https://doi.org/10.3390/rs12233976
56. Bianchini Ciampoli, L., Gagliardi, V., Ferrante, C., Calvi, A., D'Amico, F., Tosti, F.: Displacement monitoring in airport runways by persistent scatterers SAR interferometry. Remote Sens. **12**, 3564 (2020). https://doi.org/10.3390/rs12213564
57. Karimzadeh, S., Matsuoka, M.: Remote sensing X-band SAR data for land subsidence and pavement monitoring. Sensors **20**, 4751 (2020). https://doi.org/10.3390/s20174751

Intelligent Traffic Monitoring Systems

Sandra Vieira Gomes⊙

Abstract Traffic management is gaining increasing importance in solving the major problem of traffic congestion present in many countries. Traditional techniques of RADAR, LIDAR and LASAR to address this problem are no longer an efficient solution, as they are time-consuming and expensive. In this modern age of growing technology and population, challenges are being over-come with several new ways to make the traffic system smarter, more reliable and robust, considering the overall interaction of all the traffic components: vehicles, drivers and pedestrians. These systems are largely supported on Intelligent Traffic System (ITS), and their applications are mostly dedicated to traffic monitoring based on video processing, vehicle detection and tracking. Intelligent surveillance systems may also employ computer vision and pattern recognition techniques. This chapter addresses these new solutions that are being used for traffic management, their advantages and disadvantages, the challenges of their implementation, and most important, the ability to support traffic management systems in their pursue of solving critical problems of congested transportation networks.

Keywords Intelligent traffic systems · Traffic management · Vehicle detection and tracking

1 Introduction

Traffic jams are an everyday problem in any metropolitan city. It may be in fact considered an ever growing phenomenon, associated with the rise of the standard of living, which induced an increasing trend in the number of vehicles at an exponential rate.

This creates inevitable challenges for road networks which are not prepared for this level of complexity, not only in urban areas but also in rural areas; the capacity of the existing transportation networks reaches their maximum, causing severe traffic

S. V. Gomes (✉)
LNEC, National Laboratory for Civil Engineering, Lisbon, Portugal
e-mail: sandravieira@lnec.pt

© The Author(s), under exclusive license to Springer Nature Switzerland AG 2023 209
C. Chastre et al. (eds.), *Advances on Testing and Experimentation in Civil Engineering*, Springer Tracts in Civil Engineering,
https://doi.org/10.1007/978-3-031-05875-2_9

congestion. Additionally, road accidents and associated injuries are a direct consequence of the uncontrolled traffic flow and speed, raising the need for solutions to
this problem.

The constructing of additional highways cannot be considered a feasible option,
due to limited space often present in consolidated cities and their high cost of implementation, not only concerning the direct cost of construction, but also due to the
provisions necessary to keep workers safe and to maintain traffic flowing during
construction.

To respond to this problem, a lot of research is being done to manage and improve
traffic conditions, using vehicle's detection and speed monitoring in favor.

In this scope, Intelligent Traffic System (ITS) stand out with all its potentialities.
ITS are those systems which integrate both advanced control systems and wireless
communication technologies to provide innovative solutions [1]. In particular, they
refer to information and communication technologies applied to transport infrastructure and vehicles, which improves transport outcomes such as transport safety,
transport productivity, transport reliability, traveler choice, environmental performance, among others [2]. They consider a closer interaction with all the components
of a traffic including vehicles, drivers, and even pedestrians, to effectively mitigate
traffic congestion.

ITS are possible in very different technologies, from basic management systems
such as car navigation; traffic signal control systems; container management systems;
variable message signs; automatic number plate recognition or speed cameras to
monitor applications; such as security Closed-Circuit Television systems; and to more
advanced applications that integrate live data and feedback from a number of other
sources, such as parking guidance and information systems; weather information,
etc. [2].

Although the techniques may vary, they are all based on the collected data, which
is then used to perform traffic analysis to improve the use and safety of the roadway
system, either in the immediate, or for a future transportation need.

Using vehicle classification in a traffic monitoring system presents several func-
tionalities, which allows an effective traffic operation and transportation planning:

- To plan for pavement maintenance work, the use of the number of large trucks
  according to the bearing capacity of the highway is fundamental.
- For safety purposes, these tools are also very useful, as the identification of the
  vehicle types in a particular area, or community, is of high relevance for the
  application of restriction policies, or road environment changes that improve the
  safety of a particular spot or group of users.
- The geometric road design is calculated for the vehicle types that frequently utilize
  the roadway, and so, the use of specific information is of high importance.

ITS developments can be manly desegregated according to two main aspects: the
fundamental approach that is being used for performing the traffic analysis, and also
technologies used to capture the relevant data.

This chapter presents an overview of the solutions used for traffic management, either new or traditional, focusing on their advantages and disadvantages in solving critical problems of congested transportation networks.

Section 2 presents an overview of the most relevant approaches for data analysis, whilst Sect. 3 concerns the ITS technologies.

2 Fundamental Approaches for Traffic Data Analysis

Data analysis within ITS can be performed through several approaches, namely the ones related to geographical information systems, artificial intelligence or graph theory. Each of this approaches are further detailed ahead [3].

2.1 Geographical Information Systems

The integration of Geographical Information Systems (GIS) within transportation planning and operations has represented a valuable aid for its efficiency. Implementing traffic models in GIS presents several advantages, namely the easy way of handling the data (like traffic network topology, traffic network data, zone data, and trip matrices), presenting the results and controlling their quality [4]. The robustness of GIS softwares are particularly useful for modeling purposes related to the analysis of planning options and for real-time vehicle dispatch and traffic control. The visualization tools available in GIS facilitates the information interpretation within traffic management systems [5].

Several researchers have embraced this field of study, with multiple applications of GIS, either stand-alone or combined with other technologies.

The Department of Urban Studies and Planning of the Massachusetts Institute of Technology combined GIS and multimedia techniques to improve the interfaces of transport applications, including the development of prototypes for traffic control systems (referred in [6]).

The North-American Federal Transit Administration, within its National Transit Geographic Information System (GIS), developed an inventory of transit operations on the country's public transport routes, which is used to support decision-making in transport policies and planning (referred in [6]).

Nielson et al. [4] have contributed to overcome the problem of complex topology required by traffic models in GIS with the identification of several applications used to automate the process of building a traffic network topology.

Aldridge et al. [7] used GIS to analyze crashes on multilane roads, combining crash data, road data and traffic characteristics as a raster GIS plot on an aerial photograph. These authors also explored crash trends display, based on the attributes of individual crashes.

Theophilus et al. [8] used Geospatial Technology to capture, store, retrieve, analyze and display data related to global positioning to facilitate location of vehicles and incidents on roads.

Witzmann and Grössl [9] developed the tool VeGIS for the synchronization of data between traffic models and GIS.

Barman et al. [10] developed a GIS urban traffic management system to handle the roads, landmarks, sensitive areas and traffic information, solving several traffic related problems. This system works with the detection of shortest path and retrieves the information to traffic centers that have to be alerted in different situations.

2.2 Artificial Intelligence

Artificial Intelligence (AI) is particularly useful for the processing of large amounts of data. If applied to traffic systems, it allows to process and analyze different data types, including images and videos. Through the use of computer vision techniques, it is possible to recognize objects in those images and videos (even the ones captured by traffic cameras), and categorize different types of users: cars, motorcycles. It can also count vehicles, and determine if a road is congested, and re-route traffic accordingly. With enough history data, traffic AI is able to detect patterns, increasing the understanding of the traffic system, enabling to differentiate between events that may cause congestion, such as a car accident or rush hour. Additionally, when traffic AI learns that rush hours happen at certain times, it can perform predictive analysis and use the data to improve traffic flow.

Several AI techniques can be applied to different areas of transportation. The most common are Artificial Neural Networks—ANNs (computational models with several processing elements that receive inputs and deliver outputs based on their predefined activation functions), Genetic Algorithms—GAs (randomized search algorithms developed in an effort to imitate the mechanics of natural selection and natural genetics), Fuzzy Logic—FL (used to imitate human reasoning and cognition which considers 0 and 1 as extreme cases of truth but with various additional intermediate degrees of truth), and Expert Systems—ESs (is a computer system emulating the decision-making ability of a human expert, designed to solve complex problems by reasoning through bodies of knowledge, represented mainly as if–then rules rather than through conventional procedural code) [11].

Zhiyong [12] developed a traffic system that is able to react to traffic demand in real time with an updated time plan. The author summarizes the applications of intelligence methods such as fuzzy logic, neural networks, evolutionary algorithms and agent reinforcement learning to urban traffic signal control, and analyzes the superiority and inferiority of these methods in applications.

Chu et al. [13] studied the adaptive trajectory tracking control for a remotely operated vehicle (ROV) with an unknown dynamic model and the unmeasured states.

Charitha et al. [14] developed a tool, based on probe data, for estimating the travel time in signalized urban network. It is a self-learning tool, which used a Bayesian network for forecasting the travel time on a route along an arterial road.

Li et al. [15] developed a tool to reduce the time that vehicles are stopped at intersections. The system receives the time range of arriving at the intersection from vehicle-to-infrastructure communication (V2I), and a genetic algorithm determines the time that each vehicle must be delayed to arrive at the intersection. This information is then sent to each vehicle individually, allowing them to plan its own speed profile.

Bartłomiej [16] evaluated road traffic control in an on-line simulation environment, which enabled the optimization of adaptive traffic control strategies. Performance measures were computed using a fuzzy cellular traffic model, formulated as a hybrid system combining cellular automata and fuzzy calculus.

Wen [17] used an expert system approach to propose a solution for the traffic congestion problem, using a framework for dynamic and automatic light control expert system combined with a simulation model. This simulation model was composed of six submodels which adopted interarrival time and interdeparture time to simulate the arrival and leaving number of cars on roads.

Purohit et al. [18] used genetic algorithms implemented in MATLAB to optimize traffic signal timings. To represent the dynamic traffic conditions, a traffic emulator was developed in JAVA. The emulator conducts surveillance after fixed interval of time and sends the data to genetic algorithm, which then provides optimum green time extensions and optimizes signal timings in real time.

2.3 Graph Theory

The term graph in mathematics may have two different meanings [2]:

- The graph of a function or the graph of a relation.
- Related to "graph theory": collection of "vertices" or "nodal" and "links" or "edges".

Graphs are used to model transportation scenarios. They offer a convenient means of handling the topological and associated information describing a road network. Graph theoretic techniques, associated with route optimization, such as the shortest path between network nodes and spanning trees, are able to provide solutions to the definition of the functional relevance of network road segments.

To be applied, it is recurrent to define edges and nodes of a graph as components of the traffic system: the nodes represent crossings, bends or ends, and the edges as road parts in which algorithms like BFS or Dijkstra have been used to analyze road networks. This definition may be performed with the help of geospatial data.

Graph Theory is being applied in a large number of studies, associating travel time optimization to traffic control systems, road network hierarchy definition, or traffic light setting, among others.

Durgadevi et al. [2] used edge and vertex connectivity as graph theoretic tools to study the traffic control problem at intersections. The waiting time of the traffic participants can be minimized by controlling the edges of the edge connectivity (which represent the flow of traffic at an intersection), through the application of traffic sensors in the transportation network.

Galin et al. [19] developed a method for generating hierarchical road networks based an original geometric graph generation algorithm grounded on a non Euclidean metric combined with a path merging algorithm that originates junctions between the different types of roads. The geometry of the highways, primary and secondary roads, as well as the interchanges and intersections, are automatically created from the graph structure through generic parameterized models.

Shiuan-Wen et al. [20] used a graph model to represent the traffic network as the base to study the traffic light setting in order to minimize the total waiting time of vehicles. As an additional particularity, it is also mentioned that the authors also used Ant colony Optimization, Particle Swarm Optimization and Genetic Algorithms to obtain a near optimal solution.

Oberoi et al. [21] developed a qualitative model, based on graph theory, to improve the knowledge about the spatial evolution of urban road traffic. The model included several real-world objects which affect the flow of traffic, and the spatial relations between them. The mathematical formalization of graphs at different levels of granularity was also considered.

Khalil et al. [22] performed a dynamic modelling of a transportation system network based on a graph theory model. It took into consideration both static and dynamic aspects of the transportation system. The optimal path between two nodes can be found through an adaptive time estimation algorithm that takes into account the absolute positioning, relative speed and status indicators for each transportation system components (vehicles, road network, stations, amongst others).

Head et al. [23] used a similar approach to the one found in project management techniques like CPM—Critical Path Method and PERT—Program Evaluation and Review Technique. They used precedence graphs on the analysis of a series of simple operational treatments at an intersection related to traffic signal timings.

3   Traffic Monitoring Technologies

ITS applications are mostly dedicated to traffic monitoring based on video processing, and to vehicle detection, tracking and classification.

Under the scope of traffic monitoring based on video processing, technologies like Closed-Circuit Television—CCTV, Radio Frequency Identification—RFID, Global Positioning Systems—GPS, and Wireless Sensor Networks—WSN can be used for this purpose:

3.1 Closed-Circuit Television

Closed-circuit television (CCTV), commonly known as video surveillance cameras, are nowadays widely spread particularly in urban areas, and can be used for multiple purposes, one of them traffic monitoring. They are able to collect video images where manual observation can be difficult, problematic or unfeasible. By processing them, it is possible to extract useful information as speed, traffic composition, vehicle shapes, vehicle types, vehicle identification numbers and occurrences of traffic violations or road accidents.

A typical CCTV system is composed by (see also Fig. 1) [24]:

– Camera System: cameras to monitor specific areas and capture images, which can be analogue or digital (analogue cameras have lower resolution and storage limitations), automatic or non-automatic recording (the non-automatic recording cameras have no motion detectors to trigger recording and therefore spend more power and storage).
– Reviewing Process: either images are just viewed in real-time or if they are recorded, they can be viewed later on, as long as storage is provided.
– Central Controller: this involves the processing of captured images, which can be done manually or automatically with the use of computer hardware and software tools.

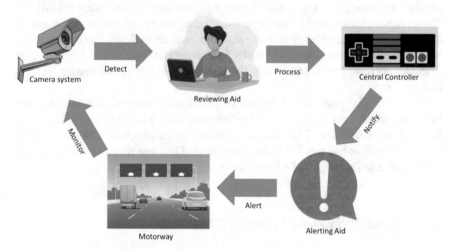

Fig. 1 Main components and workflow in a traffic control CCTV system (adapted from [24])

Fig. 2 Basic structure of an
RFID system. *Source* TT
Electronics—https://blog.tte
lectronics.com/rfid-techno
logy

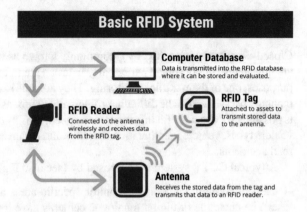

3.2 Radio Frequency Identification (RFID)

Radio Frequency Identification systems (RFID) detects changes in the electromagnetic field, which allows an automatic identification of a tag attached to an object. Two main components are necessary: a passive tag which collects energy from radio waves, and an active source reader that emits those waves (see also Fig. 2). If passive RFID tags are placed in every vehicle and active elements are installed in the road network, the location of any vehicle having a passive tag could be easily traced [25]. These authors established a real-time traffic management system with the use of RFID by deploying hardware at every possible junction in Delhi. Clear advantages were observed, particularly in the traffic jams reduction, and in the distributing of traffic to alternate paths, controlling therefore the level of traffic density in each location.

Singh et al. [26] presented an approach of RFID technology for detecting a speed violation, which presented a better performance than other solutions (like Radar Based Technology, Laser Light System, Average speed computer System or Vision Based Systems) in bad weather or light condition. They also present better accuracy, lower cost, a wider range and a better focus.

3.3 Global Positioning Systems (GPS)

Global Positioning Systems—commonly referred as GPS, can also be used for the purpose of traffic management, as it provides efficient vehicles monitoring. Vehicle speed and direction of traffic flow can be collected and uploaded in Geographical Information System (GIS) environment. If made in real-time it can provide a clear picture of the traffic-state of every route in the network. The GPS has clearly indicated the road sections where speeds are unacceptable and driver behavior is affected giving transport planners the option to choose the desired speed management technique to improve the traffic system. Owusu et al. [27] studied the use of a GPS/GIS systems for traffic monitoring in urban areas of Ghana, stating clear advantages in the identification of road sections where speeds are high, information that can be used by transport planners within their speed management actions.

3.4 Wireless Sensor Networks (WSN)

Wireless Sensor Networks (WSN) consist on a net of small sensor nodes, communicating using wireless technology to collect data. The system is composed by low-cost small vehicle detectors (powered by a battery and/or energy harvesting system) connected by a wireless network. They are easy to install and have proven to be quite accurate [28].

Pascale et al. [29] studied the application of a WSN-based traffic monitoring in California, USA. They stated that a higher-density of data collection can enhance the ability of modelling tools to reconstruct the spatial–temporal evolution of traffic flows. WSM sensors, installed in high numbers throughout the road network can therefore provide a comprehensive and high-resolution characterization of traffic dynamics.

Application of WSM is also very common in traffic lights management. Khalil et al. [30] analyzed the application of a traffic flow control system using WSN to control the traffic flow sequences. WSN is used as a tool to instrument and control traffic signals, while an intelligent traffic controller coordinates the operation of the traffic infrastructure supported by the WSN.

3.5 Vehicle Classification Systems

Particularly in what concerns vehicle classification, it is possible to systematize the available systems usually categorized into three classes depending on where the system is deployed: in-roadway-based, over roadway-based, and side roadway-based systems [31], as presented in Fig. 3.

Roadway-based vehicle classification systems. In the in-roadway-based vehicle classification systems, sensors are installed under the pavement of the road. This solution, although very common, has high costs for installation and maintenance, due to the fact that the pavement has to be cut to install the sensors. The cost increases significantly due to traffic disruption and lane closure to provide safety to road workers. Several types of sensors are available, which in general present a high accuracy because the sensors are close to the vehicles, which allows an effective capture of their presence and movement:

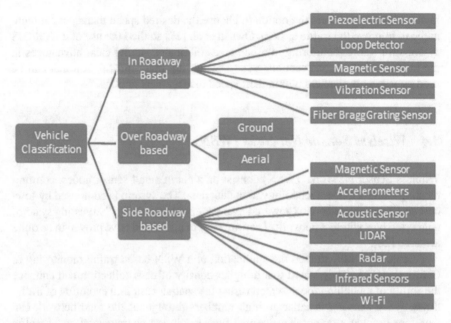

Fig. 3 Overview of available vehicle classification systems (adapted from [31])

- **piezoelectric sensors**—the physical principle of this system is based on the conversion of kinetic energy into electrical energy and some materials possess this property and are called piezoelectric materials, such as some types of polymers. The piezoelectric material consists of a crystal which produces a differential voltage when pressure is applied to its faces. The electrical signal produced by these sensors depends on the force applied, in this case the weight of the vehicles. It is a very common solution, widely used for obtaining vehicle classification. However, some common disadvantages are worth mentioning, like the ones related to pavement damage, high installation costs and a high rate of inaccurate motorcycle classification.
- **loop detectors**—which consist on a coil of wire embedded in the road pavement, which work by detecting the change of inductance when a vehicle passes over it. At this point a time-variable signal is generated, which differs by class of vehicle in shape, amplitude, statistical parameters, duration, and frequency spectrum. These different indicators create a magnetic profile of the vehicle and is the basis of the vehicle classification data-processing algorithm. This system, although very common, presents a rather low accuracy, is low, and is not very easy to install, as it requires cutting the pavement [31].
- **magnetometers**—which consist on wireless magnetic sensor nodes glued on the pavement. The sensors send data via radio to the "access point" on the side of the road, which, by turn, sends it either to a Traffic Management Center or to a roadside controller. Vehicles are detected by measuring the change in the Earth's

magnetic field caused by the presence of a vehicle near the sensor. Vehicles' classification is possible through the identification of the magnetic patterns [32]. For speed estimation two sensor nodes should be placed at a close distance.

- **vibration sensors** capture vibration patterns induced by passing vehicles due to the low elasticity of road pavement that makes vibrations well localized in time and space—see Fig. 4 [31, 33].

   In order to solve the challenges of the propagation of the waves, different systems were developed, mainly grouped in two types: (a) systems that use vibrations to count the number of axles and measure their spacing, and (b) systems based on the analysis of seismic waveforms induced by passing vehicles for vehicle classification.

- **fiber Bragg grating sensors** work by launching intense Argon-ion laser radiation into a Germania-doped fiber. The Bragg wavelength is formed due to reflected light from the periodic refraction change (see Fig. 5). A multiple sensors network is required to ensure the detection of all vehicles' axles and speeds. The related

Fig. 4 Example of a vibration sensor [33]

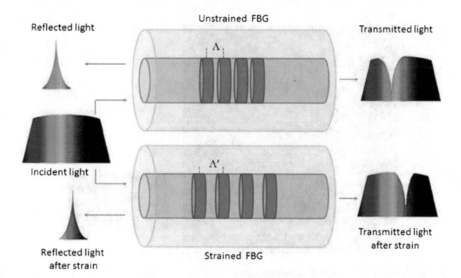

Fig. 5 Operational principle of fiber Bragg grating sensors [34]

accuracy of the system is dependent of numbers (at least two), locations (installed under the vehicle wheel path), and distances between sensors (between 2.1 and 6.1 m—7 and 20 ft.) are the major factors influencing the accuracy of a sensor network [34].

Over-roadway-based systems. In the over-roadway-based systems, sensors are placed above the roadway, allowing to capture information from several lanes at the same time. They can be differentiated by:

- Having a **ground support**, like the camera-based systems, which usually present a high classification accuracy. They are, however, affected by weather and lighting conditions. Privacy concerns also affect the choice of this solution, as people do not like to be exposed to cameras. To solve this issue, some systems adopt different types of sensors like infrared sensors and laser scanners [35]. The images that are captured go through advanced processing technologies for classifying multiple vehicles very quickly and accurately, and as so, a single camera is enough for classifying vehicles in multiple lanes—see Fig. 6.
- **Aerial devices without a ground support**, like unmanned aerial vehicles (UAV) or satellites, which serve as fixing devices for cameras, allowing to cover wide areas, like an entire roadway segment [31]. The main issue about this wide cover relates the low image resolution. Due to this issue, many of them are only used for a limited number of vehicle types that are relatively easily distinguishable such as cars and trucks.

Side roadway-based systems. In the side roadway-based systems the sensors are installed on a roadside, and as so there is no need for lane closure or construction. Some of the most common sensors used in these systems include:

Fig. 6 Example of an image processing system

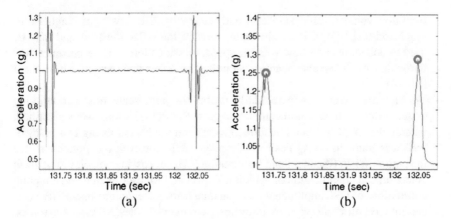

Fig. 7 Acceleration signal of a two-axle vehicle: **a** raw signal, **b** filtered signal [36]

– **Magnetometers**—already discussed in the in-roadway-based vehicle classification systems, but also possible to install on the side of the road.
– **Accelerometers**—which measure the vibrations of the pavement caused by the movement of the vehicle's wheel. Figure 7 presents the acceleration signal of a two-axle vehicle. Time between peaks is inversely proportional to vehicle speed. If we know speed, axle spacing is equal to speed × time between the two peaks [36].
– **Acoustic sensors** that capture the acoustic sounds emitted by vehicles along the road. It is a nonintrusive technique where a microphone array collects the road-side acoustic signals. Then, lane positions are automatically detected by the built-in lane detection module. Vehicles' detection creates a response curve which differs by vehicle type, allowing for their classification [37].
– **Laser Infrared Detection and Ranging (LIDAR)**, which is a device that generates a light pulse from a LIDAR gun and registers the time taken by the pulse to travel to the moving object and return to the gun (reflection). This information is then used to determine the distance between a moving object and a LIDAR gun. Through a recursive process, it is also possible to calculate the speed of moving objects. In terms of price, LIDAR devices are considered to be very costly [31].
– **RADAR**, which is a device that bounces a radio signal back from a moving object and the receiver captures the reflected signal. Vehicle speed is calculated through the difference in frequency. It is rather expensive and presents a relevant susceptibility to target identification error, although in comparison with LIDAR, their sensors provide a less accurate representation of the vehicle. Additionally, radar sensors are less vulnerable to weather and light conditions than LIDAR [31].
– **Infrared sensors** are electronic gadgets used to detect the energy generated by vehicles, road surfaces or other objects, which is then converted into electrical signals that are sent to the processing unit. These Infrared sensors may be categorized into Passive Infrared—PIR and Active Infrared—AIR. PIR detect vehicles based on emission or reflection of infrared radiation and allow to collect data

from flow volume, vehicle presence and occupancy. AIR sensors use Light Emitting Diodes (LED) or laser diodes to measure the reflection time and may be used to collect data on flow volume, speed, classification, vehicle presence, and traffic density. It can also quantify warmth of an object and recognize movement [31, 38].

– **Wi-Fi transceivers** are particularly useful for large-scale traffic monitoring systems due to their significantly low cost. Some limitations were identified, namely the simplified classification into cars and trucks, but its high accuracy of 96% was stated as a very positive advantage [39]. However, it is possible to use the unique Wi-Fi Channel State Information (CSI) of passing vehicles to perform detection and classification of vehicles. Won et al. [40] used spatial and temporal correlations of CSI amplitude with a machine learning technique to classify vehicles into five different types: motorcycles, passenger vehicles, SUV, pickup trucks, and large trucks.

4 Intelligent Traffic Systems

Intelligent traffic systems can be defined as systems powered by the latest advancements in computer, information and telecommunications technology supporting the acquisition and analysis of traffic information. Science is evolving at a high speed pace, and as so, technologies used today, may be obsolete tomorrow. Nevertheless, current technology, as presented in the previous chapters, is being used and combined with new approaches, creating new solutions for the everyday problems.

This chapter presents recent Intelligent Traffic Systems examples created under the referred philosophy that take time into consideration: either by focusing on real-time information or a priori monitoring to prevent congestion or out normal events; or take space into consideration: collecting information outside the traffic system, as vehicle tracking solutions, or by opposing, the Vehicular Ad Hoc Networks (VANETs), installed within the vehicles themselves.

4.1 Real Time Signal Control

Real-time systems applied in the scope of traffic management collects information of the current situation through video surveillance or WSN and analyzes that using some approach to control and monitor traffic.

Choudekar et al. [41] proposed a system for controlling traffic lights through image processing. Cameras installed in the traffic lights capture image sequences, which are then analyzed using digital image processing for vehicle detection, and according to traffic conditions on the road, support the traffic light control.

Wenjie et al. [42] developed a vehicle detection method using the WSN technology which can monitor the vehicles dynamically. Additionally, a new signal control algorithm to control the state of the signal light in a road intersection was also developed.

Jacob et al. [43] proposed a real time traffic lights system which involves calculating the density of traffic through a combination of ultrasonic sensors and image processing techniques. This information is processed and allows not only to control the traffic light indicators, but also to monitor traffic flow at periodic intervals.

Rana et al. [44] used fuzzy logic techniques to develop a traffic advisory system to identify the threshold capacity of a road segment. This indicator is then used to suggest improvements for road segments. Additionally, the tendency of traffic flow at different junctions is estimated through a neural network.

Sharma et al. [45] developed a smart traffic light system with sensors, microcontrollers, image processing hardware and cameras etc., to help to decide the timing of red and green light signal according to the traffic on each lane. Particularities of the system include assigning priority to important vehicles like Ambulances, Fire brigades and Police vans, GPS features allowing traffic information sharing through road users.

4.2 Traffic Load Prediction and Computation

Analysis using traffic load prediction can be considered a proactive action, as the network is monitored to prevent congestion or other abnormal events. Many researchers focus on the traffic management using load prediction and forecasting:

- Tang et al. [46] proposed a deep learning-based Traffic Load (TL) prediction algorithm to forecast future TL and congestion in network; a deep learning-based partially channel assignment algorithm to intelligently allocate channels to each link in the SDN-IoT network.
- Queen and Albers [47] monitored traffic flows in a network in order to improve traffic management efficiency. The real time data was provided by the induction loops planted on the road surfaces. The authors used a Bayesian graphical dynamic model called the Linear Multiregression Dynamic Model for forecasting traffic flow.
- Haar and Theissing [48] studied the dynamics of passenger loads in multimodal transportation networks to mitigate the impact of perturbations. The authors used an approach based on a stochastic hybrid automaton model for a transport network that allows to compute how probabilistic load vectors are propagated through the transport network, and develop a computation strategy for forecasting the network load at a certain time in the future.

4.3 Vehicle Tracking

Vehicle tracking is also a very common technology, within ITS. Although it only involves images processing and tracking vehicles, several authors have published their work on this filed, with multiple improvements to overcome particular problems of this systems, like bad lighting conditions, or coverage and precision accuracy.

Foresti and Snidaro [49] proposed a real-time traffic monitoring system for vehicle detection and tracking in bad illuminated scenarios, aiming to monitor the traffic flow, estimate the vehicle's speed or determine the state of the traffic, and detect anomalous situations, like road accidents or stopped cars.

Dangi et al. [50] presented a proposal to implement an intelligent traffic controller using real time image processing. Image sequences captured by a camera were analyzed using edge detection and object counting methods. Special attention was given to emergency vehicles, as in case of an emergency the lane is given priority over all the others.

Harjoko et al. [51] used Haar Cascade Classifier method for vehicle detection and Optical Flow method for tracking and counting vehicles within a closed detection region, ensuring a complete coverage of all vehicles.

4.4 Vehicle Routing Vehicular Ad-Hoc Networks (VANETs)

Vehicular Ad Hoc Networks (VANETs) are technologies that integrate the capabilities of wireless networks into vehicles. VANETs do not depend on a fixed infrastructure; instead, they use moving vehicles as nodes in a network to create a mobile network. The units are small, portable, and battery-powered, and they communicate among each other through radiofrequency signals, as long as they are close enough (approximately 100–300 m). As so, every car turns into a wireless router or nod, passing information from one vehicle to the others [52].

Several authors have dedicated their research to this technology, some of the most recent are presented ahead:

- Bello-Salau et al. [53] focused their research on the design of a new route metric for VANET communication, considering several parameters such as the received signal strength; transmit power, frequency and the path loss. Additionally, they presented an improved genetic algorithm-based route optimization technique (IGAROT) for an improved routing and optimal routes identification, required to communicate road anomalies effectively between vehicles in VANETs.
- Saravanan and Ganeshkumar [54] proposed a machine learning architecture using a deep reinforcement learning model to monitor and estimate the necessary data for the routing. A roadside unit was used to collect data which feeds the deep reinforcement learning process, allowing to predict the movement of the vehicle and propose routing paths. The application of deep reinforcement learning

over VANETs yields increased network performance, which provides on-demand routing information.

- Saha et al. [55] presented a detailed dissertation about wireless ad-hoc networks, not only VANETs, but also MANETs (Mobile Ad Hoc Network) and FANETs (Flying Ad-Hoc Networks) in what concerns their characteristics and protocol.

5 Final Remarks

Research and innovation within the scope of traffic monitoring has been growing at a high speed pace in the past decade. Thanks to recent advances in sensing, machine learning, and wireless communication technologies, the accuracy of these systems has improved greatly at a significantly reduced cost.

Intelligent Traffic System are no exception, and the need to improve network management emphasizes the creation of new solutions which integrate both advanced control systems and wireless communication technologies. Their main focus relies in the traffic data collection, and its subsequent communication to a central controller which uses a specific approach for the data analysis that feeds the traffic management system.

This chapter presented an overview of the current technologies adopted for the different aspects of these systems: either the approach used for the data analysis, or the technology used for the different components.

In a near future, with the rapid development of vehicle-to-everything (V2X) technology, traffic monitoring systems accuracy will be significantly enhanced. Nevertheless, they will also face technical challenges, mainly related to the creation of reliable and secure data transmission, dynamic range adjustment, interference reduction, amongst others.

Challenges related to all these technologies are diverse, but one crucial aspect is transversal to all: the interaction of all the components of a traffic including vehicles, drivers, and even pedestrians, to effectively mitigate traffic congestion is fundamental for an effective traffic management.

References

1. Sładkowski, A., Pamuła, W.: Intelligent Transportation Systems—Problems and Perspectives, vol. 32. Springer (2015)
2. Durgadevi, M., Harika, S., Anusha, D., Malleswari, D.: Use of graph theory in transportation problems and different networks. JASC J. Appl. Sci. Comput. $\mathbf{V}$(XII) (2018)
3. Gupta, P., Purohit, G.N., Dadhich, A.: Approaches for intelligent traffic system: a survey. Int. J. Comput. Sci. Eng. $\mathbf{4}$(9) (2012)
4. Nielson, O., Israelsen, T., Nielson, E.: GIS based application for establishing the data foundation for traffic models. In: ESRI User Conference (1997)

5. El-Geneidy, A., Bertini, R.L.: Integrating geographic information systems and intelligent transportation systems to improve incident management and life safety. In: 8th International Conference on Computers in Urban Planning and Urban Management Conference, Sendai, Japan (2003)

6. Pons, J., Pérez, M.: Geographic information systems and intelligent transport systems: technologies used to form new communication networks. NETCOM Netw. Commun. Stud. 53–70 (2003)

7. Aldridge, J., Harper, R., Wilson, D., Dunn, R.: Motorway crash analysis using GIS. In: IPENZ. Transport Group Conference, Auckland (2011)

8. Theophilus, A., Adekayode, F., Gbadeyan, J., Ibiyemi, T.: Enhancing road monitoring and safety through the use of geospatial technology. Int. J. Phys. Sci. 4(5), 343–348 (2009)

9. Witzmann, U., Grössl, S.: VeGIS—tool for the connectivity between traffic models and geographic information systems. In: Real Corp 2010 Proceedings, Vienna, pp. 18–20 (2010)

10. Barman, S., Sarkar, A., Roy, R., Sarkar, J.: An approach to GIS-based traffic information system using Spatial Oracle. Int. J. Spat. Temporal Multimed. Inf. Syst. 1(3), 253 (2019)

11. Machin, M., Sanguesa, J., Garrido, P., Martinez, F.: On the use of artificial intelligence techniques in intelligent transportation systems. In: 2018 IEEE Wireless Communications and Networking Conference Workshops (WCNCW), pp. 332–337 (2018)

12. Zhiyong, L.: A survey of intelligence methods in urban traffic signal control. Int. J. Comput. Sci. Netw. Secur. 7(7) (2007)

13. Chu, Z., Zhu, D., Yang, S.X.: Observer-based adaptive neural network trajectory tracking control for remotely operated vehicle. IEEE Trans. Neural Netw. Learn. Syst. (2016)

14. Charitha, D., Mark, M., Masao, K.: Travel time estimation using probe data and Bayesian network learning. Int. J. ITS Res. 6(2) (2008)

15. Li, J., Dridi, M., El-Moudni, A.: A cooperative traffic control for the vehicles in the intersection based on the genetic algorithm. In: 4th IEEE International Colloquium on Information Science and Technology (CiSt), pp. 627–632 (2016)

16. Bartłomiej, P.: Performance evaluation of road traffic control using a fuzzy cellular model. In: Corchado, E., et al. (eds.) Hybrid Artificial Intelligence Systems. Lecture Notes in Artificial Intelligence, vol. 6679, pp. 59–66. Springer Verlag, Berlin Heidelberg (2011)

17. Wen, W.: A dynamic and automatic light control system for solving the road congestion problem. Expert Syst. Appl. 34, 2370–2381 (2008)

18. Purohit, G., Sherry, A., Saraswat, M.: Time optimization for real time traffic control system using genetic algorithm. Glob. J. Enterpr. Inf. Syst. 3(IV) (2011)

19. Galin, E., Peytavie, A., Guérin, E., Beneš, B.: Authoring hierarchical road networks. Pac. Graph. 30(7) (2011)

20. Shiuan-Wen, C., Chang-Biau, Y., Yung-Hsing, P.: Algorithms for the traffic light setting problem on the graph model. In: TAAI. National Sun Yat-Sen University, Taiwan (2007)

21. Oberoi, K., Mondo, G., Dupuis, Y., Vasseur, P.: Spatial modeling of urban road traffic using graph theory. In: Proceedings of Spatial Analysis and GEOmatics (SAGEO) 2017, pp. 264–277. INSA de Rouen, France (2017)

22. Khalil, W., Merzouki, R., Ould-Bouamama, B.: Modelling for optimal trajectory planning of an intelligent transportation system. IFAC Proc. Vol. 43(16), 389–394 (2010)

23. Head, L., Gettman, D., Bullock, D., Urbanik, T.: Modelling traffic signal operations using precedence graphs. Transp. Res. Rec. J. Transp. Res. Board 2035, 10–18 (2007)

24. Kurdi, H.: Review of closed circuit television (CCTV) techniques for vehicles traffic management. Int. J. Comput. Sci. Inf. Technol. (IJCSIT) 6(2) (2014)

25. Saini, A., Chandok, S., Deshwal, P.: Advancement of traffic management system using RFID. In: 2017 International Conference on Intelligent Computing and Control Systems (ICICCS), Madurai, India, pp. 1254–1260 (2017)

26. Singh, S., Chawla, R., Singh, H.: Intelligent speed violation detection system. Int. J. Eng. Tech. Res. 2(1), 92–95 (2014)

27. Owusu, J., Afukaar, F., Prah B.: Urban traffic speed management: the use of GPS/GIS. In: Shaping the Change XXIII FIG Congress, Munich, Germany, pp. 1–11 (2006)

28. Margulici, J.D., Yang, S., Tan, C.W., Grover, P., Markarian, A.: Evaluation of Wireless Traffic Sensors by Sensys Networks. California Center for Innovative Transportation (2006)
29. Pascale, A., Nicoli, M., Deflorio, F., Dalla, C., Spagnolini, U.: Wireless sensor networks for traffic management and road safety. IET Intell. Transp. Syst. **6**, 67–77 (2012)
30. Khalil, Y., Jamal, A., Ali, S.: Intelligent traffic light flow control system using wireless sensor networks. J. Inf. Sci. Eng. **26**, 753–768 (2010)
31. Won, M.: Intelligent traffic monitoring systems for vehicle classification: a survey. IEEE Access **8**, 73340–73358 (2020)
32. Cheung, S., Ergen, S., Varaiya, P.: Traffic surveillance with wireless magnetic sensors. In: 12th World Congress on Intelligent Transport Systems, United States (2005)
33. Bajwa, R., Rajagopal, R., Varaiya, P., Kavaler, R.: In-pavement wireless sensor network for vehicle classification. In: International Conference on Information Processing in Sensor Networks (IPSN), pp. 85–96 (2011)
34. Al-Tarawneh, M.: Traffic monitoring system using in-pavement fiber Bragg grating sensors. Dissertation for the Degree of Doctor of Philosophy, North Dakota State University of Agriculture and Applied Science, Fargo, North Dakota (2019)
35. Chen, Z., Ellis, T., Velastin, S.A.: Vehicle detection, tracking and classification in urban traffic. In: International IEEE Conference on Intelligent Transportation Systems, pp. 951–956 (2012)
36. Ma, W., Xing, D., McKee, A., Bajwa, R., Flores, C., Fuller, B., Varaiya, P.: A wireless accelerometer-based automatic vehicle classification prototype system. IEEE Trans. Intell. Transp. Syst. **15**(1), 104–111 (2014)
37. Na, Y.: An acoustic traffic monitoring system: design and implementation. In: UIC-ATC-ScalCom (2015)
38. Guerrero-Ibáñez, J., Zeadally, S., Contreras-Castillo, J.: Sensor technologies for intelligent transportation systems. Sensors **18**, 1212 (2018)
39. Won, M., Zhang, S., Son, S. H.: WiTraffic: low-cost and non-intrusive traffic monitoring system using WiFi. In: International Conference on Computer Communication and Networks (ICCCN), pp. 1–9 (2017)
40. Won, M., Sahu, S., Park, K.-J.: DeepWiTraffic: low cost WiFi-based traffic monitoring system using deep learning. In: IEEE 16th International Conference on Mobile Ad Hoc and Sensor Systems (MASS) (2019)
41. Choudekar, P., Banarjee, S., Muju, M.K.: Real time traffic control using image processing. Indian J. Comput. Sci. Eng. **2**(1) (2011)
42. Wenjie, C., Lifeng, C., Zhanglong, C., Shiliang, T.: A realtime dynamic traffic control system based on wireless sensor network, parallel processing. In: International Conference on ICPP Workshops, pp. 258–264 (2005)
43. Jacob, S., Rekh, A., Manoj, G., Paul, J.: Smart traffic management system with real time analysis. Int. J. Eng. Technol. (UAE) **7**, 348–351 (2018)
44. Rana, S., Younas, I., Mahmood, S.: A real time traffic management system. In: 2nd IEEE International Conference on Computer Science and Information Technology, Beijing, pp. 193–197 (2009)
45. Sharma, S., Kumar, A., Singh, R.: Real-time auto controllable traffic management system. J. Basic Appl. Eng. Res. **1**(8), 78–82 (2014)
46. Tang, F., Fadlullah, Z.M., Mao, B., Kato, N.: An intelligent traffic load prediction-based adaptive channel assignment algorithm in SDN-IoT: a deep learning approach. IEEE Internet Things J. **5**(6), 5141–5154 (2018)
47. Queen, C., Albers, C.: Forecasting traffic flows in road networks: a graphical dynamic model approach. J. Am. Stat. Soc. Appl. Case Stud. (2008)
48. Haar, S., Theissing, S.: Predicting traffic load in public transportation networks. In: American Control Conference (ACC), pp. 821–826 (2016)
49. Foresti, G., Snidaro, L.: Vehicle detection and tracking for traffic monitoring. In: Roli, F., Vitulano, S. (eds.) Image Analysis and Processing—ICIAP 2005. ICIAP 2005. Lecture Notes in Computer Science, vol. 3617. Springer, Berlin, Heidelberg (2005)

50. Dangi, V., Parab, A., Pawar, K., Rathod, S.: Image processing based intelligent traffic controller. Undergrad. Acad. Res. J. **1**(1) (2012). ISSN: 2278-1129
51. Harjoko, A., Candradewi, I., Bakhtiar, A.: Intelligent traffic monitoring systems: vehicles detection, tracking, and counting using Haar cascade classifier and optical flow. In: Proceedings of the International Conference on Video and Image Processing—ICVIP 2017, pp. 49–55. Association for Computing Machinery, New York, USA (2017)
52. Alves Junior, J., Wille, E.: Routing in vehicular ad hoc networks: main characteristics and tendencies. J. Comput. Netw. Commun. **2018**, 10 pages. Article ID 1302123 (2018)
53. Bello-Salau, H., Aibinu, A.M., Wang, Z., Onumanyi, A.J., Onwuka, E.N., Dukiya, J.J.: An optimized routing algorithm for vehicle ad-hoc networks. Eng. Sci. Technol. Int. J. **22**(3), 754–766 (2019). ISSN 2215-0986
54. Saravanan, M., Ganeshkumar, P.: Routing using reinforcement learning in vehicular ad hoc networks. Comput. Intell. **36**, 682–697 (2020)
55. Saha, D., Wararkar, P., Patil, S.: Comprehensive study and overview of vehicular AdHOC networks (VANETs) in current scenario with respect to realistic vehicular environment. Int. J. Comput. Appl. **178**(15) (2019). ISSN 0975 – 8887

Testing and Monitoring in Railway Tracks

Eduardo Fortunato(ID) **and André Paixão**(ID)

Abstract Railway tracks may be perceived as simple physical structures, but in fact they entail a significant level of complexity as refers to the assessment and prediction of their transient and long-term behaviors. These mostly result from the intrinsic characteristics of their components and of the dynamic interaction with the successive passing trains at different speeds and with different loading characteristics. The railway industry's traditionally conservative approach to new technologies and the limited access, mostly for safety reasons, to related infrastructures somewhat hinder new developments and make it difficult to obtain further insight into the behavior of these structures. To overcome these limitations, advanced methods have been proposed for characterizing the materials that integrate the tracks and a few developments have been implemented in the monitoring of the structures under static and dynamic loading conditions. The information obtained has been essential to validate track models and to predict the transient response and the degradation behavior of the structures and their materials. This has promoted the introduction of new materials and construction methods with a view to improve the structural and environmental performances of railway infrastructures. The work presented herein provides an overview of current and advanced characterization and monitoring techniques, which are exemplified by a few applications concerning both R&D and consulting initiatives.

Keywords Railway tracks · Testing · Instrumentation · Monitoring

E. Fortunato (✉) · A. Paixão
LNEC, National Laboratory for Civil Engineering, Lisbon, Portugal
e-mail: efortunato@lnec.pt

A. Paixão
e-mail: apaixao@lnec.pt

© The Author(s), under exclusive license to Springer Nature Switzerland AG 2023   229
C. Chastre et al. (eds.), *Advances on Testing and Experimentation in Civil Engineering*, Springer Tracts in Civil Engineering,
https://doi.org/10.1007/978-3-031-05875-2_10

1 Introduction

The traditional railway track integrates both the superstructure (rails, sleepers, fastenings, and ballast) and the substructure that includes the sub-ballast and the subgrade; the upper part of the latter is usually called the capping layer or form layer (Fig. 1). The functioning of these elements is relatively complex, leading to a non-linear behavior of the structure when subjected to loading and unloading cycles caused by the passage of vehicles [1, 2].

The rails accommodate the wheel loads and distribute these loads across the sleepers. They also guide the wheels in a lateral direction. The rails may also act as electrical conductors for the signal circuit and as electrical conductors in an electrified line. In ballasted tracks, the rails are held by sleepers that attenuate and distribute the forces transmitted by the rolling stock to the ballast.

The sleepers also play an important role in the stability of the track, in its plane, because it is their own weight and the fact that they are embedded in the ballast that allows the track to withstand the lateral forces produced by the rolling stock and the forces associated with the thermal variation of long welded rails. Rail fastenings can be either rigid (typically for wooden sleepers) or flexible (for concrete sleepers) and are intended to ensure the correct positioning of the rails on the sleepers, by considering vertical, transverse, and longitudinal forces. The rail pads provide track resiliency, influencing its dynamic response, and reduce wear on sleepers.

The ballast layer is one of the most important elements of conventional tracks and plays an important role in the economic efficiency and sustainability of railway transport. The ballast layer—usually 20–40 cm thick—is composed of coarse hard rock particles of nearly uniform grain size distribution and its functions include the transmission and dissipation of the cyclic loads imposed by the passing trains (transmitted from the rails to the sleepers) to the underlying layers, by simultaneously ensuring the horizontal position of the track and providing proper drainage. Throughout its life cycle, the ballast layer densifies, mainly due to particle breakage and subgrade soil pumping, under successive impact loads, maintenance actions and climatic effects [3, 4]. Plastic deformations [5] and differential settlements in the supporting layers lead to changes in the position of the rails, to track defects and wear of components, thus reducing the performance of this infrastructure and eventually leading to the need to perform maintenance interventions to reestablish rail geometry.

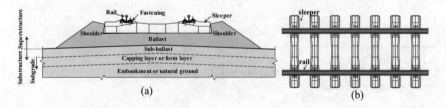

Fig. 1 Ballasted railway track: **a** schematic cross section; **b** schematic track plan view

The methods for characterizing materials and for modeling and monitoring the behavior of the railway track have evolved, allowing to design, build and maintain these structures in a more appropriate manner. In this chapter, we provide an overview of some current and advanced characterization and monitoring techniques.

2   Aspects of the Characterization of Materials

2.1   Advances in Ballast Particle Morphology

The natural aggregate for the ballast layer is one of the best controlled raw materials. Compressive particle-particle contact, particle-particle and particle-sleeper friction and imbrication are the major mechanical actions resulting from traffic and heavy maintenance interventions. Cubic shape, rough faces, and sharp edges favor imbrication, thus avoiding rolling or significant translational movements, resulting in higher layer stability and less contact forces between particles (by increasing contact points).

To ensure that the material undergoes limited degradation and preserves its optimal morphological parameters throughout its lifecycle, some mechanical and geometric properties are usually required for ballast particles, which are defined in specific regulations, by considering well-established characterization procedures [6, 7].

Though extensively validated, some of these procedures have limitations. Traditionally, particle morphology is evaluated using several classification criteria and a variety of indices or descriptors have been proposed by different authors, which were mostly presented over the last century [8]. The most practical and basic approach consists of sieving the material, but this makes only possible to capture partial particle size. Maximum and minimum particle dimensions can be determined manually using calipers, but this method is slow, subjective [9] and prone to human error [10]. Moreover, these approaches are reductive of the 3D aspects of the particles, and do not take advantage of recent knowledge and current automated methods and of image analysis approaches to fully characterize these aggregates. These new procedures allow, for example, to carry out a thorough analysis of the wear of the particles, when they are subjected to mechanical actions. Furthermore, computational advances in hardware and simulation methods led to the development of discrete element numerical models of railway tracks that demand a deep knowledge of particle shape. When properly calibrated, these models can be used to analyze the effects of traffic and mechanical maintenance actions on particle shape evolution, layer stability and track dynamic behavior [11].

Three-dimensional scanning of natural objects such as ballast particles can be performed with sub-millimeter precision, resulting in 3D models with up to a few millions of vertices and facets. Various methods are available for particle scanning and morphology analysis [8]. Some of them are expensive, such as those related to the use of X-ray computed tomography (CT) images [12]; others are either based on

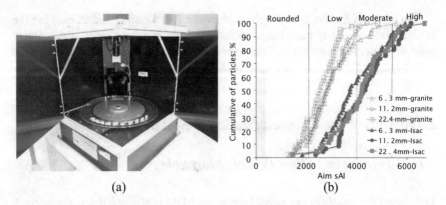

(a)   (b)

Fig. 2 Particle morphology: **a** AIMS equipment; **b** angularity [14]

2D particle morphology characterization or are only applicable to limited particle dimensions [13].

Delgado et al. [14] used the Aggregate Image Measurement System (AIMS) [15] (Fig. 2a) to quantify the sphericity, surface texture (roughness) and angularity of aggregate particles, based on the digital evaluation of a set of particles. They compared ballast particles with different nominal diameters (6.3, 11.2 and 22.4 mm) of inert steel aggregates for construction (ISAC) containing granite particles. The values obtained for particle sphericity and surface texture were similar in both materials. On the contrary, angularity was low in granite and was low/moderate to high in slag (Fig. 2b). This aspect can help explain the improved performance of slag particles when compared with that of granite aggregate, under the same loading conditions [16–18].

Several authors employed laser scanning to study ballast particle shape or to assemble particle libraries for Discrete Element Method simulations [19]. Jerónimo et al. [20] developed some studies with a portable laser scanner using a laser scanner that uses a laser emitter, 3 high definition cameras and 8 LEDs placed around the cameras (Fig. 3a). Image resolution was 0.05 mm and accuracy was at least 0.04 mm,

(a)   (b)   (c)

Fig. 3 Laser scanning procedure: **a** portable laser scanner; **b** support pedestal and screen with reflective targets; **c** triangle mesh of the particle at 1 mm resolution [20]

which made it suitable for small, complex-shaped ballast particles. The point cloud and the particle size were obtained with the aid of retroreflective targets placed on the object and/or the surrounding area (Fig. 3b). The targets were recognized by the scanner and a reference system was automatically generated. The cameras also detected ambient light reflected by the scanned surface and recorded color.

Based on digital models of a set of particles (Fig. 3c), it was possible to calculate geometric parameters and their evolution in face of particle degradation during mechanical testing (Fig. 4). Laser scanning detects and quantifies particle wear and fragmentation in a way that is not possible with traditional methods, which can only quantify volume/mass loss and qualitative shape change. In that study more sophisticated analyses were explored, such as calculation of sphericity.

Considering the capabilities of the photogrammetry and its recent advances, Paixão et al. [21] presented a cost-efficient photogrammetry method for 3D reconstruction of ballast particles (Fig. 5), as an alternative to the significantly expensive laser scanning. They compared these approaches with the laser scanning method employed in the previous study, using the same set of granite ballast particles, and considering their condition after fragmentation and wear tests. The authors compared the laser scan and photogrammetry meshes of 18 digital particles. They obtained

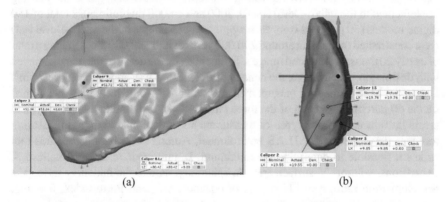

(a)   (b)

Fig. 4 Digital determination of particle dimensions: **a** L (maximum); **b** S (smallest)

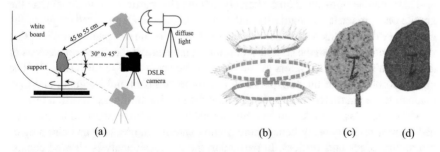

(a)   (b)   (c)   (d)

Fig. 5 Workflow: **a** setup of the photography session; **b** camera pose calculation, image matching and sparse reconstruction; **c** dense reconstruction; **d** generated mesh [21]

digital models of equivalent or higher quality with photogrammetry, namely: (i) peak differences were below 1 mm; (ii) more than 50% of the surface area showed differences less than 0.1 mm; (iii) average standard deviation of the deviations was about 0.1 mm.

According to the results, the authors considered the performance of the photogrammetry as very good, allowing for advanced and automated particle geometry analyses: sub-millimeter accuracy can be achieved using cheaper equipment and software.

Other recent applications of close-range photogrammetry have evidenced the potential of this approach for low cost detailed 3D scanning of coarse aggregates [22].

Paixão and Fortunato [23] used this method to compare the abrasion evolution of a steel slag with a granite aggregate that fulfilled the requirements for ballast in Europe. An analysis was performed on the evolution of the morphology of aggregates, which included both conventional and advanced approaches, as well as a spherical harmonic analysis of a 3D digitalization of the particles obtained by close-range photogrammetry.

Quantitative analyses were conducted on the abrasion and 3D morphology evolution of particles by micro-Deval testing. It should be noted that due to the phasing implemented in the micro-Deval test (0, 2000 and 14,000 revolutions), a total of 180 digital models (30 particles from each aggregate, in 3 phases of the micro-Deval test) were constructed (averaging around 290,000 vertices and 580,000 faces), using a total of nearly 20 thousand captured images. To elucidate on the detail of the scans achieved with this method, on average, the density of the meshes was about 38 vertices/mm^2. Due to the large number of manual measurements that would have to be performed on the 60 particles, for each of the three phases of the test, and to avoid the typical errors introduced in these types of measurements, an automated digital measuring tool was developed in MATLAB environment. Among other parameters, this tool makes it possible to determine the maximum, L, intermediate, I, and smallest, S, dimensions, volume, V, surface area, A, sphericity, ψ, and the flattening ratio, p = S/I, elongation ratio, q = I/L, degree of equancy, S/L, and Form index, F = p/q. Figure 6 presents an example of the automated analysis of a steel slag particle using this tool.

This study demonstrated quantitatively that the slag particles were richer than the granite ones, in terms of angularity and surface texture, in all analyzed stages of the micro-Deval test. This was evidenced and quantified by the morphology indices that were calculated by both the spherical harmonic analysis and the 3D mesh analysis of the particle's surfaces, as refers to surface curvature or asperity radius (Fig. 7). The initial morphology of the slag particles was more complex and irregular, and they retained these characteristics for longer than the granite during the abrasion test.

Moreover, slag particles underwent similar or less surface wear, although they had more asperities, being hence more prone to abrasion, breaking and chipping of protruding edges and vertices. In particular, the 3D mesh analysis showed consistently less surface wear in the slag aggregate than in the granite one, for different

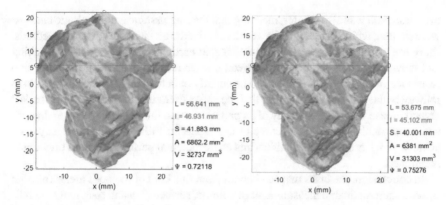

Fig. 6 Abrasion of a steel slag particle: initial (left) and final configuration (right) [23]

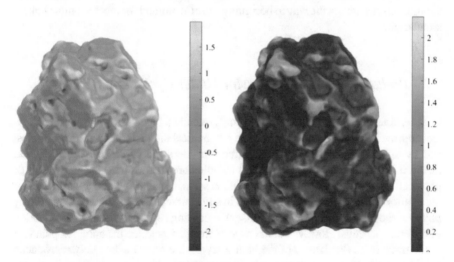

Fig. 7 Example of curvature [1/mm] (left) and wear [mm] (right) on a slag particle [23]

ranges of surface curvatures. The granite ballast showed lower micro-Deval abrasion coefficients than the slag aggregate in both the dry and the wet procedures, which could imply that the granite had greater resistance to abrasion. However, the advanced morphological analyses showed that the particles of the two materials underwent similar abrasion, in terms of both mass and volume loss. This suggests that the micro-Deval abrasion test does not fully translate the abrasion mechanism of these aggregates and may be especially disadvantageous for the slag aggregate. Regarding the traditional morphologic characterization, the results were very similar for both materials, which suggests that traditional characterization methods may not be adequate for this type of in-depth analyses. Nevertheless, the characterization of the steel slag particles in a modified Zingg diagram [9] was more uniform and their evolution path was also shorter, meaning that the slag particles morphologic changes were smaller, which agrees with the results obtained with the more advanced methods. Since the results of this work indicate that slag particles provide better interlocking and show less or comparable surface wear, thus, in terms of particle morphology and resistance to abrasion, slag appears to be equally or better suited to the application in railway ballast layers. Considering that the shear strength and stability of the ballast layer increase with the angularity of its particles, these findings contribute to demonstrate that steel slag can be a source of aggregate for railway ballast and may even achieve a higher level of mechanical and environmental performances, as previously mentioned.

In conclusion, the tasks that can benefit from this kind of methods are as follows: rigorous and replicable measurement of volume, surface area and roughness; quantitative assessment of wear and fragmentation; production of digital particle libraries, which can be shared globally or used to create digital particles for discrete element simulations that are on the way to becoming a useful method for more accurate ballast simulation.

2.2 Ballast Mechanical Behavior Under Cyclic Loading

In general terms, there are two ways to analyze ballast behavior: by a micromechanical approach, based on the physical and mechanical characteristics of the particles; and by a macromechanical approach that considers the behavior of the granular medium. In the latter case, compacted samples are characterized [17, 24, 25], or physical models are tested, in real or reduced scale, considering either some parts or the totality of the structure. Compared to in situ characterization methods, laboratory physical modelling has several advantages regarding repeatability and reproduction of different settings. However, some aspects that characterize the complex behavior of the track (e.g., the density of the ballast layer), are hardly adequately reproduced in laboratory, which may affect the validity of the results.

On a physical model recently built at National Laboratory for Civil Engineering in Portugal it was possible to perform a few tests to evaluate accumulated settlements over a relatively large number of load cycles of the ballast layer [5]. The tests comprised cyclic loading on one concrete monoblock sleeper embedded in ballast, with a conventional and modern track foundation, and an Iberian track gauge. A schematic view of the test and photos are shown in Fig. 8. Confining metallic walls were devised to minimize friction between the soils/aggregates and the walls. The subgrade and sub-ballast layers were heavily compacted during the construction phase using a compaction roller. The sub-ballast was compacted in sublayers of 0.05 m. The ballast layer was compacted with both the compaction roller and the mobile Cobra TTe hammer.

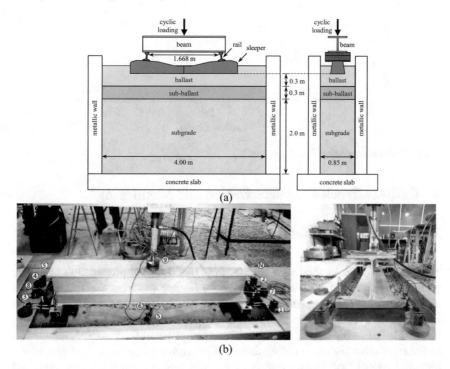

Fig. 8 Physical model: **a** scheme (front and lateral views); **b** test (front and lateral views) [5]

The loading was transmitted from one hydraulic actuator to the sleeper by a spreader steel beam. The displacements were measured with LVDTs placed at positions 1–9, as shown in Fig. 8: positions 1–4 on the rails; positions 5–8 on the sleeper and position 9 is the base of the actuator.

The simulated train loading was applied 5.5×10^5 times at a frequency of 1 Hz. The maximum load applied by the actuator was 100 kN, and the minimum load was 4 kN to avoid impacts on the sleeper. Assuming a typical maximum wheelset to sleeper load transmission of 50%, the load of 100 kN corresponded to an axle load of approximately 19.6 tones. Following an initial phase of rapid settlement measured on the sleeper (average of the four measured positions), the permanent deformation rate decreased nearly to zero, a phenomenon referred to as shakedown [26]. The representation in a logarithmic scale, in Fig. 9, shows however that for $N > 10^5$ the settlements continue progressing at an approximate log-linear rate. A small bump can be observed on the settlements curve before cycle 1×10^5, which was caused by a short interruption in the test to replace the actuator.

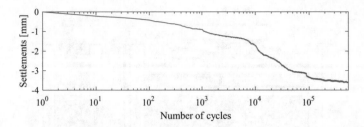

Fig. 9 Settlements measured on the sleeper [5]

3 Railway Track Characterization and Monitoring

3.1 Surface Wave Method in Testing the Railway Platform

The Surface Wave Method (SWM), which is based on the propagation of elastic surface waves along the ground, is a powerful method to assess the stiffness of the medium in depth [27]. As a geophysical method, it relies on the propagation of seismic waves across the medium with very low strain values, typically below 10^{-5}, making it possible to use the Linear Elasticity Theory to interpret the results. It allows fast and non-invasive testing, avoiding delays in the construction works and decreasing repairing costs in the tested locations; the equipment is compact, lightweight, and user-friendly.

This method was adopted in a research performed during the renewal of a Portuguese railway line platform (Northern Line), in which minimum values of the deformation modulus at the top of capping and sub-ballast layers were specified [28]. In that work, two techniques were used, i.e. the Spectral Analysis of Surface Waves (SASW) and the Continuous Surface Waves (CSW) [29]. From among the equipment used, reference is made to the following: an electro-mechanical vibrator with a controlling unit (for the CSW method, capable of generating loads up to 500 N); six 2 Hz geophones; a control unit for definition of parameters, for acquisition and analysis of load and geophones signals, as well as for data processing and data recording; hammers of different masses (for the SASW method); and a power generator. From the spectral records, the phase of the signal generated by the source in each geophone position was determined and the phase difference between the signals in each geophone and the coherence between signals was calculated. In the case of CSW, the minimum square method was used to determine the phase angle for each vibration frequency. Both the analysis and the processing of data regarding the SASW, as well as the inversion in the dispersion curve and the calculation of the variation of the shear modulus in depth, were performed with the software WINSASW 2.0 [30].

Figure 10a, b show some aspects of the tests performed on both the old and the renewed railway platforms. Figure 10c presents the profiles of the deformation modulus obtained in one place with the SASW, E_{SASW}, during the phases of renewal

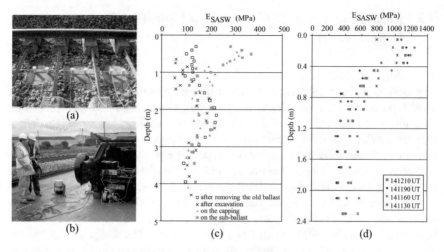

Fig. 10 Aspects of the tests: **a** SASW on the old track; **b** CSW on the renewed platform; **c** E_{SASW} in one place during the renewal; **d** E_{SASW} in some places in a newly built zone [28]

of the railway platform. The depth presented refers to the bottom of sleepers, prior to old ballast removal. The following comments can be made: the variation in the E_{SASW} is important, both in depth and between tests; following the old ballast removal, E_{SASW} values of about 140 MPa were measured from the surface, which are likely to correspond to a zone of higher compactness, due to rail traffic over the years; below and up to about 1.0–1.2 m depth, the average values of E_{SASW} ranged from 100 to 130 MPa; between that depth and around 2 m, E_{SASW} ranged between 200 and 230 MPa, reaching 100 MPa between this depth and above 3.5 m; the results obtained after excavation of a 0.35 m thickness seem to indicate that the consequent vertical stress reduction caused a decrease in the E_{SASW}, when compared with the values at the same depth before excavation, particularly up to about 1.5 m depth; after placing the 0.20 m thick capping layer of crushed limestone of well graded particle-size distribution, E_{SASW} increased to about 300 MPa in that layer; E_{SASW} in the sub-ballast layer (0.15 m) was about 300–350 MPa; the placement of that layer on the capping have increased the E_{SASW} of the latter to nearly 400 MPa.

Figure 10d shows the E_{SASW} obtained in a stretch where the railway platform was built on a new embankment. The tests were performed on the sub-ballast when the average water content was 1.4%, measured at a depth of 0.15 m, in the capping layer. We can draw the following conclusions: at the surface, corresponding to the sub-ballast and capping layers, E_{SASW} ranged from 800 to 1200 MPa; at the upper part of the embankment, down to about 0.30 m below the aggregate layers, E_{SASW} was 600–800 MPa; in the deeper layers, the values ranged from approximately 300–500 MPa.

In view of these results, it can be concluded that the SWM methods allow to estimate the deformability of the layers of the railway track substructure.

3.2   Evaluating Track Stiffness with a Light Weight Deflectometer

The falling weight deflectometer (FWD) is a non-destructive test that is traditionally used to perform stiffness assessment of road pavements. However, there are some reports of its use on railway tracks during substructure construction and on track under operation [31–33]. Paixão et al. [34] used the light falling weight deflectometer (LWD), a portable version of FWD, on sleepers to assess sleeper support conditions. These tests were carried out as part of a study that aimed to assess the influence of Under Sleeper Pads (USP) [35] on the behavior of transition zones between earthworks and civil engineering structures. Tests were performed on two similar underpasses in reinforced concrete closed frames (UP1 and UP2), located at a new single track of the Portuguese railway network. At UP1, 44 sleepers with USP were installed, while UP2 consisted of a standard track. The design established the construction of transition zones in accordance with demanding design requirements, similar to those implemented in new European high-speed railway lines [36] to avoid poor track performance [37]. Relevant aspects of the transition zones at UP1 and UP2 are presented in Fig. 11. The layers of the backfills were constructed with cement bound mixture (CBM) and unbound granular material (UGM).

The LWD applied a load impulse of 15 kN on a 0.10 m diameter plate at the center of each sleeper (Fig. 12a). A good coherence of results was observed between tests performed on sleepers with similar support conditions. Figure 12b presents the average deflection (δ_s) measured at the center of the sleepers, along two transition zones to UP1 and UP2. It can be observed that the δ_s of the sleepers without USP is about 50–100 mm. At UP1, the use of USP has increased δ_s up to about 450 mm.

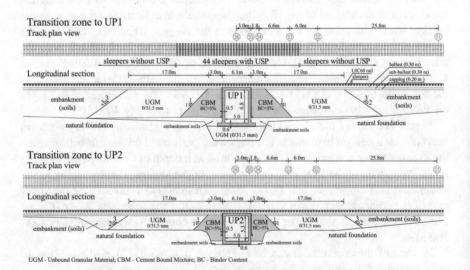

Fig. 11 Longitudinal profile of transition zones to UP1 and UP2 [34]

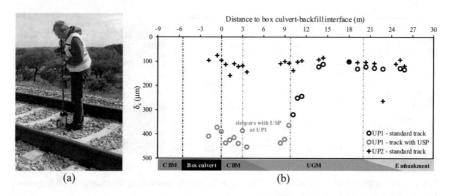

Fig. 12 Sleeper deflections obtained in LWD tests: **a** test; **b** deflection [34]

Thus, it was found that the deflections of sleepers with USP were about 4–4.5 times higher than those obtained in sleepers without USP. At the transition between sleepers with and without USP, a variation in deflections is observed, denoting a transition in stiffness.

This unconventional procedure of using the LWD test at the center of the sleeper was found to be a very simple, straightforward, and effective procedure to swiftly assess track support conditions. Results indicate that an experienced operator can obtain good testing repeatability for several drops on the same sleeper.

3.3 Receptance Tests on the Track

Receptance tests are a very practical way to assess the dynamic behavior of the track [38]. The receptance tests that have been carried out by the authors consist of the excitation of the rail using an instrumented hammer, with input frequencies of up to 450 Hz, and of the evaluation of the track response using accelerometers installed on the rail web and sleepers. Figure 13 presents an example of a sampled time recording of the measured load impulse, along with its frequency content and respective rail response.

In order to characterize the dynamic behavior of the track across both transition zones described in 3.2, receptance tests were performed on specific sections identified in Fig. 11: S1 on open track; S2 and S3 on the UGM; S4 on the CBM; S5 on the vertical alignment of the box culvert walls; and S6 at mid-span of the box culvert.

Figure 14 depicts the average receptance functions obtained in each section for frequencies between 25 and 450 Hz, corresponding to the range of impulse-acceleration signal coherence above 0.95. In general, apart from the results obtained in sections with USP, no significant differences can be observed between other receptance function curves along the transition zones. Regarding all sections, it is possible to identify a resonant frequency between 300 and 375 Hz, which corresponds to the

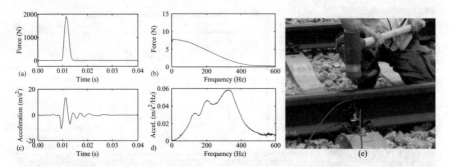

Fig. 13 Receptance test: **a** impulse time history; **b** frequency content; **c, d** respective rail responses in time and frequency domains; **e** test [34]

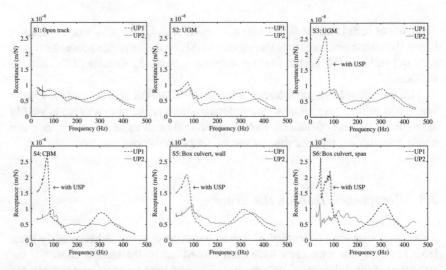

Fig. 14 Receptance functions obtained in each section [34]

vibration of the rail on the rail pads. The variability in the resonant frequency of the rail may be due to the uneven preload of the fastening system along the track, which generally influences the stiffness of rail pads [39]. The presence of USP at UP1 leads to a shift in the resonant frequency peak: the frequency of about 325 Hz in S1 and S2 (without USP) is reduced to about 308 Hz in S3 to S6 (with USP). This reduction is followed by a slight increase in the amplitude of the receptance function peak. The full track resonant frequency peaks are also visible at frequencies around 70 Hz for the track with USP (S3 to S6 in UP1) and at about 90–110 Hz without USP (the remaining sections). Higher amplitudes of the full track resonant frequency peaks are identified in sections with USP, indicating an increase in the dynamic flexibility of the track provided by these elements. It is worth noting the increased amplitude of the receptance function in lower frequencies (f < 55 Hz) in the presence of USP, which denotes an increment of about 90% in the vertical flexibility of the track. The receptance curves in S3 (on the UGM) and S4 (on the CBM) are quite similar, within each transition zone. In S6, additional peaks were identified at 44 and 40 Hz, corresponding to the resonant frequencies of UP1 and UP2 box culverts.

The USP led to a reduction of about 18% in the full track resonant frequency and to an increase of about 50% in its receptance value. The rail resonant frequency peak was reduced by 5% and an increase of about 14% in the receptance value was observed.

Considering the previous results, we can conclude that the receptance tests on the rail, performed at different positions along the transition zones, allowed assessing the dynamic flexibility of the track for different support conditions. In addition, these tests can contribute to the calibration of numerical models of the railway track [40].

3.4   Evaluating Track Stiffness with the Portancemètre

Portancemètre is a non-destructive loading equipment that was initially developed to be used in earthworks [41]. It consists of a rolling vibrating wheel that makes it possible to perform continuous stiffness measurements and to obtain the respective values of the deformation modulus (E_{V2}) [42]. For some years now, Portancemètre has been used to assess the deformability of embankment layers and railway track platforms [36, 43].

The Portancemètre has had some modifications so as to be used to assess track stiffness, measured continuously on the rail [44]. The static load of the equipment may vary between 70 and 120 kN and the maximum dynamic load amplitude may increase up to 70 kN. The stiffness can be measured by exciting the track with a frequency of up to 35 Hz. The method has good repeatability and good correlation with other methods.

Figure 15 shows an on-site measurement and track stiffness at different excitation frequencies and a running speed of 6 km/h. There are relevant differences between

(a)   (b)

Fig. 15 Portancemètre: **a** on-site measurement; **b** track stiffness values [44]

the results obtained with 25 Hz and other excitation frequencies, which may be explained by the closeness to the resonance frequency of the track. The results of 10 and 30 Hz excitations (probably lower and higher than the resonance frequency) are close to each other especially for low stiffness values.

Other methods have been developed for similar purposes. In any case, none of these methods are widely used [45].

3.5   Evaluating Track Support Conditions with the Ground Penetrating Radar

By detecting variations in the electromagnetic properties of the medium, the Ground Penetrating Radar (GPR) nondestructive technique can provide valuable and almost continuous information on the track condition. Examples of its application include detecting different materials in depth and identifying the degradation of the physical and mechanical characteristics of the track, such as ballast pockets, fouled ballast, poor drainage, subgrade settlement and transition problems [3, 46, 47].

In the last few years, the development of new GPR systems with higher antenna frequencies, better data acquisition systems, more user-friendly software, and new algorithms for calculation of materials properties have been leading to a more generalized use of GPR. As an example, Fig. 16 shows some results of the application of GPR on a railway track under renewal. The transition between the old and the renewed zones is clearly defined. As expected, the old section is more scattered and the interface between ballast and subgrade is not clear in some parts of the image, due to ballast fouling, whereas on the renewed section, a clearer identification is achieved and the two interfaces, between ballast and sub-ballast and sub-ballast and subgrade, are well revealed.

In other cases, it was possible to relate defects in track geometry with the information obtained with the GPR and to propose different rehabilitation techniques [48].

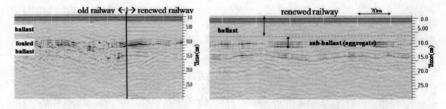

Fig. 16 GPR measurements on a transition between an old and a renewed track section (left) and on a renewed track section (right) [50]

With a view to improve GPR data analyses by an appropriate interpretation, it is necessary to obtain reliable information on the dielectric properties of tested materials. These characteristics affect GPR signal propagation, reflection, and data resolution; therefore, it is important to study which factors influence the dielectric constants and in which degree. De Chiara et al. [49] performed laboratory tests to determine the dielectric properties of infrastructure materials (clean granite ballast; silt soils; fouled ballast, as a mixture of the first two) under different conditions, in terms of both water content and fouling level, with GPR antennas of 5 different frequencies (400, 500, 900, 1000 and 1800 MHz). Fouling condition and the water content significantly influenced the dielectric values of the materials (Figs. 17 and 18). It is also known that ballast of different types of rock has different dielectric constant values [50].

Recent works using data from GPR surveys of railway infrastructures also explore the possibility of analyzing the GPR signal in the frequency domain to obtain further information about the track condition. Fontul et al. [51] presented an approach to analyze the data in both time and frequency domains, with a view to identify changes along significant lengths of railway lines that can correspond to either known track singularities or track pathologies.

The processing focuses on specific time intervals and frequency ranges of the signal that are more representative for identifying changes in the infrastructure. A

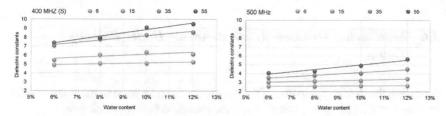

Fig. 17 Variation in the dielectric constant values of ballast with the fouling index (6–55), for the 400 MHz IDS suspended antenna (left) and the 500 MHz GSSI antenna (right) [49]

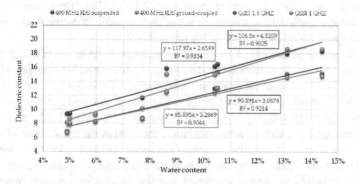

Fig. 18 Variation in the dielectric constant values of the soil with the water content [52]

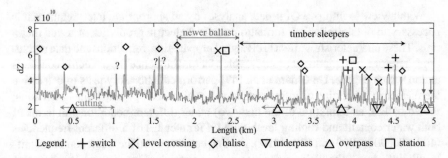

Fig. 19 Frequency domain analysis of GPR signal and correspondence with track events [51]

tool was developed to process, compare, and visualize the GPR data that can be further integrated with track geometry data and aerial photography for a user-friendly interpretation. Figure 19 depicts an example of this approach using the GPR data obtained in a 5-km section with an inspection vehicle equipped with 400 MHz GPR antennas by IDS. The figure shows the difference in signal amplitudes, dZ, in a specific frequency interval (0.7 and 2.0 GHz), calculated using a short sliding window (10 m), against the signal amplitude calculated using a wider sliding window (200 m). The results show a clear correspondence with specific track events.

3.6 Monitoring Dynamic Track Behavior Under Railway Traffic

Assessing the dynamic response of the railway track under passing trains is essential to understand its structural behavior and the evolution in its performance.

Within the scope of some research projects, the authors have used monitoring systems, including several devices, to measure variables related with the behavior of the track [53, 54]. These systems included various types of transducers to measure: (i) wheel loads; (ii) vertical displacements of the rail; (iii) rail-sleeper relative displacements (which are related to the deformation of the rail pad); (iv) vertical accelerations of the sleepers; (v) rail seat loads on the rail pads [55].

To estimate dynamic wheel loads and the reaction force at the rail seat portion of the sleeper, shear deformations on the rail can be measured between two sleepers (Fig. 20a). To compensate for any transversal eccentricity in the load applied by the train, at each measurement section of the rail, a pair of strain gauges was welded to each side of the web and connected to a full Wheatstone bridge. Each pair of strain gauges consisted of a shear rosette installed at the level of the neutral axis of the rail and with the respective sensitive patterns oriented $+ 45°$ and $-45°$ with respect to the rail longitudinal axis. At the longitudinal surface passing through the neutral axis of the rail, a pure shear state is developed, and the above orientations correspond to principal directions. Thus, deformations along these directions are equal in magnitude

Fig. 20 Example of measuring systems: **a** strain gauges; **b** LASER diode; **c** PSD; **d** LVDT; **e** piezoelectric accelerometer; **f** MEMS accelerometers [54]

and opposite in sign ($\varepsilon_1 = -\varepsilon_2 = \varepsilon$) and shear strain is $\gamma = 2\varepsilon$. Absolute rail vertical displacements can be measured by optical systems based on a diode LASER module, mounted away from the track (Fig. 20b), and by a Position Sensitive Detector (PSD) module attached to a support fixed to the rail web (Fig. 20c). Using PSD transducers, it is possible to calculate rail displacements as trains pass by, measuring variations in the position of the LASER beam in the PSD. The vertical deformation of the rail pad can be assessed by a LVDT transducer placed between the rail and the sleeper (Fig. 20d). Vertical accelerations may be measured by piezoelectric accelerometers (Fig. 20e) and Micro Electro-Mechanical System accelerometers (MEMS) placed at the sleeper ends (Fig. 20f). In some studies, rail seat loads on the rail pads, using fiber optic sensors, are also measured. Typical and portable acquisition systems comprise a laptop computer, one or more acquisition units, as well as support units for the different types of transducers (Fig. 21).

There are many other devices and systems for instrumenting sleepers, the ballast layer and the railway substructure [56–59]. The data obtained with all these systems have led to deepen the knowledge about the behavior of the track and have enabled the calibration of numerical models.

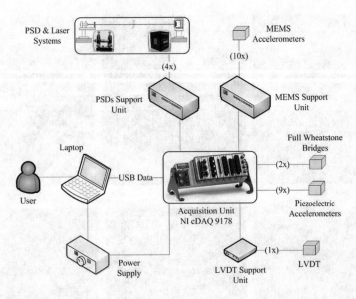

Fig. 21 Schematic representation of the data acquisition system [55]

4 On Board Monitoring with In-Service Railway Vehicles

Railway infrastructure managers run dedicated inspection vehicles to monitor the geometric quality of the track (among other aspects), to detect irregularities and to ensure safe running conditions of railway lines, in accordance with specific regulations [60]. The reconstruction of track geometry from track inspection cars is a well-established procedure and these vehicles are currently manufactured on an industrial basis by a few suppliers [61]. Modern inspection cars integrate an inertial measurement that uses the known measuring technique, by which the spatial position of a measuring sensor is determined by double integration of acceleration measurements [62]. Unfortunately, these inspections disturb the normal traffic operation, especially in networks with intensive traffic; are expensive and generally are carried out only a few times per year; and, consequently, do not provide a prompt identification of critical situations. Therefore, there is a need to develop alternative and expeditious methods to provide timely information on the performance and condition of the track, as implemented in recent years [63].

4.1 Non-dedicated Equipment

Due to the increasing sensing capabilities and cost reduction of sensors for smartphones, significant developments have been achieved as regards the application of micro-electro-mechanical systems (MEMS) to monitor transport systems. Paixão

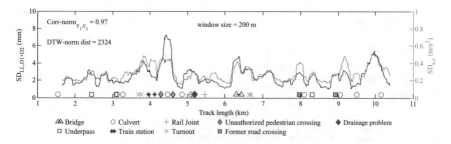

Fig. 22 $SD_{LL,D1+D2}$ and $SD_{a,z}$, considering a sliding-window size of 200 m [64]

et al. [64] presented an approach to use these technologies to perform continuous acceleration measurements on-board a passenger train with a smartphone. These measurements were analyzed and compared against the longitudinal level of the track geometry records and its degradation with time. That study also demonstrates further applicability of the approach to evaluate the structural performance of railway tracks, by assessing how different wavelengths and amplitudes of the geometric irregularities affect degradation, running safety, and ride comfort [65].

To demonstrate that the vertical accelerations measured inside the coach car are correlated to the longitudinal level of the track, Fig. 22 presents the standard deviation of the longitudinal level, $SD_{LL,D1+D2}$, regarding wavelength ranges $D1(3 < \lambda \le 25m) + D2(25 < \lambda \le 70m)$, and the standard deviation of the vertical accelerations, $SD_{a,z}$, considering a sliding-window size of 200 m. The normalized cross-correlation (Corr $-$ $norm_{y1,y2}$) and dynamic time warping distance (DTW-norm dist) values [66, 67] between $SD_{LL,D1+D2}$ and $SD_{a,z}$ are also indicated. The very good correlation between the variables is visible on an 11-km railway stretch.

In order to justify the peak values of $SD_{LL,D1+D2}$ and $SD_{a,z}$, the authors analyzed the characteristics of the track, looking for any relevant structural aspects or sudden changes in the track structure (track discontinuity or singularities). About 30 relevant discontinuities were identified, including transition zones, rail joints, turnouts, bridges, underpasses, culverts, and stations, among others. Figure 23 presents some examples of these discontinuities, and their location was also included in Fig. 22. It can be observed that, in general, these correspond to the location of peak values

Fig. 23 Discontinuities: **a** culvert; **b** turnout; **c** rail joint [64] (Imagery ©2016 Google)

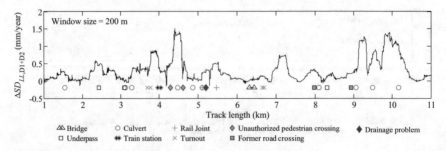

Fig. 24 $\Delta SD_{LL,D1+D2}$ considering a sliding window of 200 m [64]

of $SD_{LL,D1+D2}$ and $SD_{a,z}$. This suggests that these locations experience a higher geometric degradation and affect passenger comfort, which is in agreement with other studies [68–70].

Subsequently, to assess the actual track degradation rates, the authors analyzed the evolution of the $SD_{LL,D1+D2}$ between the surveys of 2013 and 2016, $\Delta SD_{LL,D1+D2}$. Figure 24 shows that most of the locations denoting higher degradation rates (higher $\Delta SD_{LL,D1+D2}$) generally also correspond to the locations where higher values of $SD_{LL,D1+D2}$ are observed in Fig. 22; for example, between km 3.5 and 4.5 and between km 9 and 10.5. It is also interesting to note that most $\Delta SD_{LL,D1+D2}$ peaks are also centered on the track discontinuities identified above.

4.2 Dedicated Equipment

As part of a research project, the authors contributed to the development and testing of a prototype of a monitoring system mounted on a railway maintenance vehicle [71, 72]. To interpret the results obtained with part of this system [73], a methodology was developed for determining the railway track support conditions. This methodology is based on the modal analysis of the characteristic frequencies of a 2-DoF model composed of the railway infrastructure and an instrumented vehicle moving over it. The methodology is integrated in the group of vibration-based structural damage identification methods and is focused on observing the characteristic frequencies of the combined system, which can be correlated with changes in the physical properties of the infrastructure under analysis. By performing this assessment of the railway infrastructure across its length and over time, by comparing different rides over the same railway stretch, important information can be gathered about the track support conditions.

The sensors used to provide the data for this study (Fig. 25) consisted of two high-sensitivity ±4 g triaxial accelerometers, installed symmetrically in each vehicle's cabins. Four high-range ±500 g accelerometers, mounted on the wheel boxes, were used to measure the vehicle dynamic interactions with the track and to compare their output with the cabin accelerometers. The prototype also included two velocity

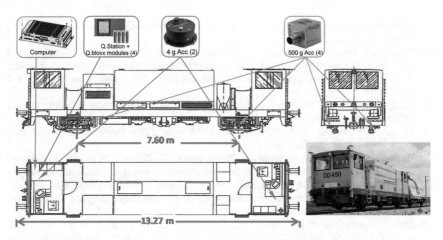

Fig. 25 Schematic view of the instrumentation locations in the railway vehicle [72]

sensors, installed on each shaft, and based on magnetic proximity sensors. The data generated by these velocity sensors, together with a GPS system, was used to obtain the position of the vehicle along the line. The data acquisition system collected data at a sampling frequency of 1000 Hz. This system comprised a data collector, some measuring modules and an industrial computer.

Preliminary validation of the developed methodology was performed by analyzing the passage of the instrumented vehicle over a transition zone between earthworks and a bridge [54, 73]. The comparison of on-board measurements with way-side measurements, allowed to conclude that the developed methodology is able to identify changes in functional and structural parameters of the railway line.

5 Final Remarks

This chapter presents some testing techniques, devices and monitoring systems that have been used for characterizing materials and for evaluating the structural behavior of the classic railway track. Some evaluation methods presented here are not yet commonly used, either due to the associated cost or to the need for qualified human resources. However, the improvement of knowledge and the technological advances in several areas will certainly allow their dissemination and widespread use.

References

1. Paixão, A., Varandas, J., Fortunato, E., Calçada, R.: Non-linear behaviour of geomaterials in railway tracks under different loading conditions. In: Advances in Transportation Geotechnics

3. Proceedings of the 3rd International Conference on Transportation Geotechnics, Guimarães, Portugal, 4–7 Sept 2016
2. Varandas, J., Paixão, A., Fortunato, E., Hölscher, P.: A numerical study on the stress changes in the ballast due to train passages. In: Advances in Transportation Geotechnics 3. Proceedings of the 3rd International Conference on Transportation Geotechnics, Guimarães, Portugal, 4–7 Sept 2016
3. Selig, E.T., Waters, J.M.: Track Geotechnology and Substructure Management. Thomas Telford, London (1994)
4. Indraratna, B., Salim, W., Rujikiatkamjorn, C.: Advanced Rail Geotechnology—Ballasted Track. Taylor & Francis (2011)
5. Varandas, J.N., Paixão, A., Fortunato, E., Zuada Coelho, B., Hölscher, P.: Long-term deformation of railway tracks considering train-track interaction and non-linear resilient behaviour of aggregates—a 3D FEM implementation. Comput. Geotech. **126**, 103712 (2020)
6. AREMA: Manual for Railway Engineering. American Railway Engineering and Maintenance-of-Way Association, Lanham, MD (2020)
7. CEN: European Standard EN 13450:2002, Aggregates for Railway Ballast. 93.100—Construction of Railways; 91.100.15—Mineral Materials and Products. CEN TC 154. Comité Européen de Normalisation, Brussels (2002)
8. Guo, Y., Markine, V., Zhang, X., Qiang, W., Jing, G.: Image analysis for morphology, rheology and degradation study of railway ballast: a review. Transp. Geotech. **18**, 173–211 (2019)
9. Blott, S.J., Pye, K.: Particle shape: a review and new methods of characterization and classification. Sedimentology **55**(1), 31–63 (2008)
10. Folk, R.L.: Student operator error in determination of roundness, sphericity, and grain size. J. Sediment. Petrol. **25**(4), 297–301 (1955)
11. Guo, Y., Zhao, C., Markine, V., Jing, G., Zhai, W.: Calibration for discrete element modelling of railway ballast: a review. Transp. Geotech. **23** (2020)
12. Quintanilla, I.D., Combe, G., Emeriault, F., Voivret, C., Ferellec, J.F.: X-ray CT analysis of the evolution of ballast grain morphology along a micro-Deval test: key role of the asperity scale. Granul. Matter **30**, 3–12 (2019)
13. Moaveni, M., Qian, Y., Boler, H., Mishra, D., Tutumluer, E.: Investigation of ballast degradation and fouling trends using image analysis. In: 2nd International Conference on Railway Technology: Research, Development and Maintenance—Railways 2014, Ajaccio, Corsica, France, 8–11 Apr 2014
14. Delgado, B.G., Viana da Fonseca, A., Fortunato, E., Goretti da Motta, L.M.: Particle morphology's influence on the rail ballast behaviour of a steel slag aggregate. Environ. Geotech. 1–10 (2019), published ahead of print
15. Al-Rousan, T.: Characterization of aggregate shape properties using a computer automated system. PhD thesis, Texas A&M University, Texas, 2004
16. Esmaeili, M., Nouri, R., Yousefian, K.: Experimental comparison of the lateral resistance of tracks with steel slag ballast and limestone ballast materials. Proc. Inst. Mech. Eng. Part F J. Rail Rapid Transit **231**(2), 175–184 (2017)
17. Delgado, B.G., Viana da Fonseca, A., Fortunato, E., Maia, P.: Mechanical behavior of inert steel slag ballast for heavy haul rail track: laboratory evaluation. Transp. Geotech. **20**, 100243 (2019)
18. Delgado, B.G., Viana da Fonseca, A., Fortunato, E., Paixão, A., Alves, R.: Geomechanical assessment of an inert steel slag aggregate as an alternative ballast material for heavy haul rail tracks. Constr. Build. Mater. **279**, 122438 (2021)
19. Suhr, B., Skipper, W.A., Lewis, R., Six, K.: Shape analysis of railway ballast stones: curvature-based calculation of particle angularity. Sci. Rep. **10**(1), 6045 (2020)
20. Jerónimo, P., Resende, R., Fortunato, E.: An assessment of contact and laser based scanning of rock particles for railway ballast. Transp. Geotech. **22**, 100302 (2020)
21. Paixão, A., Resende, R., Fortunato, E.: Photogrammetry for digital reconstruction of railway ballast particles—a cost-efficient method. Constr. Build. Mater. **191**, 963–976 (2018)

22. Zhao, L., Zhang, S., Huang, D., Wang, X., Zhang, Y.: 3D shape quantification and random packing simulation of rock aggregates using photogrammetry-based reconstruction and discrete element method. Constr. Build. Mater. **262**, 119986 (2020)
23. Paixão, A., Fortunato, E.: Abrasion evolution of steel furnace slag aggregate for railway ballast: 3D morphology analysis of scanned particles by close-range photogrammetry. Constr. Build. Mater. **267**, 121225 (2021)
24. Fortunato, E., Pinelo, A., Matos, F.M.: Characterization of the fouled ballast layer in the substructure of a 19th century railway track under renewal. J. Jpn. Geotech. Soc. Soils Found. **50**(1), 55–62 (2010)
25. Fortunato, E., Paixão, A., Fontul, S., Pires, J.: Some results on the properties and behavior of railway ballast. In: 10th International Conference on the Bearing Capacity of Roads, Railways and Airfields (BCRRA 2017), Athens, Greece, 28–30 June 2017
26. Werkmeister, S., Dawson, A., Wellner, F.: Permanent deformation behavior of granular materials and the shakedown concept. Transp. Res. Rec. **1757**, 75–81 (2001)
27. Stokoe II, K.H., John, S.H., Woods, R.D.: Some contributions of in situ geophysical measurements to solving geotechnical engineering problems. In: ICS-2 on Geotechnical and Geophysical Site Characterization, Porto (2004)
28. Fortunato, E., Bilé Serra, J., Marcelino, J.: Application of spectral analysis of surface waves (SASW) in the characterisation of railway platforms. In: International Conference on Advanced Characterization of Pavement and Soil Engineering Materials, Athens, 20–22 June 2007
29. Matthews, M.C., Hope, V.S., Clayton, C.R.I.: The use of surface waves in the determination of ground stiffness profiles. Proc. Inst. Civ. Eng. Geotech. Eng. **119**(2), 84–95 (1996)
30. Joh, S.-H.: Data Interpretation and Analysis for SASW Measurements. Chung-Ang University, Anseong, Korea (2002)
31. Burrow, M.P.N., Chan, A.H.C., Shein, A.: Deflectometer-based analysis of ballasted railway tracks. Proc. Inst. Civ. Eng. Geotech. Eng. **160**(3), 169–177 (2007)
32. Fontul, S., Fortunato, E., De Chiara, F.: Non-destructive tests for railway infrastructure stiffness evaluation. In: The Thirteenth International Conference on Civil, Structural and Environmental Engineering Computing, Crete, Greece, 6–9 Sept 2011
33. Fortunato, E., Fontul, S., Paixão, A., Cruz, N., Cruz, J., Asseiceiro, F.: Geotechnical aspects of the rehabilitation of a freight railway line in Africa. In: 3rd International Conference on Railway Technology: Research, Development and Maintenance—Railways 2016, Cagliari, Sardinia, Italy, 5–8 Apr 2016
34. Paixão, A., Alves Ribeiro, C., Pinto, N.M.P., Fortunato, E., Calçada, R.: On the use of under sleeper pads in transition zones at railway underpasses: experimental field testing. Struct. Infrastruct. Eng. **11**(2), 112–128 (2015)
35. UIC: Under Sleeper Pads—Summarising Report. Union Internationale des Chemins de Fer, Vienna (2009)
36. Paixão, A., Fortunato, E., Calçada, R.: Design and construction of backfills for railway track transition zones. Proc. IMechE Part F J. Rail Rapid Transit **229**(1), 58–70 (2015)
37. Varandas, J.N., Paixão, A., Fortunato, E.: A study on the dynamic train-track interaction over cut-fill transitions on buried culverts. Comput. Struct. **189**, 49–61 (2017)
38. De Man, A.P.: DYNATRACK: a survey of dynamic railway track properties and their quality. PhD thesis, Faculty of Civil Engineering, Delft University of Technology, Delft, 2002
39. Wu, T.X., Thompson, D.J.: Effects of local preload on the foundation stiffness and vertical vibration of railway track. J. Sound Vib. **219**(5), 881–904 (1999)
40. Alves, R.C., Paixão, A., Fortunato, E., Calçada, R.: Under sleeper pads in transition zones at railway underpasses: numerical modelling and experimental validation. Struct. Infrastruct. Eng. **11**(11), 1432–1449 (2015)
41. Quibel, A.: New in situ devices to evaluate bearing capacity and compaction of unbound granular materials. In: Gomes Correia, A. (ed.) Unbound Granular Materials—Laboratory Testing, In-Situ Testing and Modelling, pp. 141–151. A. A. Balkema, Rotterdam (1999)
42. Railways I.U.o.: IRS-70719:2020, Ed1: Railway Application—Track & Structure—"Earthworks and Track Bed Layers for Railway Lines"—Design and Construction Principles (2020)

43. Fortunato, E., Paixão, A., Fontul, S.: Improving the use of unbound granular materials in railway sub-ballast layer. In: Advances in Transportation Geotechnics II, Hokkaido University, Japan, 10–12 Sept 2012

44. Hosseingholian, M., Froumentin, M., Robinet, A.: Dynamic track modulus from measurement of track acceleration by Portancemetre. In: WCRR 2011—World Congress on Railway Research, Lille, France, 22–26 May 2011

45. Wang, P., Wang, L., Chen, R., Xu, J., Xu, J., Gao, M.: Overview and outlook on railway track stiffness measurement. J. Mod. Transp. **24**, 89–102 (2016)

46. Fontul, S., Antunes, M.L., Fortunato, E., Oliveira, M.: Practical application of GPR in transport infrastructure survey. In: International Conference on Advanced Characterization of Pavement and Soil Engineering Materials, Athens, Greece, 20–22 June 2007

47. Cruz, N., Fortunato, E., Asseiceiro, F., Cruz, J., Mateus, C.: Methodologies for geotechnical characterization in railways in operation. An experience. In: XVI European Conference on Soil Mechanics and Geotechnical Engineering, Edinburgh, United Kingdom, 13–17 Sept 2015

48. Fontul, S., Fortunato, E., De Chiara, F., Burrinha, R., Baldeiras, M.: Railways track characterization using ground penetrating radar. In: The 3rd International Conference on Transportation Geotechnics (ICTG 2016), Guimarães, 4–7 Sept 2016

49. De Chiara, F., Fontul, S., Fortunato, E.: GPR laboratory tests for railways materials dielectric properties assessment. Remote Sens. **6**(10), 9712–9728 (2014)

50. Fontul, S., de Chiara, F., Fortunato, E., Paixão, A.: Non destructive tests for evaluation of railway platforms—application of ground penetrating RADAR. In: 1st International Conference on Railway Technology: Research, Development and Maintenance (Railways 2012), Las Palmas, Gran Canaria, Spain, 18–20 Apr 2012

51. Fontul, S., Paixão, A., Solla, M., Pajewski, L.: Railway track condition assessment at network level by frequency domain analysis of GPR data. Remote Sens. **10**(4), 559 (2018)

52. Fontul, S., Fortunato, E., De Chiara, F.: Evaluation of ballast fouling using GPR. In: 15th International Conference on Ground Penetrating Radar—GPR 2014, Brussels, Belgium, 30 June–4 July 2014

53. Paixão, A., Fortunato, E., Calçada, R.: Transition zones to railway bridges: track measurements and numerical modelling. Eng. Struct. **80**, 435–443 (2014)

54. Paixão, A., Varandas, J., Fortunato, E., Calçada, R.: Numerical simulations to improve the use of under sleeper pads at transition zones to railway bridges. Eng. Struct. **164**, 169–182 (2018)

55. Paixão, A.: Transition zones in railway tracks. An experimental and numerical study on the structural behaviour. PhD thesis, University of Porto, Faculty of Engineering (FEUP), Porto (2014)

56. Aikawa, A.: Determination of dynamic ballast characteristics under transient impact loading. Electron. J. Struct. Eng. **13**(1), 17–34 (2013)

57. Jing, G., Siahkouhi, M., Edwards, J.R., Dersch, M.S., Hoult, N.A.: Smart railway sleepers—a review of recent developments, challenges, and future prospects. Constr. Build. Mater. **271**, 121533 (2021)

58. Rocchi, D., Tomasini, G., Schito, P., Somaschini, C., Testa, M., Cerullo, M., Arcoleo, G.: Ballast lifting: a challenge in the increase of the commercial speed of HS-trains. In: WCRR2016—11th World Congress on Railway Research, Milan, 29 May–2 June 2016

59. Mishra, D., Tutumluer, E., Boler, H., Hyslip, J.P., Sussmann, T.R.: Railroad track transitions with multidepth deflectometers and strain gauges. Transp. Res. Rec. J. Transp. Res. Board **2448**, 105–114 (2014)

60. CEN: European Standard EN 13848-5:2017 Railway Applications—Track—Track Geometry Quality—Part 5: Geometric Quality Levels—Plain Line, Switches and Crossings. 93.100—Construction of Railways. CEN/TC 256—Railway Applications. Comité Européen de Normalisation, Brussels (2017)

61. Esveld, C.: Modern Railway Track, Digital Edition 2016 edn. MRT-Productions, Zaltbommel, The Netherlands (2016)

62. CEN: European Standard EN 13848-2:2020 Railway Applications—Track—Track Geometry Quality—Part 2: Measuring Systems—Track Recording Vehicles. 93.100—Construction of

Railways. CEN/TC 256—Railway Applications. Comité Européen de Normalisation, Brussels (2020)

63. Weston, P., Roberts, C., Yeo, G., Stewart, E.: Perspectives on railway track geometry condition monitoring from in-service railway vehicles. Veh. Syst. Dyn. **53**(7), 1063–1091 (2015)

64. Paixão, A., Fortunato, E., Calçada, R.: Smartphone's sensing capabilities for on-board railway track monitoring: structural performance and geometrical degradation assessment. Adv. Civ. Eng. **2019** (2019)

65. Paixão, A., Fortunato, E., Calçada, R.: The effect of differential settlements on the dynamic response of the train-track system: a numerical study. Eng. Struct. **88**, 216–224 (2015)

66. Anthony, A.W.: Speech Recognition by Machine. Peter Peregrinus Ltd, London, UK (1988)

67. Berndt, D.J., Clifford, J.: Using dynamic time warping to find patterns in time series. In: KDD-94, Knowledge Discovery in Databases (AAAI-94), Seattle, WA, July–Aug 1994, pp. 359–370

68. Ubalde, L.: La auscultación y los trabajos de vía en la línea del AVE Madrid—Sevilla: análisis de la experiencia y deducción de nuevos criterios de mantenimiento. Ph.D. thesis, Universitat Politècnica de Catalunya, Barcelona, Spain, 2004

69. Kouroussis, G., Connolly, D.P., Alexandrou, G., Vogiatzis, K.: The effect of railway local irregularities on ground vibration. Transp. Res. Part D Transp. Environ. **39**, 17–30 (2015)

70. Paixão, A., Fortunato, E., Calçada, R.: A contribution for integrated analysis of railway track performance at transition zones and other discontinuities. Constr. Build. Mater. **111**, 699–709 (2016)

71. Santos, C., Morais, P., Paixão, A., Fortunato, E., Asseiceiro, F., Alvarenga, P., Gomes, L.: An integrated monitoring system for continuous evaluation of railway tracks for efficient asset management. In: 5th International Conference on Road and Rail Infrastructure (CETRA 2018), Zadar, Croatia, 16–19 Apr 2018

72. Morais, J., Santos, C., Morais, P., Paixão, A., Fortunato, E., Asseiceiro, F., Alvarenga, P., Gomes, L.: Continuous monitoring and evaluation of railway tracks: system description and assessment. In: The 3rd International Conference on Structural Integrity (ICSI2019), Funchal, 2–5 Sept 2019

73. Morais, J., Santos, C., Morais, P., Paixão, A., Fortunato, E., Asseiceiro, F., Alvarenga, P., Gomes, L.: Continuous monitoring and evaluation of railway tracks: proof of concept test. In: The 3rd International Conference on Structural Integrity (ICSI2019), Funchal, 2–5 Sept 2019

Hydraulics and Natural Resources

Laboratory Tests on Wind-Wave Generation, Interaction and Breaking Processes

María Clavero⊙, **Luca Chiapponi**⊙, **Sandro Longo**⊙, and **Miguel A. Losada**⊙

Abstract At the atmosphere-ocean interface, the interaction between the atmospheric and ocean boundary layers determines heat and momentum exchanges and can affect the global balance of substances in air and water. In addition, wave generation, transformation and breaking play key roles in many physical processes, as they are essential for the analysis and characterisation of the behaviours of artificial and natural marine infrastructures near coasts. Breaking waves are complex phenomena that are enhanced by the vorticity and generation of turbulence and their evolution, during breaking and near the seabed. The transformation of a wave train that is breaking on a slope depends on the transport of turbulent kinetic energy (TKE), which causes the advection and spread of turbulence and generates a vortex according to the type of breaking. These processes are complex and not fully elucidated. The laboratory provides powerful tools for investigating these processes more deeply by analysing the transfer between the atmospheric and oceanic boundary layers and the turbulent characteristics of the breaking waves. This chapter presents recent advances in laboratory testing that are based primarily on physical tests that were developed in a combined wave-wind flume.

Keywords Wind · Sea gravity waves · Boundary layers · Experiments

M. Clavero (✉) · M. A. Losada
Andalusian Institute for Earth System Research, University of Granada, Granada, Spain
e-mail: mclavero@ugr.es

M. A. Losada
e-mail: mlosada@ugr.es

L. Chiapponi · S. Longo
Dipartimento di Ingegneria e Architettura (DIA), Università di Parma, Parma, Italy
e-mail: luca.chiapponi@unipr.it

S. Longo
e-mail: sandro.longo@unipr.it

© The Author(s), under exclusive license to Springer Nature Switzerland AG 2023
C. Chastre et al. (eds.), *Advances on Testing and Experimentation in Civil Engineering*, Springer Tracts in Civil Engineering,
https://doi.org/10.1007/978-3-031-05875-2_11

1   Introduction

In the atmospheric marine boundary layer (ABL) and the ocean boundary layer (OBL), many small- and large-scale flow processes occur. The vertical dimensions of the ABL (of around 500 m), of the OBL (of around 50 m) and of tidal waves (of around 5 m) are small in relation to the height of the atmosphere and the depths of the oceans. However, turbulence and boundary layer and waves play important roles on a large scale since they regulate the behaviour of the Earth's climatic and meteorological systems. The coupling of atmosphere and ocean through these processes is especially important and concerns many aspects that are being studied and analysed, such as climate prediction and climate change, wave breaking analysis, bubbles, and splash generation.

Thus, an ocean-atmosphere laboratory enables the study of many phenomena that are involved in the interaction, and the consequences on ABL and OBL properties of processes like: wave generation and breaking, heat balances in boundary layers, sediment dynamics and droplet formation, wave and wind motion actions on structures (offshore platforms, wind farms and offshore wind turbines), wave power generation and the relationships between heat exchange and life development. On a broader scale, it enables the analysis of the flows of energy, chemicals and nutrients that underlie seasonal cycles and have enabled species to evolve as we observe them today.

The lines of research are numerous and varied, and each of them requires a detailed analysis that is initially only conceptual and subsequently also experimental. From this perspective, experimentation seems irreplaceable in substantiating models. It should be recalled that mathematical modelling can lead to several solutions of the same physical problem, but they are not necessarily admissible or feasible: the only admissible mathematical solution is the one that best suits the experimental evidence, the most stable solution, the solution of minimum energy, or the solution that maximises entropy. In the light of these imperatives, researchers are facing experimental challenges that are linked to the size of the real scenario. Hence, specialised laboratories are necessary, in which not only the required experimental apparatuses, such as flumes with wave and wind generators, are available, but where a cultural approach also develops, aimed at conducting the best possible investigation of previous observations and making the vision of the future of research in the field happen.

In research laboratories, an approach is required for investigating (while respecting cultural traditions) new solutions with a free and open mind, new methods of measurement and investigation, and questioning, when necessary, what science has transmitted in a simplified way, which is sometimes much too simplified.

Although wind-tunnels date back to the end of the nineteenth century and flumes for the study of the waves date back even earlier (although without wave generation but with physical models of ships towed at a known speed in water at rest), mixed tunnels with mechanical wave generators and blowers have only been relatively recently proposed. One of the reasons is the technological complexity, but this is not the most important reason. It is presumable that the frontiers of knowledge are moving towards objectives of greater interest but also of higher difficulty. Curiosity and passion are the most powerful driving forces.

2 Experimental Facilities

When experimentally investigating wind-wave generation or interaction between the atmospheric boundary layer and the oceanic boundary layer, it is essential to have an experimental facility that can faithfully reproduce all the agents involved. For this reason, it is necessary to have an advanced wave generation system (to generate swell waves) and a wind generation system under controlled conditions.

This section provides a general description of the main characteristics of wind and wave facilities and presents current examples of such facilities. The main equipment used for measuring variables is described in the next paragraphs. Finally, a section is included on the simultaneous scaling of wind and waves in reduced model tests.

2.1 Wind-Wave Flumes

Various laboratories have approached the generation of wind-waves since the mid-1950s [1]. The experimental study of wind-wave interaction and its associate processes require a facility that is equipped with at least a wave flume and a wind-tunnel. Such facilities can have various dimensions and range from small facilities that consist of only water tanks inside wind-tunnels, with tests sections of less than or approximately 5 m long [2, 3], to large facilities that are several tens of meters long, such as the wind tunnel and wave flume facility in Harbin Institute of Technology in China [4], the Delft Hydraulics large wind-wave flume [5], and the large wind-wave flume at the hydraulics laboratory of the Scripps Institution of Oceanography [6]. The most common are the medium-sized wind-wave flumes, which are 10–20 m long, such as the ASIST wind-wave flume at the University of Miami [7], the wind-wave flume at the University of Queensland [7], the air-sea-current flume of the ASI Lab, University of Delaware [8] and the ocean-atmosphere interaction flume (CIAO) of the University of Granada [9] (see Fig. 1).

CIAO is a wind-wave flume that enables the generation and coupling of mechanically generated waves (with a paddle) and wind-generated waves and may also enable the generation and coupling of currents or rain. An active absorption system that uses real-time measurements of the water level in front of the paddles is provided to obtain a desired reflection condition of regular and irregular waves. It is possible to operate a double generation system by using the two paddles on opposite sides of the flume to improve the generation and absorption of highly nonlinear waves [10]. It is also possible to recreate various reflection conditions for regular and irregular

Fig. 1 Ocean-atmosphere
interaction flume (CIAO) at
the IISTA-University of
Granada

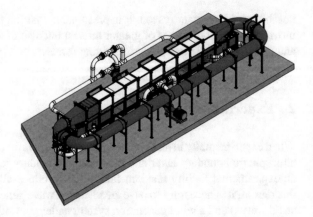

waves [9]. In the current configuration, the efficiency in controlling the reflection of wind-generated waves is limited.

In general, wind can be generated with closed- ([3, 9, 11], among others) or open-circuit [6] wind-tunnels over a wave flume. Wind is generated with one or more fans. The objective is to generate wind velocity and turbulence intensity profiles that are as similar as possible to the real profiles.

In addition to the generation of wind and waves, these facilities can generate other phenomena that also participate in the atmosphere-ocean interaction, such as:

– currents, which are usually generated by pumps;
– rain, which is generated by needles that are installed on the roof of the facility.

For a more focussed application, devices for the generation of thermal stratification can also be installed.

2.2 Other Facilities

Annular flumes to quantify the interactions among hydrodynamics, biological activity, and sediment dynamics are frequently used [12]. These are typically small facilities with limited dimensions and widths of 20–30 cm [13], although larger installations can be found [14, 15].

Facilities of this type have the advantage of presenting homogeneous conditions, in contrast to the fetch-dependent conditions in any linear facility [13]. These facilities are typically focused on research on ocean-atmosphere gas exchange.

2.3 Measurement Equipment

A large variety of variables are of interest in tests of this type; hence, a wide variety of instrumentation can be used.

One of the most important variables to measure is the water free surface displacement, from which the characteristics of the waves are obtained (e.g., height and period). The water level can be measured using various instruments, and the most common are the resistance (impedance)-wire wave gauges [6, 16] (among others). Another option is to use an ultrasonic distance meter (UDM) [9, 17], which has the advantage of being non-intrusive. Its main disadvantage is the loss of the signal if the reflective surface is too steep. This is remedied by increasing the size of the emission cone to obtain a larger footprint and, consequently, to reduce the probability of the echo being lost. The acquisition frequency is typically less than 100 Hz, which is more than sufficient for measurements of a free surface, where the surface tension limits the maximum frequency to a few hertz. The accuracy that is claimed by the manufacturers is a fraction of millimetres, but it is meaningless to assume values of less than 0.5 mm due to (i) the limitation on the wavelength of ultrasound in air (ultrasound at 100 kHz has a wavelength of 3 mm) and (ii) the finite size of the footprint, namely, a few centimetres. In summary, the estimated level is a space-average value over an area of several square centimetres.

The free surface geometry can also be identified with optical tools, as the colour imaging slope gauge (CISG), which was originally described in [18]. The imaging of the wave slope is based on light refraction at the water surface. The same principle is already used for laser slope gauges, and it enables the two-dimensional wavenumber spectrum estimation of the free surface elevation. Simpler tools have been developed for estimating the free surface profiles from images that have been captured from the lateral glass of the flumes: a laser sheet that intersects the free surface enables measurements in the mid-section of the flume, but satisfactory results can also be obtained simply by recording the profile at the lateral glass, although the meniscus effect reduces the accuracy [19].

To determine the wind characteristics, it is necessary to measure the velocity. For this purpose, equipment of various capacities can be used. The simplest is the Pitot-static tube [3, 6], but it is more frequently utilized to measure only mean velocity data since if the principal objective is to acquire turbulence, the corrections that are required to obtain accurate Pitot tube measurements render it unsuitable [20]. It is also possible to use hot-wire or hot-films [15]; this approach is based on the heat transfer from a sensing element. Hence, hot-wire and hot-films are highly sensitive to ambient variations in the temperature. Therefore, hot-wire anemometry is not usually recommended for the measurement of mean flow properties, but it is irreplaceable for investigations of rapidly varying flows and especially turbulence [21]. In fact, the main developments in turbulence analysis have been carried out with hot-wire anemometry in air flows because the technique offers a continuous signal with a flat frequency response of up to tens of kHz or even hundreds of kHz, but with distortion [22]. Its use in the presence of water has various limitations because the water droplets can attach to the sensor and possibly damage the wire or the film [23]. X-probe with constant temperature anemometers (CTAs) can be used to measure the Reynolds stress that is applied to the water surface [24]. The air flow can also be measured using a one-, two- or three-dimensional ultrasonic anemometer [15].

Water velocities can be measured with various instruments. If the intrusiveness of the equipment is not a problem for measurement, the velocity in the water can be measured with an acoustic Doppler velocimeter (ADV) [7] or an ultrasonic Doppler profiler [25–27]. The latter technique offers velocity measurements at a limited data rate (less than 100 Hz depending on the configuration and on the number of probes) in several gates along the axis of the ultrasonic cone; it enables data elaboration in the space domain, without the need to adopt the frozen turbulence hypothesis to convert data in the time domain into data in the space domain. Acoustic instruments require the seeding of particles (TiO_2 particles of a few micrometres are highly effective), and reliance on only particles that are naturally present in tap water or micro-bubbles yields poor results with frequent absence of a validated signal.

The high spatial and time resolutions required in turbulence measurements, can be provided by laser based equipment, such as a laser Doppler velocimeter (LDV), which provides high-accuracy data at a single point on one, two or three components (1C, 2C or 3C) [9, 16, 28]. The data rate depends on the concentration of seeding particles ("tracers") and on the system of data elaboration. The most effective is the burst spectrum analyser (BSA), although frequency loop-based system can yield satisfactory results. Higher accuracy is achieved with counters, although the number of validated particles decreases dramatically if the tracer concentration exceeds a threshold, and the time series has an uneven time step. Under optimal conditions, a data rate of up to a few kHz can be obtained. The tracers can be powder of suitable size or droplets. Seeding with water droplets that were generated by a spray gun was tested in a small wind-tunnel to measure the wind velocity profile [28]. The gun was located in the wind-tunnel inlet section before the honeycomb, and a strong airflow helped reduce the droplet size before reaching the measurement section and favoured sufficient mixing for the achievement of a data rate of tens of hertz, also in the trough sections of the wind-generated waves.

Particle image velocimetry (PIV) systems measure the instantaneous global velocity field in a planar region (2C-2D) of a flowing fluid by using a laser for the generation of a plane sheet of light, and one or more cameras to acquire a pair of images of the seeding particles. Stereoscopic PIV, also known as 3D-PIV, is used to obtain the three-component velocity field in the planar region (3C-2D, three-component velocity vectors in the plane of laser light), and two cameras are needed. Finally, the volumetric PIV system offers the possibility of measuring three components in a volume of the fluid (3D-3C) [17, 29, 30]. Data elaboration for extracting the velocity can be based either on fast Fourier transform (FFT) oriented algorithms of the pair of images, or particle tracking. The data rate depends on the characteristics of the laser generator, of the frame grabber and of the computational unit. Commercial instruments capture a few frames per second with 2048×2048 pixel^2-resolution cameras, and experimental high-performing instruments reach tens of kHz with much lower resolutions [31]. The huge data flow and data storage severely limit the performance of these systems. PIV and LDV have the advantage of being non-intrusive measurement methods, but it is essential to seed the water with seeds of a suitable size for the instrument and the data rate.

Additional variables that can be of interest are the temperature, relative humidity and pressure, for which classical instruments with sufficient accuracy are available. For experiments in which the salt flux is of interest, the salinity (a proxy for density) can be measured by sampling the fluid and analysing it with a pycnometer, a densimeter, a refractometer, or a conductivity probe, with eventual movement in the vertical direction to obtain the salinity profiles (see, e.g., [32] for measurements of this type). The shadowgraph and Schlieren optical techniques are also used for internal wave measurements.

The reader can refer to the previously cited or other articles for additional information on the discussed instruments.

2.4 Simultaneous Scaling of Wind-Wave Experiments

In most applications, dimensional analysis is used to develop similarity criteria that cannot be completely satisfied at the geometric scale of the laboratory, which is mainly because there are constraints in the choice of the fluid and there are obvious limitations in changing the gravitational acceleration. A partial similarity is adopted, with relevant dimensionless groups that assume different values in the model and in the prototype. The selection of the dominant groups depends on the type of phenomenon that is under study: for sea gravity waves, the Froude number is the preferred choice since most balances are between gravitational forces and convective inertial forces. However, in the presence of fluids with substantially different properties, such air and water, the scales that intervene in the process differ, and similarity must be planned by considering separately the processes in the two fluids. This is the case for wind that is blowing in a wind-tunnel at contact with water that, in turn, has a free surface that is populated by gravity waves.

Dimensional analysis for wind-waves and scaling

The correct scaling between laboratory data (the model) and field data (the prototype) in the wave flumes requires an analysis of the rules of similarity. The problem is well known, and although its principles are simple, the criteria for the development of similarity are complicated for complex experiments that involve airflow and wave motion.

For water, the physical process is described in terms of nine variables: a scale of velocity, a scale of length, time, density, fluid viscosity and surface tension, pressure, compressibility and acceleration of gravity. A group of three of the variables are independent, and using Buckingham's theorem, the problem is formulated as a function of six non-dimensional groups, namely, Reynolds, Froude, Weber, Strouhal, Euler, and Mach numbers [33].

The governed variables include the height and period of the wave, wavelength, and spectral characteristics, among others, which become dimensionless at a suitable scale with the fundamental quantities. Complete similarity requires equal values of each of these groups in the model and prototype; therefore, three degrees of freedom remain. However, using water in laboratory experiments and in the presence of gravitational acceleration, five constraints are added since the viscosity, density, surface tension, and compressibility have the same values in the model and prototype, and complete similarity is not possible. Partial similarity is obtained by (i) neglecting the Reynolds number under the hypothesis that it has minor effects, (ii) neglecting the Weber number under the hypothesis that the curvature of the air-water interface is limited, and (iii) neglecting the Mach number under the hypothesis that the compressibility of water (water is often mixed with air bubbles) is not relevant. It is assumed that aerated breaking or very fast breakers do not occur in the field, hence the reproduction of these effects in the laboratory is not required. Breaking or very fast breakers are conditions under which a transonic state may occur locally rendering relevant the bulk compressibility. Under these conditions, a Froude similarity can be adopted, with length and pressure scales that are equal to λ and speed and time scales that are equal to $\lambda^{1/2}$.

Strouhal and Euler similarities are also satisfied. Since the similarities that are related to the Reynolds, Weber and Mach numbers are not satisfied (the ratios between the values in the model and the prototype are $r_{Re} = \lambda^{3/2}$, $r_{We} = \lambda^2$ and $r_{Ma} = \lambda^{1/2}$ for the Reynolds, Weber and Mach numbers, respectively), scale effects are expected in the conversion of experimental measurements in the laboratory into field values, with higher distortion as the length scale ratio λ decreases: a reduced Reynolds number in the model could lead to less effective transport of the momentum in the model than in the prototype [34]; a reduced Weber number in the model increases the role of surface tension effects in the model, with a reduction, for example, in the inclusion of air; and a reduced Mach number in the model hides possible shock effects [35, 36].

Any attempt to modify the characteristics of the fluid to also satisfy, for example, the Weber similarity is prone to failure: if $\lambda = 1/10$ then $r_{We} = 1/100$, but there is no additive that can reduce the water-air tension surface (approximately 0.070 N m^{-1}) to $1/100$; by adding olive oil, we would obtain a mere 0.030 N m^{-1}, and adding ethyl alcohol results in 0.022 N m^{-1}, which represents a reduction of less than one order of magnitude. Similarly, we cannot find safe fluids with a reduction of the kinematic viscosity according to $\lambda^{3/2}$ or approximately $1/30$ for $\lambda = 1/10$: Ammonia has a kinematic viscosity of 0.27 mm^2 s^{-1}, which is less than $1/5$ of the kinematic viscosity of salt water, namely, 1.2 mm^2 s^{-1}.

When the experiments are focused on the boundary layer above the water, the large-scale pressure and velocity pulsations present in a real wind that is blowing over the sea can be neglected, along with the compressibility of the air because the real wind flow is isochoric. Various time scales are related mainly to turbulence, and a highly relevant time scale is the period of the small-scale fluctuating pressure, which is considered important in wave growth [37]. In the initial phase of wave generation, a time scale that is related to the shear rate in the air boundary layer is also considered relevant. Toba [38] reported a scenario in which both the air and water boundary layer have a structure similar to the turbulent boundary layer on a rough solid wall; hence, the classical boundary layer scales can be chosen: friction velocity, apparent surface roughness, density and viscosity, and a time scale that varies during the process. Three convenient dimensional groups are the Reynolds number $Re_a = u_a^* z_0 / v_a$, the Euler number $Eu_a = \Delta p / (\rho_a u_a^{*2})$, and the Strohual number $St_a = z_0 / (u_a^* t)$, where the subscript "a" refers the variable to the air. z_0 is the scale of roughness, defined as $z_0 = \alpha u_a^{*2} / g$, with Charnock's parameter α [39]. The flow regime is considered aerodynamically rough (fully turbulent and not dependent on the air viscosity) if Re_a > 2.5 and smooth if $Re_a \leq 0.13$ (see, e.g., [40]).

For the wave flow, a Reynolds number that is based on the amplitude and on the orbital velocity of the waves [41] is $Re_w = aV / v_w = a^2 \omega / v_w$, where a is the amplitude, $V = \omega a$ is the orbital velocity, ω is the pulsation, and v_w is the kinematic viscosity of water. Substitution of the dispersion relation yields $Re_w = (2g\pi)^{1/2} a^2 / v_w L^{1/2}$, where L is the local wavelength. The critical Reynolds number is $Re_{w,cr} \sim 3000$, and since the amplitude motion decays exponentially with the depth, a similar decay is expected for Re_w, with a possibly turbulent flow near the free surface and a laminar flow beneath. This effect can be highly distorted in the laboratory, where the critical condition for turbulence is seldom reached also near the free surface, whereas it is a common condition in the field near the free surface and for a significant fraction of the water column. Turbulence of this type is not related to wave breaking or to water drops that are accelerated by the wind before impacting the free surface [42]. Once generated, turbulence is diffused downwards by several other phenomena, and in the presence of currents, it can also be generated by the shear well beneath the free surface.

In consideration of the limitations of laboratory experiments with respect to the field process, it is generally assumed that the numerous uncertainties regarding wave and wind characteristics, reflection conditions in the field (near the coast), and all other parameters and variables far outweigh the uncertainties of the scale effects, which can easily be ignored.

Dimensional analysis for wind-waves and currents and scaling

The process of wave generation due to wind action in the presence of a current is considered. Hereafter, we will use the notation adopted in Longo [43]. The typical function can be expressed as $f(H_{rms}, T_p, u_a^*, F, t_d, g, u_c, h) = 0$, where H_{rms} is the root mean square wave height, T_p is the peak period, u_a^* is the friction velocity of the wind, F is the fetch length, t_d is the wind duration, g is the gravitational acceleration, u_c is a velocity scale of the current, and h is the local depth. Tension surface effects are being left out in this study. The problem is purely kinematic with a dimensionality of two, and upon the selection of g and u_a^* as fundamental variables, dimensional analysis suggests a maximum of six non-dimensional groups, and the previous equation reduces to $f(gH_{rms}/u_a^{*2}, gT_p/u_a^*, u_c/u_a^*, gF/u_a^{*2}, gt_d/u_a^*, gh/u_a^{*2}) = 0$. Although the friction velocity is considered a correct scale for growing wave characteristics (see, e.g., [44]), the group celerity of the waves c_g seems more suitable for the current effects, with the group u_c/u_a^* substituted by u_c/c_g. This last group can be introduced via the dispersion relation for linear waves, which can be expressed as $f(c_g/u_a^*, gh/u_a^{*2}, gT_p/u_a^*) = 0$. As a consequence of the choice of the velocity scale, the typical function becomes $f(gH_{rms}/u_a^{*2}, gT_p/u_a^*, u_c/c_g, gF/u_a^{*2}, gt_d/u_a^*, gh/u_a^{*2}) = 0$. If the waves are in deep water, the group gh/u_a^{*2} can be eliminated. Similarly, if the duration of the wind is sufficient to saturate the specified fetch, a stationary generation condition is satisfied, and the group gt_d/u_a^* is not relevant. Overall, the general function in deep water and under the stationary generation condition can be expressed as $f(gH_{rms}/u_a^{*2}, gT_p/u_a^*, u_c/c_g, gF/u_a^{*2}) = 0$. The approximate similarity is based on the Froude number, which forces the velocity and time scales to be equal to $\lambda^{1/2}$, where λ is the geometric scale.

To correctly extend the laboratory results to the field, all the information that is presented above must be considered. It is expected that the laboratory experiments will yield underestimates of the turbulence levels for the air and for the water and, consequently, underestimates of the diffusivities of chemicals, gases, and heat, with limited spray generation and possibly with a generation of currents with velocity scaling that is not proportional to $\lambda^{1/2}$ and with a different horizontal velocity profile along the vertical than the actual one.

3 Data Analysis Techniques

Measurements result in time series in a single point, in several points in a plane or a volume with generally periodic behaviour or with known statistical properties. The classical tools for data analysis are used according to the source of information.

Free surface level measurements are local measurements (sometimes spatial if optical tools are used) and are analysed in the time domain with a zero-crossing analysis, if the periodic wave content is dominant and detectable. Frequency domain analysis is typically adopted for signals with non-zero bandwidth. The spectra are in the wave number domain in the case of spatial measurements of the elevation or of the inclination of the free surface. To estimate the phase velocity of the waves, the cross-correlation of the signal between two measuring probes that are close to each other at a distance Δx can be used to detect the time delay τ, namely, $c = \Delta x/\tau$. The wave group celerity can be estimated by calculating the Hilbert transform of the cross-correlation function of the measured water levels, which yields the envelope of the function with a peak in a time delay τ_g that, for narrow band signals, can be substituted into $c_g = \Delta x/\tau_g$ (see [45] for the theory and [28, 34] for the application).

Velocity measurements are subject to the same type of analysis: the classical spectra of turbulence are obtained in the time domain (macro-turbulence for measurements with a limited data rate and a non-small measurement volume), and the spatial correlation and the length scales of the process are extracted directly for ultrasonic Doppler profiler measurements and PIV [46], and by adopting the frozen turbulence hypothesis for single-point measurements with LDV and ADV. To examine the internal structure of the velocity signal and to separate the mean velocity, wave-induced velocity and fluctuating velocity, the principal orthogonal decomposition

(POD) can be used [17], along with wavelet decomposition [47, 48] and quadrant analysis of the fluctuating components of the velocity [23, 49]. Wavelet decomposition is also a suitable tool for the analysis of bursts and of intermittency of turbulence [48], and for eddy detection in turbulence fields that interact with a free surface [50].

4 Recent Advances on Wind-Wave Interaction

In recent decades, many studies on the interactions between airflow and water flow were published, and many models have been developed for investigating and predicting the dynamics of flows and wave generation mechanisms. The air-sea interface covers more than 70% of the Earth's surface, and even low-speed winds transfer the momentum to the water, thereby generating waves, currents and turbulence. Wave generation is affected by turbulence not only on the air side, but also on the water side. In addition to the transfer of momentum in wave and current generation, air-water gas exchange has also environmental implications, and is a critical factor for determining the fate of pollutants or other anthropogenic materials. The global balances of various substances have been calculated on the basis of parametric models, but the exchange process at the air-water interface is poorly understood. Turbulence and surface tension are important in determining the gas transfer rate through the interface, and their contributions differ in terms of order of magnitude. For example, a small drop of a surfactant can reduce the gas transfer rate by up to 60% without affecting the turbulence structure [51]. Therefore, most of the process occurs in a thin layer, in which the turbulence is suppressed, and the molecular dynamics is dominant.

On the water side, the structure of the flow field below the interface depends on the wind speed and fetch. In the laboratory, for a fixed wind speed in the first zone immediately downstream, there are tiny ripples at the air-water interface; in the second zone, waves grow, with rounded crests and sharp troughs.

To clarify various aspects of these phenomena, we report on a series of tests conducted in a wind-tunnel that was equipped with a water tank. Gravity waves were generated entirely by wind. The experiments were conducted in a small open wind-tunnel at a low speed in the Centro Andaluz de Medio Ambiente, CEAMA, University of Granada, Spain. The boundary layer wind-tunnel has a PMMA structure with a test section of 3.00 m in length with a cross-section of 360 mm × 430 mm. Incidentally, this small wind-tunnel is a replica of a large wind-tunnel and it was built to control the overall quality of the flow inside the large tunnel. The adoption of small-scale experimental devices typically introduces disturbances and interference of various types, along with scale effects in the extrapolation of data from model to prototype. However, the relevant phenomena, such as turbulent mixing, wind-wave growth, and micro-breaking, are reproduced correctly with the enormous advantage of the use of small devices, namely, all changes in geometry and experimental conditions require much less time and less work than with large or full-scale experimental devices.

Wind-generated waves show the typical growth trends of the fetch and wind speed and are also asymmetrical, with more pronounced crests than troughs. The phase and group celerity of the waves are calculated using a technique of cross-correlation of the water levels that are measured in several spaced sections: the phase celerity, due to the current in the tank flowing in the direction of wave propagation, exceeds the theoretical velocity in the absence of the current, and is influenced by wind drift and the Stokes current, which are variable in space. The group celerity is influenced similarly, and a model that considers the relative variations of the phase speed and the group speed is developed and includes a dependence of the drift velocity on the steepness of the wave.

The grouping of the waves is also detected at this laboratory scale, with statistics on the runs and total runs close to those that were measured for the real sea waves. The correlation coefficient between two successive wave heights is smaller than the theoretical coefficient due to the asymmetry of the waves, which suggests that the crest envelope and trough envelope should be treated separately.

The air-flow boundary layer above the water waves was compared to the boundary layer on a flat, smooth, rigid wall. The average velocity shows a logarithmic profile. The apparent roughness is related to the amplitude of the waves, and the flow is turbulent with a typical friction coefficient value for a fully developed rough flow, except for sections with limited fetch. The transition to turbulence occurs at an earlier stage than the transition of a boundary layer on a smooth, flat, rigid wall. The thickness of the boundary layer above the water waves increases much faster than for the boundary layer on a smooth, solid wall.

The Reynolds stress for the boundary layer above the water waves exceeds the Reynolds stress for the boundary layer on a wall, with streamwise fluctuations that are highly dominant compared to vertical fluctuations. A rapid decline in turbulence was recorded in the domain where water was periodically present. A constant Reynolds shear stress layer was detected, with a friction velocity revealing an excess value of approximately 20%, which is approximately twice the excess value that was calculated for a flat rigid wall.

At least 50% of the boundary layer structure has been strongly influenced by Reynolds wave-induced stresses, although H_{rms} is a better scale for the vertical distribution of these stresses. The wave-induced Reynolds stresses have become negligible for $z > 5H_{rms}$. The distribution of the intermittence factor in the boundary layer on water waves was similar to that of a boundary layer on a rigid flat wall, with several differences near the interface. Here, the presence/absence of water dampens the turbulence.

Quadrant analysis, which is a tool for discriminating the contribution to the turbulent momentum flux, showed that ejection and sweep events were dominant and more concentrated. The joint probability density function (p.d.f.) of the fluctuating velocities, namely, U' and V', showed circular isolines in the upper region of the boundary layer ($z/\delta > 0.7$), which became elliptical in the lower region near the water interface. At small fetches, large negative perturbations in the flow direction were preferably lifted. For larger fetches, ejections and sweeps were concentrated near the water interface. The intensity of the momentum transfer from the wind flow to gravity waves, is much higher for the short waves than for mature waves.

Turbulence energy production peaked at $z/\delta = 0.2$ and had a distribution that was similar to that observed for a self-preserving boundary layer with a strong adverse gradient pressure. The quadrant analysis contribution to energy production showed that ejections still dominated the equilibrium, and that production was spatially modulated in the wind direction, with a couple of cells and with a minimum in the area of the free surface wave height reduction.

In all these experiments, the average velocity profiles are logarithmic, and the flows are hydraulically rough. The friction velocity for the water boundary layer is one order of magnitude smaller than that for the wind boundary layer. The turbulence level increased immediately below the interface due to the micro breakings, and this reflects that the Reynolds shear stress is of the order u_w^{*2}. It is consistent with numerous models in the literature, which support the presence of a layer immediately below the free surface where energy and momentum are added due to the waves breaking (e.g., [52]).

The vertical components of turbulent fluctuations assume common values of $v'_{rms} = (0.32, 0.35, 0.43)u_s$ at the still water level, where u_s is the root mean square of the free-surface vertical velocity. Larger values correspond to larger fetches, which is similar to the case in a canal with a Crump weir.

The autocorrelation function in the vertical direction shows the characteristics of typical anisotropic turbulence, with a wide range of wavelengths. The macro- and micro-scales, namely, Λ and λ (the latter not to be confused with the geometric scale), respectively, increase with depth in a region below the free surface. Their maximum values are $\Lambda_{max} = 0.4$ |z| and $\lambda_{max} = 0.06$ |z|. Permanent vortexes are detected by analysing the autocorrelation functions.

The ratio between the micro-scale and macro-scale can be expressed as $\lambda/\Lambda = a\,Re_\Lambda{}^n$, with the exponent n differing slightly from $-1/2$, which is the value for the case with turbulence production and dissipation in equilibrium. The negative value of coefficient a suggests that the pressure work and average turbulent energy transport by turbulent motion act as sinks of turbulent energy.

In the categorisation of free surface flows on the basis of a length scale and a turbulent velocity scale, current experiments fall into a wavy free-surface regime if the wavelength is selected as the length scale. The integral turbulent scale on the water-side alone underestimates the degree of disturbance on the free surface, and a correction can be made to include the contribution of air turbulence. However, this velocity scale correction is insignificant and does not change the classification of the flow regime at the interface in this study.

4.1 Long Waves Plus Reflected Long Waves and Short Wind-Generated Waves

Long waves in the presence of reflection are modified by wind-generated short waves. This scenario is ubiquitous in coastal regions in shallow water, where incident swell combines with reflected waves and where wind action forces short wave generation. A series of experiments were designed and conducted to analyse the degree of interaction between the long wave field and the short waves. Long waves are mechanically generated waves (MGWs), and short waves that are induced by a wind blowing in the opposing direction of incident waves (equivalently, in the same direction of reflected long waves since experiments are executed in a flume) are named wind-generated waves (WGWs).

The observed effect of a wind blowing against wave motion is a reduction in the wave height, which is consistent with the wave attenuation that has been reported in the literature [53]. The results also demonstrate new aspects of WGW interactions, especially a reduction in the reflection coefficient and a reduction in the phase shift between incident and reflected waves, which is new and cannot be quantitatively interpreted due to the dissipation of the energy that is generated by the wind. The general behaviour of the WGW height is an initial growth due to wind energy input, which is followed by attenuation due to possible micro breakings and non-linear transfer to longer waves. The peak period of the WGW increases monotonically due to the increase in wavelength, while the steepness of the wave shows a varied spatial evolution.

The effects of reflection conditions are evident for two of the three quantities: (i) the increase in reflection conditions produces a reduction in wave height, which suggests reduced efficiency in WGW energy transfer or increased non-linear energy transfer; (ii) the WGW period is almost unchanged with reflection; and (iii) the WGW steepness is lower with an increased reflection coefficient.

Further analysis was conducted by separating the WGW according to its position with respect to the MGW phase. As an initial approach, the WGW propagates over the wavy form of the MGW with linear interactions, and different characteristics of the WGW in the crest and trough of the MGW, and in the downwind and upwind sides. A phase analysis (phase refers to that of the MGW) in which four subdomains were considered for the WGW, provided much more information on the growth of the WGW and the evolution, with the estimation of the WGW length, height, period and steepness.

The following insights were obtained from the experiments on WGWs:

- As the reflection coefficient is increased, H_{rms} decreases with the length of the fetch on the upwind side and increases on the downwind side, trough and crest.
- The average period T_m increases with the length of the fetch, with a variegated trend overall and without identifiable trends in the four subdomains.
- As the reflection coefficient is increased, the steepness $k\,H_{rms}$ increases with the length of the fetch only on the downwind side.

The propagation of the WGW along the flume is also examined by estimating the phase and group celerity. The effects of sheared horizontal currents are included through a perturbation model [54], which shows satisfactory overlap with the experimental phase celerity when analysed in bulk. The celerity that is resolved in phase is best interpreted by including in each phase the orbital horizontal velocity that is induced by waves, with variable effects that also depend on the spatial position of the measurement section (the variability is due to the node-antinode sequence of the MGW plus the reflected long wave field). A further correction includes the apparent vertical acceleration of the non-inertial reference frame (the MGW) of the WGW: this is an apparent body force that adds and subtracts from gravity in the troughs and crests, respectively, thereby affecting the results of the dispersion relationship. The overall results provide higher consistency between theory and experiments, although differences remain. Additional ad hoc experiments are required to shed light on these aspects, all of which are relevant for improving conceptual models and increasing the accuracy of numerical models.

4.2 Effects of Currents on Short Wind-Waves

The interaction between wind-waves and currents is frequent in many natural and artificial environments, such as lakes, lagoons and reservoirs. In these zones, currents are often generated by tides, estuary inlets, thermal and density effects, while sea and land breezes cyclically generate waves. A frequent scenario is that of a wind blowing and generating free gravity waves on the surface (sea waves) in the absence of swell and in the presence of a current. A series of experiments that were conducted in a wind-wave tunnel with co- and counter-currents have revealed peculiar aspects of the complex flow field. The complexity arises (i) from the non-homogeneity, with energy progressively transferred by the wind; (ii) from the shear action of the current; (iii) from the breaking; and (iv) from the grouping and a mix of scales that govern the process.

A co-current reduces the growth of the wave height as measured in the absence of a current, with effects that are proportional to the speed of the current. This is a consequence of the reduction in the relative speed (between air and water), which is accompanied by reductions in the friction factor and energy transfer efficiency. A counter-current interrupts monotonous wave growth with the fetch length, even if the wave height is larger (in the same section and at the same wind speed) than for a co-current of equal absolute speed. Part of the wave energy is dissipated by breaking (microbreaking is always expected at high wind speed), but the transfer of energy from the wind is facilitated; hence, the energy balance remains positive. The section in which the wave height fall occurs depends on the wind speed and the current speed.

The steepness of the wave is maximal for weak counter-currents and decreases both when the co-current is present and when the counter-current becomes more intense. For the co-current, the explanation is intuitive since a decreasing wavelength is accompanied by an increasing wavelength; for the counter-current, the interpretation is based on the wavelength increasing faster than the wave height. Therefore, in the latter configuration, an unexpected stabilising effect occurs, which favours the transfer of energy to longer waves. This transfer of energy is also evident from the spectral shape, of which the evolution is influenced both by the energy that is introduced by the action of the wind and by the presence of a counter-current. A double-peak spectrum also develops for small values of u_c/c_g in the counter-current case.

The phase celerity is strongly influenced by the presence of a co-current, and the stronger the current, the stronger the effects, while a counter-current has no appreciable effects. The co-current condition always induces a strong increase in phase celerity, even compared to theoretical models that include the presence of a current. We suspect that the discrepancy can be attributed to the continuous energy input that is due to the blowing wind. We recall that these experiments are conducted under highly inhomogeneous conditions and that an increase in the fetch length indicates that more energy is transferred to the waves.

In addition, the celerity of the groups exhibits a strong variation compared to that in the absence of a current. First, in the absence of a current, the group celerity of the wind-waves increases faster than the phase celerity (the increments are calculated with respect to the group celerity and the phase celerity for the equivalent swell), with a ratio of $(c_g - c_{g0})/(c - c_0) > 1.2$. The current acts by reducing this value, with a gradual decrease for co-current conditions and with dispersed data for counter-current conditions. A counter-current significantly reduces the energy flow along the path, thereby resulting in a rapid increase in the wave height. From this perspective, the counter-current has a shoaling effect.

An analysis of the grouping shows that the average length of the groups is minimally affected by a co-current (unless the current is very strong) and is subject to non-monotonic variations in the presence of a counter-current. A minimal value of the length is observed in the counter-current domain, where a significant change in the wave field occurs. A higher counter-current speed favours longer wave groups.

The experiments were conducted on a small scale, where, for example, the Weber number was neglected because the surface effects of tension were considered negligible. Furthermore, the finite size of the flume induced an extra-circulation, which was included in bulk in the analysis, without a detailed separation of the wind drift, imposed current and secondary circulation. However, the results are clear with respect to the profound difference between a co-current and a counter-current. Due to the complex and sometimes unexpected phenomena, the overall scenario is sufficient to justify further investigation with the objective of generalising the present results.

5   Recent Advances on Wave Breaking Flow Measurements and Breaker Types

Breaking almost always occurs in shallow water, where it is induced by bathymetry. The breaking occurs in random spots, even if the waves are periodic and regular, as in a laboratory flume. If there is a natural or artificial submerged bar, the breaking section for regular waves is almost deterministic because the bar forces its instability. However, even if the bar forces the regular appearance of breakers, the phenomenon retains intrinsic non-deterministic variabilities in its flow structure and overall geometry.

Breaking waves are rarely used as a flow field to model turbulence due to the instability and complexity of the phenomena that occur, which are enhanced by the vorticity and the generation and evolution of turbulence during the breaking and near the bottom. While the phenomena at the bottom are relevant in shallow water, the phenomena near the free surface are always relevant, although with a different level of importance that depends on the stage of breaking. The vorticity is often accompanied by turbulence, even in cases where the classic air entrapment of the breaking wave is not observed. The energy balance of incoming waves is not dominated by dissipation at a viscous level (the final stage of the action of turbulence occurs at the molecular scale, with the generation of heat), but is also based on transfer to large scale characteristics, such as (i) long-shore and cross-shore currents, (ii) free (unbounded) long waves that escape offshore and edge waves that are trapped in the nearshore, (iii) and macro eddies at the wavelength scale. In this regard, the analysis of the structure of turbulence in breaking waves facilitates understanding of all these new flow field structures. Numerous parameters can be evaluated, and numerous models can be tested if fully experimental 3D velocity time series are available. A 3D system provides sufficient temporal velocity and spatial resolution for measuring complex phenomena, such as wave breaking. As a general consideration, the need for an adequate synthesis of the results that are obtained with 3D measurements arises from the inability to easily and adequately display the enormous amount of obtained data.

A series of experiments were conducted in the wave flume at IISTA-University of Granada, where an artificial slope of 1:10 was constructed with a berm of stones over it to ensure wave breaking. The objective was to analyse the flow field in the breaker by measuring the flow in the three directions.

5.1   Momentum Balance

The tests showed that the average transverse phase velocity (alongshore component) is two orders of magnitude smaller than the transverse and vertical velocities and still exhibits periodicity with alternating negative and positive flows, which is due to a systematic asymmetry that is induced by the bathymetry of their experiments and the presumed formation of transverse recirculation cells (a laboratory effect). The cross-shore and the vertical phase-averaged velocities (the wave components) are almost in quadrature but remains a small correlation. As a result, a transfer of momentum due to Reynolds' wave stresses enters the balance. The momentum is positive during breaking (transfer towards the crest) and negative immediately after breaking (transfer from the crest towards the underlying flow field). Its profile in the vertical direction (cycle average phase) is positive (negative in the crest for some experiments), but the behaviour depends on the characteristics of the incoming wave. The inconsistencies are attributed to the presence of the bar, which modulates the flow field of the breaker according to the characteristics of the incoming wave (height and period).

The balance of the cross-shore momentum shows the important role of local inertia, while the advection is of comparable intensity only immediately after breaking. Turbulent stresses act similarly to inertia in the post-breaking phases and require the addition of a net pressure gradient to the net pressure gradient that is required to accelerate the fluid. The measured acceleration that is experienced in the breaker of the present experiments is a fraction of gravity, but it is expected to increase for stronger breakers. Its experimental value is limited, in terms of accuracy, by the absence of exact measurements during breaking, and is smoothed out by the spatial averages that are used to present the data synthetically. The mean (vertical) pressure gradient during breaking affects the profile of the breaker, which is steeper for a rapidly increasing pressure gradient, and reaches a higher peak value. The peak value is delayed compared to the crest of the breaker, except for less energetic breakers. Steepening of the breaker front is correlated to the cross-shore net pressure gradient term, and smoothing of the breaker profile is correlated to the vertical net pressure gradient term. The pressure gradient also works against turbulent stresses, which, in some phases, behave as inertial terms.

5.2 Turbulent Kinetic Energy Balance

The turbulent kinetic energy is essentially in the transverse and vertical directions. The average values that were measured in the experiments indicate a distribution among the three directions similar to those observed in other turbulent flow fields, with a transverse contribution to the dominant cross-shore contribution and with significant variations during the wave cycle. The turbulence structure is far from isotropic, with a one-component behaviour only at the crest, and a structure that evolves towards the bottom. The distance to isotropy should be carefully considered because the available experimental data have insufficient spatial and temporal resolutions for micro-scale flow detection in the most dissipative and the most isotropic cases. Various estimates of the dissipation rate, in comparison to the common hypothesis of the dissipation rate of isotropic turbulence, indicate that in many cases, this hypothesis overestimates the actual dissipation. Thus, the same warning is issued due to limited spatial and temporal resolutions. The main terms of the balance are advection and transport that are due to the turbulence. The production can be "negative", namely, during some phases, the turbulence transfers energy to the average (wave) flow field. The hypothesis of "frozen turbulence", which is adopted to estimate spatial gradients according to the local temporal gradient, yields better results if the advection velocity is the velocity of the wave component rather than, for example, the wave celerity.

5.3 Cross-Shore Variability and Vorticity Dynamics

The vorticity coexists with a wave flow (in principle, a potential flow) and evolves to fill the domain with vorticity that is associated with shear or coherent structures. Vortexes, which have both an alongshore axis and a diagonal axis in cross-shore vertical planes, and vortex tubes with a tortuous path have been detected by visual observation and velocity measurement analysis, respectively. These vortex tubes behave similarly to vortex tubes in other complex flow fields, and engage in a variety of mutual interactions that can be related to transition phenomena.

The main component of the vorticity is directed alongshore. The temporal evolution of the vorticity in the cross-shore direction is controlled by vortex stretching that is due to the average flow, especially during flow reversal. In the breaking phase, the stretching of the vortex is also associated with the fluctuating strain rate and the flux of the fluctuating vorticity due to turbulence. In the alongshore direction, the equilibrium is due mainly to the fluctuating vorticity flow, which is induced by turbulence; the remaining terms are less relevant. In the vertical direction, all terms are of the same magnitude.

The measured values of the vorticity are smaller than the values without the submerged berm that have been documented in previous experiments. The topologically induced vorticity that is concentrated near the free surface is reduced in favour of a more uniform vorticity in the water column.

5.4 Turbulent Stress Invariant Analysis and Relation with the Dissipation Tensor

The anisotropy was quantified by the invariants and was relatively high during all phases and more evident for the more energetic events. For low-energy breakers, the anisotropy was lower in the dissipation tensor than in the Reynolds stress tensor. For high-energy breakers, the dissipation tensor still had a sufficient level of isotropy, and the Reynolds stress tensor was polarized towards anisotropy.

The relationship between the main components of the two tensors is a simple proportionality for the low-energy breakers. Hence, various classic turbulence models are still applicable. For high-energy breakers, the relationship seems much more complex, and no simple model can be suggested.

The unique availability of the experimental data that are provided by 3D volumetric velocimeters enabled an in-depth analysis by following the same path as was adopted for the numerical simulation data. However, the dissipation tensor is expected to be affected by the limited spatial resolution of the data, and both tensors are affected by the limited time resolution; hence, the results should be evaluated in the light of these limitations.

Beyond the application of well-known models for turbulence estimation, a huge experimental data set is available to calibrate these models, which is an important novelty. The availability of this data set is a step towards the detailed interpretation of the fundamental mechanisms of turbulence generation, evolution, and dissipation.

6 Conclusions

After several decades of models and experiments, the processes at the air-water interface are becoming better understood, and phenomena have been identified that, at first glance, seemed inexplicable. The co-presence of molecular diffusion and advection renders the models more complex and requires multi-scale analyses that can consider the distinct contributions of these two processes: a diffusive process can last for a very long time, with effects that are not dissimilar to those of a convective process, namely, intense but short effects. The same diffusive process can be generated by differences in temperature, salinity, and chemical substances; hence, double or triple diffusivity models are required.

The availability of more accurate and more reliable measuring instruments, with a greater supply of information, has enabled deeper investigations that have yielded more accurate results. This has required the development of data analysis techniques for managing the new and very large amount of information. For example, PIV 3D provides up to 100 000 speed vectors for each frame in a modest measurement volume; pre-processing of the frames and post-processing of the results require time and high computing power.

From the perspective of experimentation, a major current limitation is the scale effects, which become more relevant as the number of process variables increases. Hence, experiments must be designed such that the scale effects for the variables of interest are reduced.

There was a time when simple level measurements enabled the experimental validation of even highly complex wave models. Today, measurements have proliferated, and it is necessary to refine the interpretation of the data to guarantee the coherence of the model and equal accuracy between experimental data and theoretical schemes.

7 Present and Future Challenges

The perspectives of research in the sector are numerous, and it is inconceivable to list them all. Among the major problems that remain open are the global budgets, which are invoked regularly whenever climate change is analysed or discussed, but without specifying the extent of our ignorance. The balances of heat, chemical substances, and gases are difficult to determine because they involve combinations of phenomena that occur on a planetary scale: small errors in the values of the involved variables can change the directions of the evolutionary trends of the system. Among the drivers of the exchanges, it should be included the breaking and reflection of the waves in coastal areas. The former occurs practically everywhere there is a water surface, as can be observed by those who engage in aerial flight or sail on the sea: whitish ruffles are visible even when the wind blows with low intensity. The latter, namely, reflection, is almost always partial reflection, and it is omnipresent in coastal areas, where it combines with density currents in a shallow water environment, with strong interaction with the bottom.

The present wind generation of waves can be extended to include the effects of bursts in the air side, which enhance the wind action on the water surface by inducing pressure fluctuations on a large scale.

The wide use of laboratories for research motivates the development of new techniques for the control of currents. Currents that are generated by waves in flumes are artefacts due to boundaries, and currents in real environments behave differently. Hence, the control of the current velocity profile and its flow rate is necessary for advancing knowledge and avoiding interferences and lab effects.

References

1. Wu, J.: Laboratory studies of wind–wave interactions. J. Fluid Mech. **34**(1), 91–111 (1968)
2. Chiapponi, L., Longo, S., Bramato, S., Mans, C., Losada, A.M.: Free-surface turbulence, wind generated waves: laboratory data. Technical report on Experimental Activity in Granada, University of Parma (Italy), CEAMA (Granada, Spain) (2011)
3. Zavadsky, A., Liberzon, D., Shemer, L.: Statistical analysis of the spatial evolution of the stationary wind wave field. J. Phys. Oceanogr. **43**(1), 65–79 (2013)
4. Guo, A., Liu, J., Chen, W., Bai, X., Liu, G., Liu, T., Chen, S., Li, H.: Experimental study on the dynamic responses of a freestanding bridge tower subjected to coupled actions of wind and wave loads. J. Wind Eng. Ind. Aerodyn. **159**, 36–47 (2016)
5. van Vliet, P., Hering, F., Jähne, B.: Delft hydraulic's large wind-wave flume. In: Air-Water Gas Transfer, pp. 505–516. Aeon Verlag, Hanau, Germany (1995)
6. Veron, F., Melville, W.K.: Experiments on the stability and transition of wind-driven water surfaces. J. Fluid Mech. **446**, 25 (2001)
7. Olfateh, M., Ware, P., Callaghan, D.P., Nielsen, P., Baldock, T.E.: Momentum transfer under laboratory wind waves. Coast. Eng. **121**, 255–264 (2017)
8. ASI Lab Homepage. http://www1.udel.edu/ASI-Lab/facilities/index.html
9. Addona, F., Chiapponi, L., Clavero, M., Losada, M.A., Longo, S.: On the interaction between partially-reflected waves and an opposing wind. Coast. Eng. **162**, 103774 (2020)

10. Andersen, T.L., Clavero, M., Frigaard, P., Losada, M., Puyol, J.I.: A new active absorption system and its performance to linear and non-linear waves. Coast. Eng. **114**, 47–60 (2016)

11. Rhee, T.S., Nightingale, P.D., Woolf, D.K., Caulliez, G., Bowyer, P., Andreae, M.O.: Influence of energetic wind and waves on gas transfer in a large wind-wave tunnel facility. J. Geophys. Res. Oceans **112**(C5) (2007)

12. Pope, N.D., Widdows, J., Brinsley, M.D.: Estimation of bed shear stress using the turbulent kinetic energy approach—a comparison of annular flume and field data. Cont. Shelf Res. **26**(8), 959–970 (2006)

13. Schmundt, D., Münsterer, T., Lauer, H., Jähne, B.: The circular wind/wave facilities at the University of Heidelberg. In: Air-Water Gas Transfer, pp. 505–516. Aeon Verlag, Hanau, Germany (1995)

14. Krall, K.E.: Laboratory investigations of air-sea gas transfer under a wide range of water surface conditions. Dissertation, University of Heidelberg. Available at: http://www.ub.uni-heidelberg.de/archiv/14392 (2013)

15. Toffoli, A., Proment, D., Salman, H., Monbaliu, J., Frascoli, F., Dafilis, M., Stramignoni, E., Forza, R., Manfrin, M., Onorato, M.: Wind generated rogue waves in an annular wave flume. Phys. Rev. Lett. **118**(14), 144503 (2017)

16. Takagaki, N., Komori, S., Suzuki, N., Iwano, K., Kurose, R.: Mechanism of drag coefficient saturation at strong wind speeds. Geophys. Res. Lett. **43**(18), 9829–9835 (2016)

17. Clavero, M., Longo, S., Chiapponi, L., Losada, M.A.: 3D flow measurements in regular breaking waves past a fixed submerged bar on an impermeable plane slope. J. Fluid Mech. **802**, 490–527 (2016)

18. Jähne, B., Riemer, K.S.: Two-dimensional wave number spectra of small-scale water surface waves. J. Geophys. Res. **95**(C7), 11531–11546 (1990)

19. Longo, S., Ungarish, M., Di Federico, V., Chiapponi, L., Petrolo, D.: Gravity currents produced by lock-release: theory and experiments concerning the effect of a free top in non-Boussinesq systems. Adv. Water Resour. **121**, 456–471 (2018)

20. Bailey, S.C.C., Hultmark, M., Monty, J.P., Alfredsson, P.H., Chong, M.S., Duncan, R.D., Fransson, J.H.M., Hutchins, N., Marusic, I., McKeon, B.J., Nagib, H.M., Örlü, R., Segalini, A., Smits, A.J., Vinuesa, R.: Obtaining accurate mean velocity measurements in high Reynolds number turbulent boundary layers using Pitot tubes. J. Fluid Mech. **715**, 642–670 (2013)

21. Bruun, H.H.: Hot-wire anemometry: principles and signal analysis. Meas. Sci. Technol. **7**, 024 (1996)

22. Hutchins, N., Monty, J.P., Hultmark, M., Smits, A.J.: A direct measure of the frequency response of hot-wire anemometers: temporal resolution issues in wall-bounded turbulence. Exp. Fluids **56**(1), 18 (2015)

23. Longo, S., Losada, M.A.: Turbulent structure of air flow over wind-induced gravity waves. Exp. Fluids **53**, 369–390 (2012)

24. Makin, V., Branger, H., Peirson, W., Giovanangeli, J.P.: Modelling of laboratory measurements of stress in the air flow over wind-generated and paddle waves. J. Phys. Oceanogr. **37**, 2824–2837 (2007)

25. Longo, S.: The effects of air bubbles on ultrasound velocity measurements. Exp. Fluids **41**(4), 593–602 (2006)

26. Longo, S.: Experiments on turbulence beneath a free surface in a stationary field generated by a Crump weir: free surface characteristics and the relevant scales. Exp. Fluids **49**, 1325–1338 (2010)

27. Longo, S.: Experiments on turbulence beneath a free surface in a stationary field generated by a Crump weir: turbulence structure and correlation with the free surface. Exp. Fluids **50**, 201–215 (2011)

28. Longo, S.: Wind-generated water waves in a wind-tunnel: free surface statistics, wind friction and mean air flow properties. Coast. Eng. **61**, 27–41 (2012)

29. Chiapponi, L., Cobos, M., Losada, M.A., Longo, S.: Cross-shore variability and vorticity dynamics during wave breaking on a fixed bar. Coast. Eng. **127**, 119–133 (2017)

30. Longo, S., Clavero, M., Chiapponi, L., Losada, M.A.: Invariants of turbulence Reynolds stress and of dissipation tensors in regular breaking waves. Water **9**(893), 1–12 (2017)
31. Willert, C.E.: High-speed particle image velocimetry for the efficient measurement of turbulence statistics. Exp. Fluids **56**, 17 (2015)
32. Petrolo, D., Longo, S.: Buoyancy transfer in a two-layer system in steady state. Experiments in a Taylor-Couette cell. J. Fluid Mech. **896**(A27), 1–31 (2020)
33. Hughes, S.A.: Physical Models and Laboratory Techniques in Coastal Engineering, vol. 7. World Scientific (1993)
34. Chiapponi, L., Addona, F., Diaz-Carrasco, P., Losada, M.A., Longo, S.: Statistical analysis of the interaction between wind-waves and currents during early wave generation. Coast. Eng. **159**, 103672 (2020)
35. Peregrine, D.H.: Water-wave impact on walls. Annu. Rev. Fluid Mech. **35**(1), 23–43 (2003)
36. Bredmose, H., Bullock, G.N., Hogg, A.J.: Violent breaking wave impacts. Part 3. Effects of scale and aeration. J. Fluid Mech. **765**, 82–113 (2015)
37. Teixeira, M.A.C., Belcher, S.E.: On the initiation of surface waves by turbulent shear flow. Dyn. Atmos. Oceans **41**(1), 1–27 (2006)
38. Toba, Y.: Similarity laws of the wind wave and the coupling process of the air and water turbulent boundary layers. Fluid Dyn. Res. **2**(4), 263 (1988)
39. Charnock, H.: Wind stress on a water surface. Q. J. R. Meteorol. Soc. **81**(350), 639–640 (1955)
40. Kraus, E.B., Businger, J.A.: Atmosphere-Ocean Interaction, vol. 27. Oxford University Press (1994)
41. Babanin, A.V.: On a wave-induced turbulence and a wave-mixed upper ocean layer. Geophys. Res. Lett. **33**(20) (2006)
42. Longo, S., Liang, D., Chiapponi, L., Aguilera Jiménez, L.: Turbulent flow structure in experimental laboratory wind-generated gravity waves. Coast. Eng. **64**, 1–15 (2012)
43. Longo, S.: Principles and applications of dimensional analysis and similarity. Springer International Publishing, Cham (2021)
44. Janssen, P., Komen, G., De Voogt, W.: Friction velocity scaling in wind wave generation. Bound.-Layer Meteorol. **38**(1–2), 29–35 (1987)
45. Bendat, G.S., Piersol, A.G.: Random Data Analysis and Measurement Procedures. Wiley, New York (2000)
46. Longo, S., Chiapponi, L., Clavero, M.: Experimental analysis of the coherent structures and turbulence past a hydrofoil in stalling condition beneath a water-air interface. Eur. J. Mech./B Fluids **43**, 172–182 (2014)
47. Longo, S.: Turbulence under spilling breakers using discrete wavelets. Exp. Fluids **34**(2), 181–191 (2003)
48. Longo, S.: Vorticity and intermittency within the pre-breaking region of spilling breakers. Coast. Eng. **6**, 285–296 (2009)
49. Longo, S., Chiapponi, L., Clavero, M., Mäkelä, T., Liang, D.: Study of the turbulence in the air-side and water-side boundary layers in experimental laboratory wind induced surface waves. Coast. Eng. **69**, 67–81 (2012)
50. Longo, S., Domínguez, F.M., Valiani, A.: The turbulent structure of the flow field generated by a hydrofoil in stalling condition beneath a water-air interface. Exp. Therm. Fluid Sci. **61**, 34–47 (2015)
51. McKenna, S.P.: Free-surface turbulence and air–water gas exchange. PhD thesis, MIT, 2000
52. Kudryavtsev, V., Shrira, V., Dulov, V., Malinovsky, V.: On the vertical structure of wind-driven sea currents. J. Phys. Oceanogr. **38**, 2121–2144 (2008)
53. Peirson, W.L., Garcia, A.W., Pells, S.E.: Water wave attenuation due to opposing wind. J. Fluid Mech. **487**, 345–365 (2003)
54. Swan, C., James, R.L.: A simple analytical model for surface water waves on a depth-varying current. Appl. Ocean Res. **22**(6), 331–347 (2000)

Advances in Wave Run-Up Measurement Techniques

Diogo Mendes⑩, Umberto Andriolo⑩, and Maria Graça Neves⑩

Abstract Wave run-up is defined as the elevation of wave uprush on the beach profile or coastal structures above the still water level. Common applications of wave run-up measures include the prediction of flood events during storms, the design of coastal structures and the assessment of vulnerability in coastal management plans. This chapter gives a general overview of the techniques adopted to measure wave run-up. Traditional techniques, such as resistance wires, wave gauges, pressure sensors and ultrasonic sensors have been used in the field and laboratory. The advent of shore-base video monitoring systems have significantly improved the measurement on beaches in the last two decades.

Keywords Beach · Coastal structures · Hydrodynamics · Coastal management

1 Introduction

Wave run-up is defined as the elevation of wave uprush on the beach profile or coastal structures above the still water level [1] (Fig. 1). The wave run-up is a complex hydrodynamic phenomenon that depends on the local water level, the incident wave conditions (height, period, steepness, direction, and bandwidth), the bottom slope

D. Mendes (✉)
Geosciences Department and Centre of Environmental and Marine Studies (CESAM), University of Aveiro, Aveiro, Portugal
e-mail: dsmendes@ua.pt

HAEDES, Casais do Arrocho, Azoia de Cima, 2025-452 Santarem, Portugal

U. Andriolo
INESC Coimbra, Department of Electrical and Computer Engineering, Polo 2, 3030-290 Coimbra, Portugal
e-mail: uandriolo@mat.uc.pt

M. G. Neves
NOVA School of Science and Technology, NOVA University Lisbon, Lisbon, Portugal
e-mail: mg.neves@fct.unl.pt

© The Author(s), under exclusive license to Springer Nature Switzerland AG 2023
C. Chastre et al. (eds.), *Advances on Testing and Experimentation in Civil Engineering*, Springer Tracts in Civil Engineering,
https://doi.org/10.1007/978-3-031-05875-2_12

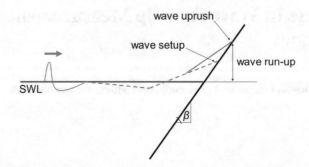

Fig. 1 Schematic representation of wave run-up over a plane bottom (black line) with slope (tan β). The wave run-up is the sum between the time-averaged wave setup (dashed dark blue line) and the individual wave uprush (dark blue line) above the still water level (SWL, light blue line). The blue arrow indicates the direction of wave propagation

and the nature of the beach or structure (e.g., reflectivity, crest height, permeability, roughness).

The physical processes that control wave run-up are the wave-induced setup and the swash [2] (Fig. 1). Under natural conditions, wave run-up is the result of the action of several waves on a beach or coastal structure slope, each one with a different wave period (i.e. irregular waves) over a time span. Therefore, wave run-up is calculated based on a statistical analysis of several wave run-up events, such as the $R_{2\%}$, the run-up exceeded by 2% during a sea state, and the R_{max}, the maximum run-up during a sea state.

The wave-induced setup increases the water level near the shoreline on a time scale longer than individual waves, but shorter than tidal oscillations. On beaches, the swash is associated with the water excursions (i.e., wave uprush) promoted by individual waves on the foreshore. Both wave-induced setup and swash are dependent on the wave breaking type, which is commonly assessed through the surf similarity parameter (ξ), as function of the wave steepness (either offshore, at the breaking point, or at the toe of the structure) and the bottom slope. Pioneering attempts to relate wave run-up with ξ were successfully performed by Hunt in 1956 [3], for monochromatic waves (i.e. one wave period only) interacting with dikes, seawalls and revetments ($R/H_0 = K\xi$, where H_0 is the offshore wave height and K is a constant factor obtained using measured data). Later, Holman in 1986 [4] used field data to evaluate the contribution of the wave setup in wave run-up ($R/H_0 = K\xi + C$, where C is a constant factor associated with the contribution of wave setup). These previous studies highlighted how wave run-up depends on the wave-induced setup and, on beaches, on the swash physical processes. A recent thorough review on wave run-up formulations can be found in Gomes da Silva et al. [5].

On beaches, the wave run-up formulation is important for numerous coastal engineering and management applications [6]. For instance, wave run-up is one of the most critical parameters contributing to coastline flooding and shoreline change [7], and it determines the landward boundary of the area affected by the wave action during coastal storms impact [6]. Regarding coastal structures, wave run-up is fundamental for the design of dikes, revetments, and rubble-mound breakwaters. As overtopping occurs when wave run-up reaches the crest of the structure and passes over it, the run-up is used for the structural design of the crest height. Besides, for rubblemound breakwaters, the prediction of wave run-up is needed to support the estimation of overtopping volumes, overtopping velocities, and flow thicknesses [3], and is a key parameter for estimating forces over crown wall structures [7]. Weggel [8] was the first to develop an empirical model to predict overtopping based on run-up. Since then, other formulae were presented for overtopping as a function of run-up, as Mase et al. [9].

Maximum wave run-up estimations during storm conditions, along with operational real-time measurements, are essential for the safety assessment of coastal structures in what concerns inundation, as flooding can potentially cause structural failures and losses of property and lives [4]. Forecasting systems of run-up during storm events are critical to mitigate possible coastal disasters, and to provide warning of possible coastal hazards [5]. This is particular true for the occurrence of tsunamis events [2, 6].

Given its importance, wave run-up has been receiving much attention by the engineering and scientific communities. Scopus database shows that the number of publications related to wave run-up increased linearly in time until 2000, and exponentially between 2000 and 2010 (Fig. 2). This distinct behavior coincides with the scientific developments and deployments of shore-based video monitoring systems (see Sect. 3), which have been widespread along the world's shorelines to support coastal studies from the 90s.

The quality of measurements obtained in the laboratory and in the field has a great influence on the accuracy of the developed empirical formulae. For both laboratory and field observations, reliable data depends on the instrumentation and/or techniques used to collect data. This chapter focuses on describing and giving a general overview on the available technological capabilities to measure wave run-up. Future perspectives on wave run-up measurement techniques closes the present chapter.

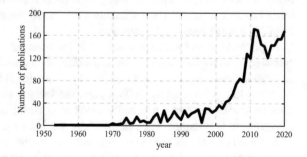

Fig. 2 Number of publications per year found in Scopus where Title, Abstract or Keywords include "wave run-up" or "wave run-up"

2 Description of Wave Run-Up Measurement Techniques

In 1950, Hunt used resistance run-up gauges in the laboratory to conduct the first wave run-up measurements [3]. In 1982, Guza and Thornton [2] used resistance wires to obtain field observations of wave run-up. In 1986, Holman [4] used a video camera to acquire field observations of wave run-up. In 1991, Nielsen and Hanslow [10] used a set of markers to measure wave run-up in the field. In 2008, Turner et al. [11] used ultrasonic sensors (similar to acoustic wave probes) to obtain both wave run-up and bed level measurements on a wave-by-wave time scale. More recently, in 2018, Dodet et al. [12] used a transect of pressure sensors to obtain measurements of wave run-up in the field. Here below, these techniques are described. Due to its importance to wave run-up measurements, the video technique applied on beaches will be described in detail in Sect. 3.

Guza and Thornton [2] used resistance wires (conductance probes) to obtain field measurements of wave run-up on a 1/20–1/30 sloping beach. The voltage registered by the wires can be converted to vertical height reached by the waves. Raubenheimer et al. [13] used run-up wires that were stacked vertically with the bottom wire about 5 cm above the bed, and the other wires 5 cm apart. Although the method was successfully used by previous authors, installing run-up wires on beach profiles has several operational issues. Guza and Thornton [2] reported problems with seaweed, sand level changes, rain, calibration procedures, among others. Although instrumenting a beach with run-up wires can be a challenging task by itself, some recent studies used this method successfully [14]. Nevertheless, operational issues are always present as "the run-up wire was maintained at approximately 0.02–0.03 m above the bed, which required continual adjustment by observers" [14].

Wave run-up on a coastal structures is usually measured using conductance probes, usually referred as run-up gauges or wave gauges, that measure the water surface elevation on the structure slope (Fig. 3). As referred above, these probes operate by measuring the current that flows between two stainless steel wires that are immersed in water. This current is converted to an output voltage that is directly proportional to the immersed depth. Since it is very difficult to measure the run-up on a very thin water layer in small-scale experiments with run-up gauges, water tongue less than 2 cm thick is not measured accurately.

In physical model tests wave run-up is becoming to be measured using images collected from digital video cameras, together with a scaled ruler fixed on the front slope to increase the accuracy of the run-up estimates [15]. For each test condition, usually with approximately 1000 waves, the images collected from digital video cameras during the test are visually monitoring frame by frame and an individual wave run-up is obtained and confirmed by the information obtained in the scaled ruler. Based on that, maximum run-up and $R_{2\%}$ can be calculated.

The deployment of a set of markers consists in counting the number of waves that cross each set of markers with known elevations (x) over a time period, to obtain a time-series of wave run-up measurements (x, t). Nielsen and Hanslow [10] used wooden stakes to study wave run-up distributions in the field. This method has,

Fig. 3 Resistance type wave gauge used to measure run-up on a coastal structure at laboratory scale

however, some drawbacks, as: (i) the accuracy of the wave run-up time-series depends on the cross-shore distance between markers; (ii) it requires intense human effort in the field; (iii) it is not adequate during energetic wave conditions. Nevertheless, this method has been still used recently coupling portable video systems (Fig. 4) for improving the observations (e.g., Manno et al. [16]).

Another method to obtain measurements of wave run-up is through ultrasonic sensors (similar to acoustic wave probes). Analogous to a set of markers, a cross-shore transect of several ultrasonic sensors will enable the measurement of sea-surface elevation through space and time, as explained by Turner et al. [11]. Consequently, such a deployment will allow wave run-up measurements with a high-frequency

Fig. 4 Example of field measurements of wave run-up using markers (orange marks) and a video-camera mounted on a tripod (from Swenson [17])

Fig. 5 Pressure sensors deployed at Banneg Island (France) on a rocky cliff (from Dodet et al. [12])

resolution. One of the main advantages, besides the sea-surface elevation which enables a detailed wave-by-wave analysis, is to record bottom elevations [11, 18]. This technique has, however, similar drawbacks as the set of markers deployment because it depends on the available number of ultrasonic sensors and the intensive labour work to instrument a surf zone, especially in macro-tidal regimes.

A third method for wave run-up measurements resides on the use of pressure sensors. A cross-shore transect of several pressure sensors is generally deployed on beach profile or on the mound slope of a coastal structure (Fig. 5). The procedure consists of analysing the time moments in which the transect indicates wet and dry conditions (i.e. swash). The measured pressure can be then converted into a vertical height. This method was used in the field on a very complex coastal setting characterized by steep rocky cliffs by Dodet et al. [12]. Although it can be used under extreme wave conditions, the method is difficult to be applied on beaches with non-cohesive sediments, where the beach profile changes occur over short-time scale.

3 Video Monitoring Technique

3.1 History of Video Monitoring Technique

Coastal video monitoring systems are being proven to be a powerful tool to monitor nearshore morpho- and hydrodynamics. When video camera devices are installed on a fixed, shore-based platform composing a coastal video monitoring station, they offer the possibility of collecting both short- and long-term observations. Despite the lower coverage in space in comparison with the other remote sensing technologies (e.g., satellite imagery and X-Band radar), they offer the advantages of autonomous and continuous collection of high-resolution images, at a relatively low cost, which are fundamental for wave run-up measurements.

The pioneer ARGUS monitoring programme [19, 20] was the first scientific programme to install a shore-based video monitoring system aiming at supporting coastal studies through video-derived observations. The system was developed by the Coastal Imaging Lab at the Oregon State University in the early 90s, and it has been providing coastal image data worldwide for the last three decades [19, 21].

In the 2000s, the expansion of commercial video systems (e.g., CoastalComs, www.coastalcoms.com, Erdman, www.video-monitoring.com) and the development of open-source image processing tools (e.g., SIRENA [22], COSMOS [23], Beach-keeper plus [24], ULISES [25]) promoted the installation of video monitoring stations for scientific purpose exploiting the relatively cheap Internet Protocol (IP) video cameras, overcoming the expensive installation and purchase of Argus system.

A video-monitoring station is usually composed of one (or more) video camera connected to a personal computer, which has the purpose of controlling the image acquisition and storing the video images. The optical device is usually installed stable at an elevated position looking at the emerged beach and the nearshore. Data sampling interval can vary between 1 and 10 Hz (image per second) depending on the aim of the study, camera properties, and storage space capability. Besides the coastal video monitoring stations, the portability of IP cameras has also promoted the temporary use of optical devices to support oceanographic fieldwork (e.g., Senechal et al. [26]) and the video monitoring application on physical modelling (e.g., Schimmels et al. [27], Vousdoukas et al. [28]).

Three types of images are commonly generated by coastal video monitoring stations: Timex, Variance and Timestack images. Timex images are digitally averaged image intensities, representing the time-averaged of all the frames collected within a period of sampling, generally, 10 min, while Variance images represent the standard deviation for the same time [19, 29]. Conversions between image coordinates and ground coordinates are usually made using the collinearity equations after corrections for lens distortions (e.g., Taborda and Silva [23], Simarro et al. [25], Sánchez-García et al. [30]).

Although some attempts were done in using Timex and Variance for wave run-up measurements [31], the main advances given for these measurements on beaches were due to the use of Timestack images.

3.2 Timestack Images

Timestack images are generated by sampling a single array of pixels, representing a beach transect, from each frame over the acquisition period. Timestack is, therefore,

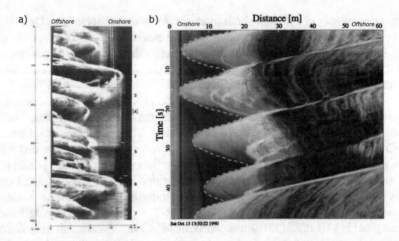

Fig. 6 First appearances of Timestack images: **a** from Aagaard and Holm [33]; **b** from Holland and Holman [34]

a time-space image composed by pixel intensity time series over a given image sequence (e.g., Andriolo [32]).

Historically, Timestacks were specifically produced to measure wave run-up, as the camera acquisitions allowed the monitoring of the high-frequency waterline oscillation on the beach slope. The first appearance of a Timestack image (Fig. 6a) is dated to 1989 by Aagaard and Holm [33]. A clearer image was published in Holland and Holman [34] in 1993 (Fig. 6b), originated from Argus images collected at Duck during the DELILAH experiment.

On Timestack, the swash excursions time series is typically visible as series of cuspates, which describe the water-level fluctuation (uprush and backwash) of broken waves on the beach slope. The shoreward peaks of such curvatures express the maxima swash excursion points of each individual wave swash front. Operationally, each swash maxima are marked by the user on Timestack to obtain the swash maxima time-series over time and space (Fig. 7). To compute the wave run-up (swash elevation), the space-dimension of the Timestack is interpolated with the related beach transect elevation, usually surveyed by RTK-GNNS on the field.

The typical statistic values used for run-up formulas, such as the maximum (R_{max}) and the run-up exceeded by 2% of the waves ($R_{2\%}$), are computed considering the whole time series used for producing the Timestack. Previous studies [36, 37] found the run-up distribution to be consistently represented by a normal distribution, suitable for retrieving $R_{2\%}$.

Few image processing methodologies have been proposed to automate the wave run-up detection on Timestacks. Vousdoukas et al. [38] adopted Otsu's thresholding method, while Almar et al. [39] introduced and validated a method based on the Radon Transform, better performing than a color contrast-based image processing technique [40]. Nevertheless, the manual digitalization of swash fronts has been chosen by most Timestack-based wave run-up studies due to the complex nature

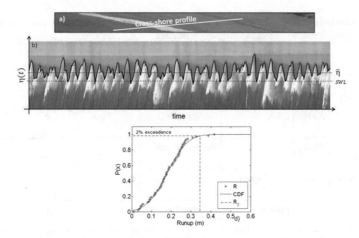

Fig. 7 Timestack-based method for wave run-up measurements: **a** portion of the image acquired by the video system; white line represents the cross-shore transect selected to extract pixels time series and generate the Timestack; **b** Timestack image with swash excursion time series; red crosses represent the individual digitized wave run-up, black line is the limit between wet and dry areas, the horizontal lines represent the still water level (SWL, dashed line) and the wave setup (solid line); **c** cumulative distribution function of individual wave run-up obtained from digitalization on Timestack. Adapted from Gomes da Silva et al. [5, 35]

both of wave run-up signal on the foreshore (high-frequency) and of Timestack images.

3.3 Advances in Wave Run-Up Measurements from Timestacks

Over the last two decades, the use of Timestack images have extensively improved the knowledge of wave run-up on coasts, promoting new parameterizations of wave run-up formula (Table 1).

Sampling several transects alongshore to produce a series of Timestacks, Ruggiero et al. [41] showed that alongshore variability of wave run-up on a dissipative beach was associated with alongshore variability of the foreshore slope. The work of Stockdon et al. [37] represented the new potentiality offered by the long-term collection of wave run-up measurements for advancing the parameterization. Timestacks produced by the Argus programme at 10 different sites (east and west coast of US and in the Netherlands) were considered. Although the bulk of the data (91%) were collected at intermediate to reflective conditions at Duck beach, Timestack-based measurements included a variety of beach characteristics and wave conditions and were statistically analyzed, to finally propose the $R_{2\%}$ formula that is currently the most used worldwide [5]. Following the framework proposed by Stockdon et al. [37],

Table 1 Chronological significant works devoted to testing and parameterization of wave run-up formulae through video Timestack imaging on beaches

Publication	Year	Country	Ocean/sea	Wave run-up formulae testing	Wave run-up parameterization
Holland and Homan [34]	1993	USA	Pacific Ocean		x
Ruggiero et al. [41]	2004	USA	Pacific Ocean		x
Stockdon et al. [37]	2006	USA	Pacific Ocean		x
Salmon et al. [42]	2007	New Zealand	Pacific Ocean	x	
Senechal et al. [43]	2011	France	Atlantic Ocean	x	x
Vousdoukas et al. [38]	2012	Portugal	Atlantic Ocean		x
Paprotny et al. [44]	2014	Poland	Baltic Sea	x	x
Poate et al. [45]	2016	England	Atlantic Ocean		x
Atkinson et al. [36]	2017	Australia	Pacific Ocean	x	x
Di Luccio et al. [46]	2018	Italy	Ligurian Sea (Mediterranean Sea)	x	
Valentini et al. [47]	2019	Italy	Ionian Sea (Mediterranean Sea)	x	
Didier et al. [48]	2020	Canada	Atlantic Ocean	x	x

other works (Table 1) aimed at the local wave run-up parameterization using medium and long-term Timestack dataset on Portuguese [49], British [45], Australian [36], Canadian [48] and Polish [44] coasts. Several wave run-up formulae have also been tested and proposed from wave run-up measured during storm wave conditions at New Zealand [42], Italian [46, 47] and French [26] coasts. Combining the Timestack-based data collected at the different above-cited coasts, more advanced wave run-up formulae were proposed with the use of Gene expression programming [50], machine learning [51] and model-of-models [36] techniques.

While the use of Timestack images has been widely used for wave run-up measurements on beaches, the field application on coastal structure is still scarce (e.g., Lee et al. [52], González-Jorge et al. [53]). This is due to the logistical constraints of video-cameras installation, which rarely allow to obtain the seaward side of breakwaters and jetties within the camera field of view.

4 Perspectives on Future Advances

The main limitation of Timestack application for wave run-up studies resides in the requirement of the beach profile elevation. Being Timestack a time-space image, the cross-shore horizontal swash excursion visible on images needs to be converted to elevation with the use of the surveyed beach slope. However, the frequency of ground surveys is often not sufficient to describe the high variability of the intertidal area. Video monitoring technique can infer the foreshore slope by tracking the edge of the waterline in Timex and/or Variance images over the tidal cycle [49, 54–57]. Nevertheless, the video-based techniques for intertidal beach slope estimation are affected by inaccuracies introduced by the photogrammetry procedures (e.g., camera calibration and image georectification), nearshore hydrodynamics (e.g., wave setup) and shoreline detection (e.g., Uunk et al. [54]). Future work may improve the video measurements of intertidal beach profile, to integrate it to hydrodynamic observations and improve wave run-up studies.

When waves enter shallow waters, they interact with the bathymetry as they propagate to the coast, and are gradually transformed to skewed and breaking waves, before they eventually trigger wave run-up (e.g., Andriolo [32], Vousdoukas et al. [38]). Numerous authors pointed out that nearshore wave transformation (i.e. wave deformation by bar-trough beach profiles) is neglected in wave run-up formulas, as only offshore wave parameters (e.g., Hs, Tp) have been considered (e.g., Gomes da Silva et al. [5] and references therein). In this regard, the application of video monitoring techniques dedicated to the estimation of breaking wave height [58–61] may be integrated into the use of Timestacks for wave run-up measurements, to propose new parameterizations based on the actual wave properties near the shore.

Future improvements on video-based run-up studies may also be provided by the exploitation of online streaming webcams and surfcams [29, 62, 63] to expand coastal monitoring around the world's coastlines.

A relatively recent alternative to measure wave run-up is the use of a laser-scanner. This type of system can measure high-frequency (more than 2 Hz) water motions over a high-resolution $O(cm)$ cross-shore transect. As an example, Almeida et al. [64] deployed a laser-scanner on top of a 5.2 m height aluminum tower and were able to cover a 90 m cross-shore transect with an average along the slope resolution of 0.06 m. Another advantage of the laser-scanner is the continuous monitoring of the beach morphology, which is needed for the bottom slope quantification and subsequent wave run-up estimation. Laser-scanners have been used in the laboratory [65, 66] and in the field [e.g., 67–69], collecting wave run-up measurements also under very extreme offshore wave conditions [70]. The deployment of a set of laser-scanners [71] can provide detailed information on the complex inner surf zone hydrodynamics and the associated wave run-up. These recent studies have shown that laser-scans allows the integration of sea-surface elevation, wave motion, beach profile evolution and swash elevation measurements, for further improving the knowledge of the complex hydrodynamic processes involved in wave run-up.

Acknowledgements Acknowledgements are due to the Portuguese Foundation for Science and Technology (FCT) and the European Regional Development Fund (FEDER) through COMPETE 2020—Operational Program for Competitiveness and Internationalization (POCI) in the framework of UIDB/00308/2020 and the research project UAS4Litter (PTDC/EAM-REM/30324/2017).

References

1. U.S. Army Corps of Engineers: Coastal Engineering Manual (EM 1110-2-1100) (2002)
2. Guza, R.T., Thornton, E.B.: Swash oscillations on a natural beach. J. Geophys. Res. **87**, 483–491 (1982). https://doi.org/10.1029/JC087iC01p00483
3. Hunt, I.A.: Design of seawalls and breakwaters. J. Waterw. Harb. Div. **85**, 123–152 (1956)
4. Holman, R.A.: Extreme value statistics for wave run-up on a natural beach. Coast. Eng. **9**(6), 527–544 (1986). https://doi.org/10.1016/0378-3839(86)90002-5
5. Gomes da Silva, P., Coco, G., Garnier, R., Klein, A.H.F.: On the prediction of runup, setup and swash on beaches. Earth-Sci. Rev. **204**, 103148 (2020). https://doi.org/10.1016/j.earscirev.2020.103148
6. Douglass, S.: Estimating Runup on Beaches: A Review of the State of the Art. USACE Report AD-A229 516. Washington, DC (1990)
7. Kobayashi, N.: Wave runup and overtopping on beaches and coastal structures. In: Advanced Series in Coastal and Ocean Engineering, pp. 95–154. World Scientific (1999). https://doi.org/10.1142/9789812797544_0002
8. Weggel, J.R.: Wave overtopping equation. Coast. Eng. Proc. **1**(15), 2737–2755 (1976)
9. Mase, H., Tamada, T., Yasuda, T., Hedges, T.S., Reis, M.T.: Wave runup and overtopping at seawalls built on land and in very shallow water. J. Waterw. Port Coast. Ocean Eng. (2013). https://doi.org/10.1061/(asce)ww.1943-5460.0000199
10. Nielsen, P., Hanslow, D.J.: Wave runup distributions on natural beaches. J. Coast. Res. **7**, 1139–1152 (1991). https://doi.org/10.2307/4297933
11. Turner, I.L., Russell, P.E., Butt, T.: Measurement of wave-by-wave bed-levels in the swash zone. Coast. Eng. **55**, 1237–1242 (2008). https://doi.org/10.1016/j.coastaleng.2008.09.009
12. Dodet, G., Leckler, F., Sous, D., Ardhuin, F., Filipot, J.F., Suanez, S.: Wave runup over steep rocky cliffs. J. Geophys. Res. Oceans **123**, 7185–7205 (2018). https://doi.org/10.1029/2018JC013967
13. Raubenheimer, B., Guza, R.T., Elgar, S., Kobayashi, N.: Swash on a gently sloping beach. J. Geophys. Res. **100**, 8751–8760 (1995). https://doi.org/10.1029/95JC00232
14. Hughes, M.G., Moseley, A.S., Baldock, T.E.: Probability distributions for wave runup on beaches. Coast. Eng. **57**, 575–584 (2010). https://doi.org/10.1016/j.coastaleng.2010.01.001
15. Pillai, K., Etemad-Shahidi, A., Lemckert, C.: Wave run-up on bermed coastal structures. Appl. Ocean Res. **86**, 188–194 (2019). https://doi.org/10.1016/j.apor.2019.02.006
16. Manno, G., Lo Re, C., Ciraolo, G.: Uncertainties in shoreline position analysis: the role of run-up and tide in a gentle slope beach. Ocean Sci. **13**, 661–671 (2017). https://doi.org/10.5194/os-13-661-2017
17. Swenson, M.: Wave runup. http://homepages.cae.wisc.edu/~chinwu/GLE401/web/Mike/Wave%20runup.htm. Accessed 1 Mar 2021
18. Masselink, G., Russell, P., Turner, I., Blenkinsopp, C.: Net sediment transport and morphological change in the swash zone of a high-energy sandy beach from swash event to tidal cycle time scales. Mar. Geol. **267**, 18–35 (2009). https://doi.org/10.1016/j.margeo.2009.09.003
19. Holman, R.A., Stanley, J.: The history and technical capabilities of Argus. Coast. Eng. **54**, 477–491 (2007). https://doi.org/10.1016/j.coastaleng.2007.01.003
20. Splinter, K.D., Harley, M.D., Turner, I.L.: Remote sensing is changing our view of the coast: insights from 40 years of monitoring at Narrabeen-Collaroy, Australia. Remote Sens. **10**, 1744 (2018). https://doi.org/10.3390/rs10111744

21. Andriolo, U.: Nearshore hydrodynamics and morphology derived from video images. Ph.D. Thesis, University of Lisbon 224 pp. (2018)
22. Nieto, M.A., Garau, B., Balle, S., Simarro, G., Zarruk, G.A., Ortiz, A., Tintoré, J., Álvarez-Ellacuría, A., Gómez-Pujol, L., Orfila, A.: An open source, low cost video-based coastal monitoring system. Earth Surf. Process. Landforms 35, 1712–1719 (2010). https://doi.org/10.1002/esp.2025
23. Taborda, R., Silva, A.: COSMOS: a lightweight coastal video monitoring system. Comput. Geosci. 49, 248–255 (2012). https://doi.org/10.1016/j.cageo.2012.07.013
24. Brignone, M., Schiaffino, C.F., Isla, F.I., Ferrari, M.: A system for beach video-monitoring: beachkeeper plus. Comput. Geosci. 49, 53–61 (2012). https://doi.org/10.1016/j.cageo.2012.06.008
25. Simarro, G., Ribas, F., Álvarez, A., Guillén, J., Chic, Ò., Orfila, A.: ULISES: an open source code for extrinsic calibrations and planview generations in coastal video monitoring systems. J. Coast. Res. 335, 1217–1227 (2017). https://doi.org/10.2112/JCOASTRES-D-16-00022.1
26. Senechal, N., Coco, G., Bryan, K.R., Holman, R.A.: Wave runup during extreme storm conditions. J. Geophys. Res. Oceans 116 (2011). https://doi.org/10.1029/2010JC006819
27. Schimmels, S., Vousdoukas, M., Wziatek, D., Becker, K., Gier, F., Oumeraci, H.: Wave run-up observations on revetments with different porosities. In: Lynett, P., Smith, J.M. (eds.) Coastal Engineering Proceedings, pp. 1–14 (2012). https://doi.org/10.9753/icce.v33.structures.73
28. Vousdoukas, M.I., Kirupakaramoorthy, T., Oumeraci, H., de la Torre, M., Wübbold, F., Wagner, B., Schimmels, S.: The role of combined laser scanning and video techniques in monitoring wave-by-wave swash zone processes. Coast. Eng. 83, 150–165 (2014). https://doi.org/10.1016/j.coastaleng.2013.10.013
29. Andriolo, U., Sánchez-García, E., Taborda, R.: Operational use of surfcam online streaming images for coastal morphodynamic studies. Remote Sens. 11(1), 1–21 (2019). https://doi.org/10.3390/rs11010078
30. Sánchez-García, E., Balaguer-Beser, A., Pardo-Pascual, J.E.: C-Pro: a coastal projector monitoring system using terrestrial photogrammetry with a geometric horizon constraint. ISPRS J. Photogramm. Remote Sens. 128, 255–273 (2017). https://doi.org/10.1016/j.isprsjprs.2017.03.023
31. Simarro, G., Bryan, K.R., Guedes, R.M.C., Sancho, A., Guillen, J., Coco, G.: On the use of variance images for runup and shoreline detection. Coast. Eng. 99, 136–147 (2015). https://doi.org/10.1016/j.coastaleng.2015.03.002
32. Andriolo, U.: Nearshore wave transformation domains from video imagery. J. Mar. Sci. Eng. 7, 186 (2019). https://doi.org/10.3390/jmse7060186
33. Aagaard, T., Holm, J.: Digitization of wave run-up using video records. J. Coast. Res. 5, 547–551 (1989). https://doi.org/10.2307/4297566
34. Holland, K.T., Holman, R.A.: The statistical distribution of swash maxima on natural beaches. J. Geophys. Res. 98, 271–278 (1993). https://doi.org/10.1029/93JC00035
35. Gomes da Silva, P., Medina, R., González, M., Garnier, R.: Infragravity swash parameterization on beaches: The role of the profile shape and the morphodynamic beach state. Coast. Eng. 136, 41–55 (2018). https://doi.org/10.1016/j.coastaleng.2018.02.002
36. Atkinson, A.L., Power, H.E., Moura, T., Hammond, T., Callaghan, D.P., Baldock, T.E.: Assessment of runup predictions by empirical models on non-truncated beaches on the south-east Australian coast. Coast. Eng. 119, 15–31 (2017). https://doi.org/10.1016/j.coastaleng.2016.10.001
37. Stockdon, H.F., Holman, R.A., Howd, P.A., Sallenger, A.H.: Empirical parameterization of setup, swash, and runup. Coast. Eng. 53, 573–588 (2006). https://doi.org/10.1016/j.coastaleng.2005.12.005
38. Vousdoukas, M.I., Wziatek, D., Almeida, L.P.: Coastal vulnerability assessment based on video wave run-up observations at a mesotidal, steep-sloped beach. Ocean Dyn. 62, 123–137 (2012). https://doi.org/10.1007/s10236-011-0480-x
39. Almar, R., Blenkinsopp, C., Almeida, L.P., Cienfuegos, R., Catalán, P.A.: Wave runup video motion detection using the Radon Transform. Coast. Eng. 130, 46–51 (2017). https://doi.org/10.1016/j.coastaleng.2017.09.015

40. Huisman, C.E., Bryan, K.R., Coco, G., Ruessink, B.G.: The use of video imagery to analyse groundwater and shoreline dynamics on a dissipative beach. Cont. Shelf Res. **31**, 1728–1738 (2011). https://doi.org/10.1016/j.csr.2011.07.013

41. Ruggiero, P., Holman, R.A., Beach, R.A.: Wave run-up on a high-energy dissipative beach. J. Geophys. Res. Oceans **109**, 1–12 (2004). https://doi.org/10.1029/2003JC002160

42. Salmon, S.A., Bryan, K.R., Coco, G.: The use of video systems to measure run-up on beaches. J. Coast. Res. 211–215 (2007)

43. Senechal, N., Abadie, S., Gallagher, E., MacMahan, J., Masselink, G., Michallet, H., Reniers, A., Ruessink, G., Russell, P., Sous, D., Turner, I., Ardhuin, F., Bonneton, P., Bujan, S., Capo, S., Certain, R., Pedreros, R., Garlan, T.: The ECORS-Truc Vert'08 nearshore field experiment: presentation of a three-dimensional morphologic system in a macro-tidal environment during consecutive extreme storm conditions. Ocean Dyn. **61**, 2073–2098 (2011). https://doi.org/10.1007/s10236-011-0472-x

44. Paprotny, D., Andrzejewski, P., Terefenko, P., Furmańczyk, K.: Application of empirical wave run-up formulas to the Polish Baltic Sea coast. PLoS ONE **9**, 1–8 (2014). https://doi.org/10.1371/journal.pone.0105437

45. Poate, T.G., McCall, R.T., Masselink, G.: A new parameterisation for runup on gravel beaches. Coast. Eng. **117**, 176–190 (2016). https://doi.org/10.1016/j.coastaleng.2016.08.003

46. Di Luccio, D., Benassai, G., Budillon, G., Mucerino, L., Montella, R., Pugliese Carratelli, E.: Wave run-up prediction and observation in a micro-tidal beach. Nat. Hazards Earth Syst. Sci. **18**, 2841–2857 (2018). https://doi.org/10.5194/nhess-18-2841-2018

47. Valentini, N., Saponieri, A., Danisi, A., Pratola, L., Damiani, L.: Exploiting remote imagery in an embayed sandy beach for the validation of a runup model framework. Estuar. Coast. Shelf Sci. **225**, 106244 (2019). https://doi.org/10.1016/j.ecss.2019.106244

48. Didier, D., Caulet, C., Bandet, M., Bernatchez, P., Dumont, D., Augereau, E., Floc'h, F., Delacourt, C.: Wave runup parameterization for sandy, gravel and platform beaches in a fetch-limited, large estuarine system. Cont. Shelf Res. **192**, 104024 (2020). https://doi.org/10.1016/j.csr.2019.104024

49. Vousdoukas, M.I., Ferreira, P.M., Almeida, L.P., Dodet, G., Psaros, F., Andriolo, U., Taborda, R., Silva, A.N., Ruano, A., Ferreira, Ó.M.: Performance of intertidal topography video monitoring of a meso-tidal reflective beach in South Portugal. Ocean Dyn. **61**, 1521–1540 (2011). https://doi.org/10.1007/s10236-011-0440-5

50. Power, H.E., Gharabaghi, B., Bonakdari, H., Robertson, B., Atkinson, A.L., Baldock, T.E.: Prediction of wave runup on beaches using Gene-Expression Programming and empirical relationships. Coast. Eng. **144**, 47–61 (2019). https://doi.org/10.1016/j.coastaleng.2018.10.006

51. Passarella, M., Goldstein, E.B., De Muro, S., Coco, G.: The use of genetic programming to develop a predictor of swash excursion on sandy beaches. Nat. Hazards Earth Syst. Sci. **18**, 599–611 (2018). https://doi.org/10.5194/nhess-18-599-2018

52. Lee, S.-C., Choi, J.-Y., Park, K.-S., Kim, S.-S., Kim, S.-J., Jun, K.-C.: Use of optical video imagery to improve wave run-up prediction accuracy. J. Coast. Res. **85**, 1271–1275 (2018). https://doi.org/10.2112/SI85-255.1

53. González-Jorge, H., Díaz-Vilariño, L., Martínez-Sánchez, J., Riveiro, B., Arias, P.: Wave run-up monitoring on rubble-mound breakwaters using a photogrammetric methodology. J. Perform. Constr. Facil. **30**, 04015075 (2016). https://doi.org/10.1061/(asce)cf.1943-5509.0000822

54. Uunk, L., Wijnberg, K.M., Morelissen, R.: Automated mapping of the intertidal beach bathymetry from video images. Coast. Eng. **57**, 461–469 (2010). https://doi.org/10.1016/j.coastaleng.2009.12.002

55. Aarninkhof, S.G.J., Turner, I.L., Dronkers, T.D.T., Caljouw, M., Nipius, L.: A video-based technique for mapping intertidal beach bathymetry. Coast. Eng. **49**, 275–289 (2003). https://doi.org/10.1016/S0378-3839(03)00064-4

56. Valentini, N., Saponieri, A., Damiani, L.: A new video monitoring system in support of Coastal Zone Management at Apulia Region, Italy. Ocean Coast. Manag. **142**, 122–135 (2017). https://doi.org/10.1016/j.ocecoaman.2017.03.032

57. Andriolo, U., Almeida, L.P., Almar, R.: Coupling terrestrial LiDAR and video imagery to perform 3D intertidal beach topography. Coast. Eng. **140**, 232–239 (2018). https://doi.org/10.1016/j.coastaleng.2018.07.009

58. Shand, T., Bailey, D., Shand, R.: Automated detection of breaking wave height using an optical technique. J. Coast. Res. **28**, 671–682 (2012). https://doi.org/10.2112/jcoastres-d-11-00105.1

59. Gal, Y., Browne, M., Lane, C.: Long-term automated monitoring of nearshore wave height from digital video. IEEE Trans. Geosci. Remote Sens. **52**, 3412–3420 (2014). https://doi.org/10.1109/TGRS.2013.2272790

60. Andriolo, U., Mendes, D., Taborda, R.: Breaking wave height estimation from Timex images: two methods for coastal video monitoring systems. Remote Sens. **12**, 204 (2020). https://doi.org/10.3390/rs12020204

61. Almar, R., Cienfuegos, R., Catalán, P.A., Michallet, H., Castelle, B., Bonneton, P., Marieu, V.: A new breaking wave height direct estimator from video imagery. Coast. Eng. **61**, 42–48 (2012). https://doi.org/10.1016/j.coastaleng.2011.12.004

62. Mole, M.A., Mortlock, T.R.C., Turner, I.L., Goodwin, I.D., Splinter, K.D., Short, A.D.: Capitalizing on the surfcam phenomenon: a pilot study in regional-scale shoreline and inshore wave monitoring utilizing existing camera infrastructure. J. Coast. Res. **165**, 1433–1438 (2013). https://doi.org/10.2112/SI65-242.1

63. Bracs, M.A., Turner, I.L., Splinter, K.D., Short, A.D., Lane, C., Davidson, M.A., Goodwin, I.D., Pritchard, T., Cameron, D.: Evaluation of opportunistic shoreline monitoring capability utilizing existing "surfcam" infrastructure. J. Coast. Res. **319**, 542–554 (2016). https://doi.org/10.2112/JCOASTRES-D-14-00090.1

64. Almeida, L.P., Masselink, G., Russell, P.E., Davidson, M.A.: Observations of gravel beach dynamics during high energy wave conditions using a laser scanner. Geomorphology **228**, 15–27 (2015). https://doi.org/10.1016/j.geomorph.2014.08.019

65. Blenkinsopp, C.E., Matias, A., Howe, D., Castelle, B., Marieu, V., Turner, I.L.: Wave runup and overwash on a prototype-scale sand barrier. Coast. Eng. **113**, 88–103 (2016). https://doi.org/10.1016/j.coastaleng.2015.08.006

66. Hofland, B., Diamantidou, E., van Steeg, P., Meys, P.: Wave runup and wave overtopping measurements using a laser scanner. Coast. Eng. **106**, 20–29 (2015). https://doi.org/10.1016/j.coastaleng.2015.09.003

67. Brodie, K.L., Raubenheimer, B., Elgar, S., Slocum, R.K., McNinch, J.E.: Lidar and pressure measurements of inner-surfzone waves and setup. J. Atmos. Ocean. Technol. **32**, 1945–1959 (2015). https://doi.org/10.1175/JTECH-D-14-00222.1

68. Bergsma, E.W.J., Blenkinsopp, C.E., Martins, K., Almar, R., de Almeida, L.P.M.: Bore collapse and wave run-up on a sandy beach. Cont. Shelf Res. **174**, 132–139 (2019). https://doi.org/10.1016/j.csr.2019.01.009

69. Almeida, L.P., Almar, R., Blenkinsopp, C., Senechal, N., Bergsma, E., Floc'h, F., Caulet, C., Biausque, M., Marchesiello, P., Grandjean, P., Ammann, J., Benshila, R., Thuan, D.H., da Silva, P.G., Viet, N.T.: Lidar observations of the swash zone of a low-tide terraced tropical beach under variable wave conditions: the Nha Trang (Vietnam) COASTVAR experiment. J. Mar. Sci. Eng. **8**, 302 (2020). https://doi.org/10.3390/JMSE8050302

70. Fiedler, J.W., Brodie, K.L., McNinch, J.E., Guza, R.T.: Observations of runup and energy flux on a low-slope beach with high-energy, long-period ocean swell. Geophys. Res. Lett. **42**, 9933–9941 (2015). https://doi.org/10.1002/2015GL066124

71. Martins, K., Blenkinsopp, C.E., Power, H.E., Bruder, B., Puleo, J.A., Bergsma, E.W.J.: High-resolution monitoring of wave transformation in the surf zone using a LiDAR scanner array. Coast. Eng. **128**, 37–43 (2017). https://doi.org/10.1016/j.coastaleng.2017.07.007

Recent Advances in Instrumentation and Monitoring Techniques Applied to Dam Breach Experiments

Sílvia Amaral⬮, Teresa Viseu⬮, Gensheng Zhao, and Rui M. L. Ferreira⬮

Abstract The flow through an eroding dam is three-dimensional and exhibits several length scales. Laboratory experiments are fundamental to characterize these flows. Instrumentation and monitoring techniques in laboratory environment must be able to capture this complexity and scale amplitude and should be, preferentially, non-intrusive, particularly if the failure event is related with earth dams, where erosion patterns may be affected by instrument intrusiveness. Regarding embankment dam failures, the processes usually subjected to great scrutiny require the measurement of the breach effluent flow (BEF), the breach morphological evolution (through longitudinal and transversal profiles as well as 3D reconstructions) and the flow velocity over the dam and through the breach. This chapter presents advances in instrumentation to be deployed in dam breach laboratory tests and novel measuring methods made possible with such instrumentation. Emphasis is placed on the description of the critical components of the instruments, with emphasis on digital imagery, and on image analysis and post-processing techniques. A detailed exposition of existent methods, and respective evaluation of its suitability in the context of laboratory work, is presented.

Keywords Dam failure and breaching · Imaging techniques and post processing · Velocity fields (LSPIV, PTV) · 3D reconstruction (Kinect, laser scan, photogrammetry)

S. Amaral (✉) · T. Viseu
National Laboratory for Civil Engineering (LNEC), Lisbon, Portugal
e-mail: samaral@lnec.pt

T. Viseu
e-mail: tviseu@lnec.pt

G. Zhao
Nanjing Hydraulic Research Institute (NHRI), Nanjing, China
e-mail: gszhao@nhri.cn

R. M. L. Ferreira
Instituto Superior Técnico (IST), Universidade de Lisboa (NHRI), Lisbon, Portugal
e-mail: ruif@civil.ist.utl.pt

© The Author(s), under exclusive license to Springer Nature Switzerland AG 2023
C. Chastre et al. (eds.), *Advances on Testing and Experimentation in Civil Engineering*, Springer Tracts in Civil Engineering,
https://doi.org/10.1007/978-3-031-05875-2_13

1   Introduction

Dam failure events have the potential to cause loss of human lives and catastrophic environmental impacts in the downstream valley, justifying major investments in dam breach studies. Embankment dams are of particular interest as these are most of the large dams worldwide [1–3] presenting the highest percentage of accidents, many of which led to the structure failure. A recent example is the accident with two earthen dams in Michigan (USA) that, in the sequence of a flood, suffered a cascade collapse in 2020, expelling billions of cubic meters of water and sending them downstream in a powerful rush of destruction. Fortunately, the urgent evacuation of 10,000 people from communities downstream prevented the loss of lives but the remaining damage in the downstream valley still caused around $175 million of losses.

Several approaches are available to conduct research studies on embankment dam failures. In recent years, an increasing trend towards the improvement of experimental based approaches does exist. The success of experimental studies to investigate the hydraulic and geotechnical aspects of dam breaching processes requires an adequate specification of instrumentation to measure and monitor the most important variables as discharge rates, water surface elevations, flow velocities, scour, etc.

This chapter presents advanced instrumentation solutions, its main components and measuring methods that can be used in experimental dam breach studies. This includes the presentation of a monitoring scheme specifically developed to acquire data near or at the breach site ('local measurements'), allowing for estimates of variables such as the flow rate through the breach and erosion rates. Emphasis is placed on post-processing techniques and analysis of digital imagery, for which there is detailed exposition of existent methods and respective evaluation of its suitability for several purposes.

Concerning the present chapter organization, Sect. 2 claims the reader attention for the *relevance of using recent instrumentation, improving the measuring methods*, and optimizing laboratorial procedures when conducting dam breach experiments. Section 3 gives a description on the *experimental facility* where the dam breach experiments were carried out including a *general overview of the global instrumentation constituting the measurement and monitoring systems*. Section 4 is devoted to the full characterization of the *instrumentation and measuring methods applied on dam breach tests* and Sect. 5 offers final remarks concerning the instrumental layout presented and its suitability for each purpose.

2   Relevance of Using Recent Instrumentation and Improved Measuring Methods in Dam Breach Experiments

Before the mid-1980s, experimental investigation on dam breach could be regarded as mostly qualitative [4]. However, attempts were made to measure the breach effluent flow and the breach evolution process albeit with indirect methods. Technologic difficulties that did not allow observing/characterizing the breach at a local scale, i.e., technical difficulties on acquiring variables in an evolving breach, hindered the understanding of some key phenomena.

The variables usually measured in dam breach tests are the water levels (in the upstream reservoir and downstream the dam) and the breach outflow discharges. The velocity fields at the breach vicinity (commonly at the surface and sometimes across the water column), the dam breach erosion (transversal and longitudinal profiles of the breach as well as 3D reconstruction), the average pressures and the air concentration can also be measured.

For a long time, the laboratory measurement and recording of these variables was performed using the available measuring devices, such as:

(i) *limnimeters* with a nonious, to measure water levels; (ii) *piezometers* connected to tubular manometers, to measure average pressures along flow boundaries; (iii) *current meter* and *micro-current meter*, to measure flow velocities and (iv) *Bazin spillways*, for flow measurements.

Although all devices combine facility of using and good accuracy, they are mostly based on manual data acquisition procedures and require numerous discrete measurements to define the variables' fields, thus involving important human resources. Knowing that recent advances in dam breach tests are essentially due to the use of advanced technological instrumentation, measuring techniques and powerful computing, the present chapter focuses on this topic. It acts as a summary of the recent instrumentation (technologic advanced devices and measuring techniques) that can be currently used to measure and register hydraulic variables in dam breach tests with great accuracy.

Specifically, key instruments and techniques for measuring the flow rate, water levels, internal pressures and for the characterization of the flow kinematics and spatial surface velocimetry as well as the breach erosion (morphology and scour, through 3D reconstruction) will be presented in the next Sub-chapters. Generically, the following will be introduced: (i) instrumentation to continuously monitor *water levels* and *inflow*; (ii) methodologies to characterize the *surface flow velocimetry* (PIV and PTV) to obtain surface velocity fields in the breach vicinity and accurate velocity vectors at specific locations, respectively; and (iii) approaches to estimate the *breach morphology and scouring* evolution during failure (breach detection and 3D reconstructions of the failed dam).

Also presented in the next Sub-chapters, are some application examples of the developed methodologies in solving specific problems of dam breach experiments, as the calculation of *estimates of the breach effluent flow (BEF), non-locally and locally* (respectively based on the measurement of local and distant variables) and the understanding of the processes involved with occurrence of *mass detachment from the dam body* during the dam failure evolution.

Although most of the instrumentation and measuring techniques introduced in this chapter are presented under the context of earthen dam breach experiments, most can be used to other types of hydraulic experiments. In fact, some equipments presented along the following section were used in dam breach experiments but could have also been applied to other types of hydraulic experiments.

3 Experimental Facility and Instrumental Layout

Most of the advances in instrumentation and experimental procedures were devised, tested and ameliorated in successive dam breach campaigns carried out in a medium-scale laboratory facility at the Portuguese National Laboratory for Civil Engineering (LNEC). The reservoir is 31.5 m long and 6.60 m wide—Fig. 1. Scaled 0.45 m high earthen dams were subjected to overtopping. To estimate the breach effluent flow (non-locally and locally) and the local flow velocities, as well as to characterize the breach evolution process, the following quantities were monitored: (1) flow discharge entering the reservoir; (2) water levels in the reservoir, in the downstream settling basin and near the breach; (3) surface flow velocity fields near the breach; (4) breach geometric evolution, with both 2D and 3D cameras.

The instrumental layout required to characterize the latter variables was composed by the following devices, whose main characteristics and relative position are presented in Table 1 in Fig. 2, respectively:

- high-speed video cameras (CCD monochromatic);
- HD video cameras;
- 3D cameras (Kinect sensor);
- floodlights to enhance the contrast between the tracers (white) and the flow surface (black);
- high-power laser coupled to a cylindrical lens to create a laser sheet—Quantum Finesse; $P = 6$ W (continuum laser beam); $r = 1$ mm; $\lambda = 532$ nm (green light);
- mechanical dispenser of tracers to seed the flow (perforated pipes coupled to low intensity engines, filled with Styrofoam beads—$d_m = 3$ mm; $\rho = 1.05$ g/cm^3);

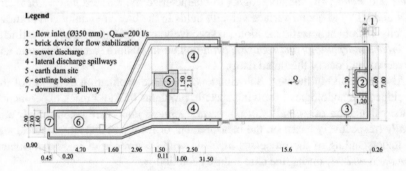

Fig. 1 Plan view of the experimental facility (dimensions in meters)

Table 1 Main characteristics of the instrumental layout

Video cameras	Lens/vision angle (°)	Resolution (px^2)	Acquisition rate (fps)	
HD Sony (DCR-SX53E)	/22.5°	720 × 576	25	
HD Canon	/28.5°	1440 × 1080	25	
Mikrotron	50 mm/F1,8–16/46°	1680 × 1710	4 frames (at 150fps every 1/4 s)	
Photonfocus	75 mm/F1,3/20°	1024 × 1024	1	
Kinect sensor				
– 1 colour sensor	/43° (vertical)	1920 × 1080	Variable	
– infrared depth sensor	/57° (horizontal)			
Laser model	Emitted light	Max theoric power	Operation temperatures	Cooling system
Finesse	λ = 532 nm	6 W	External: (15–32) °C Internal: (20–38) °C	Yes
Water level probes	Manufacturer	Model	Scanning range	Accuracy
Ultrasonic	Baumer	U500	100–1000 mm	<0.5 mm
Resistive	LNEC product	Wave height probes	100–1200 mm	<0.3 mm

- water level sensors (ultrasonic level sensors—Baumer U500 and resistive sensors) for water levels monitoring, distributed across the reservoir, around the dam breach site and inside the downstream settling basin.

A digital flowmeter was used to perform real time acquisition of the flow discharge at the reservoir inlet (inflow). This was manually controlled by a gate valve, placed at the entrance pipe (ø350 mm), Fig. 3.

Further details on the instrumental layout and its components can be found in [5–7].

4 Instrumentation and Measuring Methods

4.1 Water Levels and Inflow

Water levels and piezometric heights (average pressures) were traditionally measured by piezometers or limnimeters (Fig. 4a and b).

The measurement principle of piezometers is based on recording the water height attained inside the tube that works by the principle of the communicating vessels.

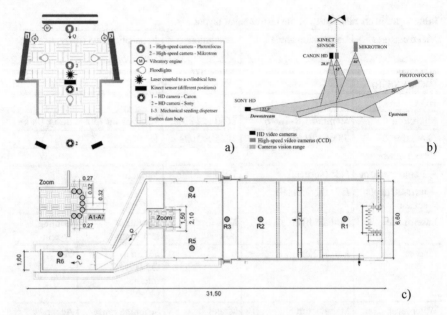

Fig. 2 Instrumental layout. **a** Top view; **b** lateral view of relative positions of the high-speed cameras and HD cameras; **c** water level sensors distributed inside the reservoir, around the dam and in the downstream settling basin (green—resistive sensors; yellow—ultrasonic level sensors)

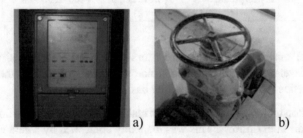

Fig. 3 Monitoring the flow discharge at the reservoir inlet during the failure tests: **a** digital flowmeter; **b** gate valve

Fig. 4 Water levels measurements: **a** piezometer; **b** limnimeter; **c** resistive level sensor; **d** ultrasonic level sensors

More recently, water levels have been preferentially measured with ultrasonic or resistive level sensors (Fig. 4c and d), being the former non-intrusive and the latter intrusive, as they operate under water, i.e., have to be submerged.

Ultrasonic level sensors use ultrasonic technology to generate a high-frequency sound wave to measure the time for the echo to reflect from the target and return to the ultrasonic liquid level sensor. The distance from the level sensor to the fluid is then calculated based on the speed of sound and the profile programmed into the sensor.

Resistive level sensors (also called conductive sensors) are used for point-level sensing conductive liquids such as water (or highly corrosive liquids). The working principle is based on the use of two metallic probes of different lengths (one long, one short) submerged in the flow. The long probe transmits a low voltage; the second shorter probe is cut so the tip is at the switching point. When the probes are inside the liquid, the current flows across both probes to activate the switch. One of the benefits of these devices is that they are safe due to their low voltages and currents. They are also easy to use and install but regular maintenance checks must be carried out to ensure there is no dust on the probe.

Although both ultrasonic and resistive level sensors allow the acquisition of accurate water level data, they still require denoising and depicting post-processing. Ultrasonic level sensors can be preferable when non-intrusiveness is required. This was the case of the dam breach experiments performed in the facility described in Sect. 3. Note that the goal of acquiring water levels around the breach zone would be the characterization of the surface curvature that should, preferably be obtained with non-intrusive measurements. This was possible by distributing a set of ultrasonic level sensors around the dam model as represented in Figs. 2c and 4d.

Other non-intrusive technologies for water level measurements are currently being tested at LNEC. These rely mostly on the application of image post-processing techniques applied to image data acquired with HD digital video cameras [8, 9].

4.2 Characterization of the Surface Flow Velocimetry

Flow velocities were traditionally measured by current and micro-current meters (Fig. 5a).

Doppler velocity sensors technology, namely Ultrasonic Velocity Profiler (UVP, Fig. 5b) and Acoustic Doppler Velocimetry (ADV, Fig. 5c), have been used in the last decades and showed many advantages when compared to more traditional measurements. In addition to not requiring any type of calibration, it makes an instantaneous measurement of the velocity value. On the other hand, as it detects the direction of the flow, the velocity vector is directly measured. This technology also presents some drawbacks: (i) it does not provide continuous observations; (ii) the slowness of the process, which requires the repositioning of the equipment at each of the measuring points; (iii) it requires the use of seeding; and (iv) it uses intrusive devices although the UVP, due to their small size, generate only a small disturbance.

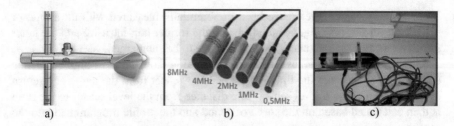

Fig. 5 **a** Current meters; **b** ultrasonic velocity profiler; **c** acoustic Doppler velocimetry

4.2.1 PIV—Particle Image Velocimetry to Obtain Breach Vicinity Velocity Fields

Velocity measurements based on PIV are a nonintrusive technique for flow visualization that has been successfully used in a wide range of flow conditions [10]. Generally, PIV is based on cross-correlation techniques from pairs of consecutive images [11] and it defines the velocity based upon the average particle motion in the possible space (interrogation areas), i.e. the Eulerian velocity [12]. To characterize the velocity field in depth, a coupled laser and a highly seeded flow would be required [13]. A PIV analysis typically includes the steps [14–16]: (i) image pre-processing (to increase quality and eliminate noise—and; (ii) image evaluation obtained from a cross-correlation method solved by either a direct cross correlation (DCC) or a discrete Fourier transform (DFT) approach; (iii) post-processing (to eliminate spurious velocity vectors, recalculate those by interpolation with valid neighbours and to calibrate the velocity map). The surface velocity field and the flow tendency near the breaching dam were measured in the dam breach experiments by using an algorithm of Particle Image Velocimetry (PIV). Surface velocity maps in the breach vicinity were based on the HD images acquired with a high-speed CMOS video camera (Mikrotron), placed perpendicularly at an appropriate distance of the flow surface so that the entire area of interest was caught (Table 1, Figs. 2a, b and 6), to monitor the tracers displacement (seeding between consecutive images). Such camera was chosen because of its high spatial resolution, low-light capacity, and camera shutter variability (down to 4 μs), which allows to optimize the images quality for PIV applications.

To provide an appropriate seeding concentration to the surface flow at a constant rate, a mechanical dispenser was manufactured, Fig. 2a. To increase the image contrast between the white of the styrofoam particles and the black of the surface flow (aiming the creation of high contrast images suitable for post processing analysis) and to avoid light reflections in the water surface that could induce spurious correlations, three spotlights 500 W were strategically placed above the dam, illuminating the surface flow in the breach vicinity, Fig. 2a. Seeding particles used as tracers consisted in styrofoam beads ($\Phi_{mean} = 3$ mm; $\gamma_{styrofoam} = 1.05$ gcm^{-3}).

The base images for the PIV application—Fig. 6a—were filtered for quality enhancement, namely for contrast heightening, noise reduction and peak removal

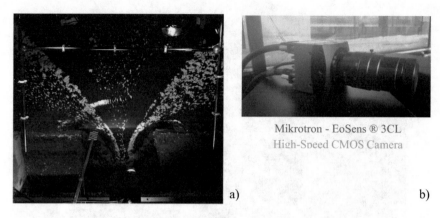

Mikrotron - EoSens ® 3CL
High-Speed CMOS Camera

a)   b)

Fig. 6 **a** Example of HD image of the breach flow with seeding particles for PIV application; **b** Mikrotron high-speed CMOS camera

[14]—*Step 1, image pre-processing.* The filtered images were the base for the estimation of the surface velocity maps with the free-use software PIVlab [15].

Step 2, image evaluation. The velocity maps in metric units were obtained by considering calibration images that took into account the distance variations between the camera position and the water surface—*Step 3, image post-processing*, this last being extracted from the data of the ultrasonic sensors placed around the breach and considering the curvature of the breach approach flow.

Figure 7 illustrates a velocity map obtained with a PIV application. It shows the base images with the definition of the calculation domain (after image pre-processing), the velocity map calculated with PIVlab software, with exclusion of low seeded areas or with high light reflections (image evaluation) and the final velocity map converted to metric units after calibration and interpolation processes (velocity maps post-processing).

4.2.2 PTV—Particle Tracking Velocimetry to Obtain Velocity Vectors at Specific Locations

Local velocities at specific locations were obtained in the dam breach experiments with a Particle Tracking Velocimetry (PTV) algorithm. This method was especially useful for the characterization of local velocity vectors where great accuracy was required. For example, the analysis of the velocity vector in the local marked with a red square, Fig. 8a, to perceive its variation during the detachment of a mass of soil from the dam body, does not require the analysis of the entire image because it possesses too much information. Instead, the focus of analysis can be reduced to the rectangle illustrated in Fig. 8b.

Figure 9 illustrates the application of a PTV analysis. Figure 9a results from the application of noise floor removal, Gaussian filter passage and image enhancement of

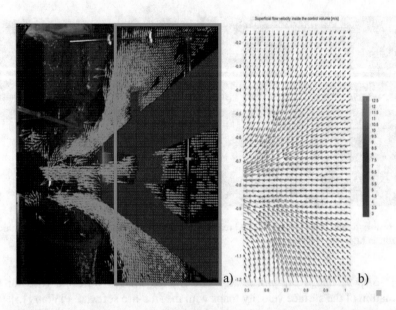

Fig. 7 PIV application example: **a** definition of the calculation domain of interest (blue rectangle) and calculated velocity vectors with excluded areas (masked in red); **b** final velocity vectors (after calibration and interpolation)

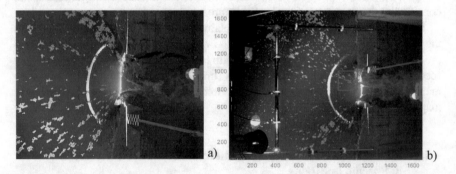

Fig. 8 Top view images of the breach approach flow: **a** block fall on the left with representation of the analysis position (red square) of the velocity vector; **b** focus of analysis and application of the PTV algorithm (red rectangle)

grey scale, showing the tracer's motion in detail. Figure 9b displays the peaks' location in consecutive images (A and B). Figure 9c shows the peaks' motion, between red and blue positions, during Δt. Note that this time interval (Δt) must be very small so that the nearest neighbor is still inside the consecutive image (B = A + Δt) and hence, this technique can be applied.

For this purpose, a high framerate was used to assure peaks' displacements among images were small. Figure 9d illustrates the surface velocity map that resulted from

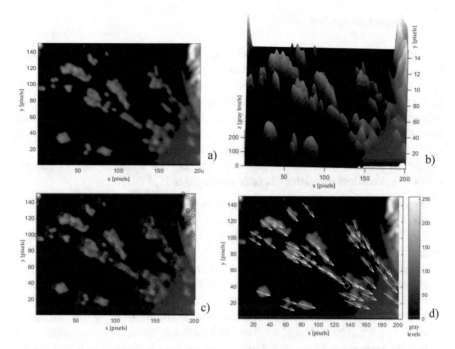

Fig. 9 PTV steps; **a** post-processed image **b** peaks' identification **c** peaks' location—images A (red circles) and B (blue circles); **d** PTV surface velocity map in image B

the PTV analysis. Conversion of the PTV surface velocity maps from pix/s to m/s was performed using the same approach described for the PIV analysis.

4.3 Characterization of the Breach Morphology and Scouring

Over decades of research, scouring has been surveyed by limited point-wise measurements at specific locations, by using either gauges and probe sensors or scales recorded by cameras inside transparent walls of experimental facilities. To the authors' best knowledge, few works have used non-intrusive techniques and high-resolution topographic measurements capable of characterizing the evolution of the breaching in embankment's dams, namely the use of image-based techniques to build three-dimensional models with significant spatial resolution for continuous monitoring of the failure process.

4.3.1 Breach Detection

Breach detection consists in determining the breach geometry, position and area during the experiments. The two different approaches for breach detection developed in this work consisted in the application of optical techniques and image analysis procedures to the digital images acquired with both monochromatic high-speed video cameras—Photonfocus and Mikrotron (Fig. 2a and b) with the aid of a laser sheet (Fig. 10) to detect the breach boundaries. Because the laser sheet is projected vertically, both approaches lead to vertical breach areas.

The breach area definition of each approach is represented in Fig. 11 and the respective procedure of calculus is fully described in [5, 17]:

Approach A. It corresponds to a vertical plane area contained between the laser sheet trace in the emerged dam body and the free surface; the section where the breach area is measured was located right upstream the initial dam crest; Figs. 11a and 12c shows this definition;

Approach B. It assumes the existence of a curved breach crest line with a constant elevation separating the dam upstream slope from the breach channel (as observed by [18, 19]) over which the flow passes at an approximately critical state (Fr = 1). Accepting the submerged crest line with an almost fixed height (dashed line—Fig. 11b), it can be collected from the intersection of the vertical laser plane with the dam body (grey zone—Fig. 11b). The water depth is dictated by a comparable thinking; in this approach B the transversal breach cross-section is the curved bright line in Fig. 13a, correspondent to the boundary where the level gradient starts to change (Fig. 13b), which was determined by image analysis through a thresholding algorithm.

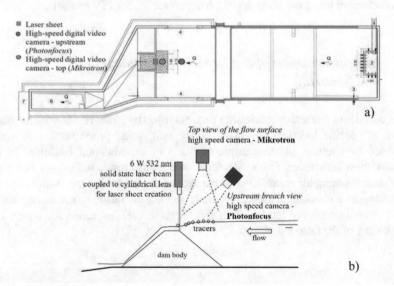

Fig. 10 Instrumentation for breach detection, **a** plant of the positions required for the data acquisition; **b** lateral view of relative positions of the high-speed cameras and laser

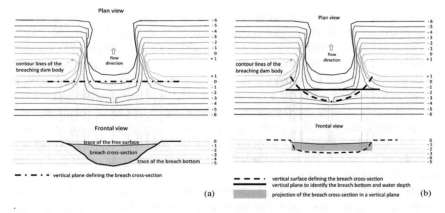

Fig. 11 Breach detection definition: **a** approach A; **b** approach B

Fig. 12 **a** Laser sheet projected in the dam, right upstream the crest; **b** image acquired with the upstream high-speed camera (Photonfocus—Fig. 10) illustrating the laser traces projected at the dam body and at the free-surface; **c** breach area determined by approach A

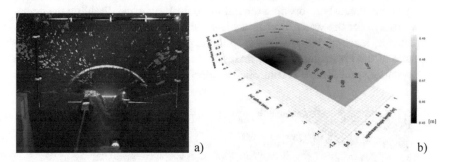

Fig. 13 **a** Image acquired with top high-speed camera (Fig. 10) illustrating the flow passing over the crest line; **b** water surface upstream the breach illustrating the crest line position (beginning of the gradient increasing)

4.3.2 3D Reconstruction of the Failed Dam—Kinect Sensor

Non-intrusive breach bottom measurements were performed with the aid of a motion sensor. The 3D reconstruction of the failed dams was based on RGB-D images acquired with a motion-sensing device—Microsoft Kinect v2. It encompasses a *Colour Sensor* for RGB images acquisition (640 × 480 pixels—30 fps) and an *IR Depth sensor* for depth images acquisition within a range of 800–4000 mm (640 × 480 pixels), Fig. 14. The latter determines the depth by emitting infrared light with modulated waves and detecting the shifted phase of the returning light allowing to construct a 3D object map. The accuracy and precision of the depth images are influenced by the environment temperature, camera distance and scene colour. Reliable results were obtained with a 25 min pre-heat of the device and the accuracy level was maintained by adopting a constant offset [20]. The scene colour influence was considered in the filtering approach since some of these rely on coinciding colour and depth changes.

The use of this sensor allowed characterizing the breach temporal evolution based on the 3D reconstructions of the failed dam obtained with the acquisition of depth images before, during and after the dam failure. For this acquisition, the following steps were considered:

1. a flow cut was required for each depth image acquisition, since breach is not detected by the device when the flow passes above it, mostly because of turbidity issues due to emulsified sediments that disturb underwater measurements;
2. breach morphologic characterization was achieved by placing the Kinect sensor at different focal points to carry out a complete sweep of the monitoring area (Fig. 2a); for alignment and georeferencing purposes it was assured that at least three target points with known coordinates are visualized from every sensor position (Fig. 14b);
3. after each image acquisition, the failure test was re-started and continued until the next stop considered significant for 3D reconstructing; test was restarted by simultaneously re-opening the inflow control valve and disabling the lateral spillways for the water level to achieve the pre-stop value; while not instantaneous, reaching the previous magnitude conditions depend on the advance state of the breaching process, taking 2–5 min; the breach erosion during the re-starting procedure was controlled by placing a metallic plate to avoid the flow

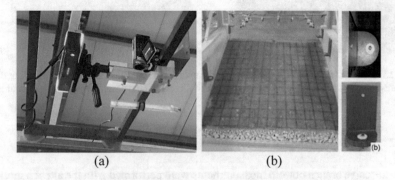

(a)   (b)

Fig. 14 **a** Kinect sensor above the dam; **b** position of the targets with known coordinates

to pass through the breach previously the pre-stop breach approach conditions were achieved;

4. several flow stops were performed during the period of interest of the experimental tests (i.e., during the occurrence of the largest mass detachment episodes and of the highest outflows); this stopping and re-starting procedure demonstrated not to affect the breach erosion process.

The 3D reconstructions of the failed dam were based on the Kinect sensor data. First, the Kinect depth images were converted into point clouds and second, these were processed in a 3D point cloud and a mesh processing software (Cloud Compare). The point clouds post-processing steps can be resumed in:

– *removal of outliers* (i.e., the points that clearly do not correspond to the embankment dam) and *identification of target points* in each point cloud;
– *georeferenciation*—assignment of the real coordinates in the target points identified in each cloud, hence assuring the same referential for all clouds;
– *alignment and merging of points clouds*—identification and matching of common points for a finer alignment between clouds and for covering the entire dam so that a unique georeferenced point cloud is achieved;
– *filtering*—passing of a noise filter to remove the outliers, i.e., points that are clearly measurement errors.

The application of Kinect sensor to perform a 3D reconstruction of the failure process of an embankment dam as here described allowed a deeper understanding of the embankment's erosion evolution, showing the failure to occur predominantly due to headcut erosion with infra-excavation, with occasional appearance of lateral erosion cavities (Fig. 15). Note that these infra-excavated structures (cavities) would be difficult to survey with only one top acquisition. At the same time, for tests to be able to continue after the stops, this type of information can only be obtained with a non-intrusive and fast acquisition equipment as the one used in this work.

Laser scanning. Another non-intrusive device proper for 3D reconstruction of dam breach tests is a Laser scan. In this sub-section an application of a laser scanning successfully applied to dam breach experiments is presented (Fig. 16).

A set of dam breach experiments were conducted in a flume of 60 m × 3 m × 3 m in the Changjiang River Scientific Research Institute, Changjiang River Water Resources Commission, China (Fig. 17). The measurement of the breach and downstream valley morphologic variations were successfully achieved with the use of a 3D Laser Scanner (Fig. 18)—in [21].

A complete sweep of the area of interest, in this case, the entire flume, was performed every 5 min until the end of the test. This method allowed observing a scour hole and the breach channel development during dam breach test (Fig. 18).

In general, both depth devices are suitable for monitoring dam breach tests, namely the progression of erosion and the breaching processes, as the headcut erosion and underscouring. Nevertheless, it should be remarked that the laser scanning is a very expensive equipment, with a high precision level, and is, hence, more suitable for detailed analysis; for the measurements with lower accuracy levels, Kinect sensor is a good economical alternative that suits the purpose.

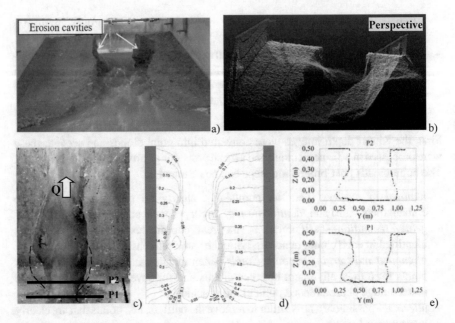

Fig. 15 Example of an embankment dam failure test (final stage): **a** detail of erosion cavities; **b** point cloud of the failed dam surface; **c** plan view of the georeferenced point clouds post-processed; **d** topographic contour map of the failure final stage frontal; **e** representation of cross-sections P1 and P2 (enhancing the existence of erosional cavities)

Fig. 16 Experimental flume for dam breach experiments located at the Changjiang River Scientific Research Institute. **a**) flume overview; **b**) lateral view near the dam site

Fig. 17. 3D scanner used in dam-breaching tests

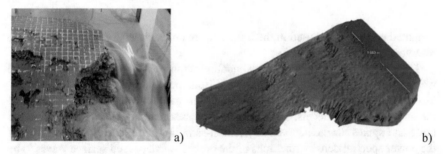

Fig. 18. Example of the 3D reconstruction of the dam breaching obtained from 3D scanner acquisition

4.4 Application Examples of the Developed Measuring Methods in Solving Specific Problems of Dam Breach Experiments

4.4.1 'Non-local' Estimates of the Breach Effluent Flow

The breach effluent flow (BEF) in dam breach studies is usually based on a mass balance, through the application of the continuity equation within the upstream reservoir (finite differences scheme below), or based on rating curves in known sections, usually downstream the dam site.

$$Q^{n+1} = Q_{in}^{n+1} - A^n \frac{z^{n+1} - z^n}{\Delta t} \tag{1}$$

where Q_{in} is the water inflow; Q is the breach effluent flow; $\frac{z^{n+1} - z^n}{\Delta t}$ is the time variation of the water level in the reservoir and A^n is the superficial reservoir area (a

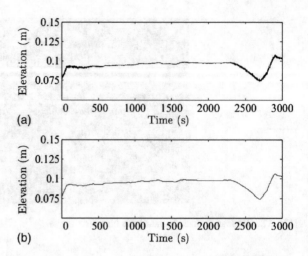

Fig. 19 Example of the signal of an ultrasonic level sensor located inside the reservoir: **a** unfiltered; **b** filtered

weighted contribution based in the influence area of each level sensor placed in the reservoir has been considered).

The application of the continuity equation requires the reservoir inflow and the water levels variations within the reservoir. Both digital flowmeter and level probes, distributed across the reservoir, allowed real-time data acquisition, i.e. time series of the inflow and water levels. The acquired data presented high levels of background noise and spikes for the adopted acquisition-sampling rate (10 Hz). The analysis of the power spectral density functions of the probes signal proved surface waves to be associated to 0.5–2.0 Hz frequencies.

To eliminate the high frequency electric noise and the small amplitude surface waves (to sufficiently smooth the signal for it to be numerically differentiated) a signal filtering based on the application of a Butterworth low-pass filter process was applied followed by the application of a 'moving average' (Fig. 19). Despiking process consisted in the application of a threshold criterion applied to detrended data (trend was after added back). The reliability of the acquired data was improved by performing a sensitivity analysis of the signal filtering (denoising) and applying a local average technique ('moving average'). The order and cut-off frequency of the filter were chosen according to the combination that best suited the filtering goals.

4.4.2 'Local' Estimates of the Breach Effluent Flow

Two novel methods for the estimation of the breach effluent flow based on local data, i.e., data collected near or at the breach site, were developed. These consisted, generically in the product of flow velocities with the breach area, as illustrated in the following equation:

$$Q = \int_A u_s \cdot n dS \approx U_n A \tag{2}$$

where u_s is the surface velocity at the breach section, n is the unit normal vector pointing out of the reservoir, A is the breach flow area and U_n is the depth-averaged flow velocity normal to the breach section.

As referred, both these methods were based on data acquired near the dam site, differing only in the breach area approach—A or B (Sect. 4.3, Breach detection). The breach effluent flow was gathered at the entrance of the breach channel, using the definition of a flux tube discharge. As input data these methods required (i) accurate predictions of the breach flow area (as described in Sect. 4.3, Breach detection); and (ii) surface flow velocity maps in the breach vicinity (as described in Sect. 4.2, PIV).

4.4.3 Example of Both Estimates of the BEF ('Non-local' and 'Local')

Figure 20 shows an example of 'non-local' estimates (black and grey continuous lines) and 'local' estimates (black, thick and thin-dashed lines) of the breach effluent flow calculated as described in Sect. 4.4 'Non-local' and 'Local' estimates of the breach effluent flow. A red dashed line symbols the occurrence of one mass detachment from the dam body as the one illustrated in Fig. 21. Note that this kind of occurrences integrate the main failure mechanisms present in the failure evolution of cohesive embankment dams as those used in the experimental tests program.

Consistency between novel estimates (A and B—Fig. 20), as well as analogous evolution between these and the 'non-local' ones was found, Fig. 20b. Method B, in particular suits well non-local estimates in both progression and quantity. This adequate fit was considered as a validation of both instrumentation and experimental methods used in this study for the calculation of the breach geometry and flow velocimetry.

In particular, even during mass detachments from the dam body, both types of estimates were consistent in what concerns the flow behaviour, showing a continuous progression with no occurrences of local flow peaks. This was interpreted as a fact, that the discontinuous erosion really does not cause any flow discontinuities (i.e., sudden peaks), enhancing the credibility of both estimates and allowing to assume that both give achievable results.

4.4.4 Breach Hydrodynamics and Morphologic Evolution

As referred in the example of both estimates of the BEF, Sect. 4.4 the breach discontinuous erosion originated by the mass detachments do not induce discontinuities in the flow (sudden peaks). As this is counterintuitive, it raised the question: '*If BEH is*

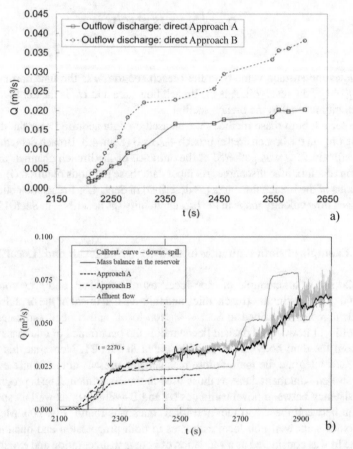

Fig. 20 Breach effluent flow: **a** 'local' estimates—based on approaches A and B of the calculation of the breach flow area (Sect. 4.3.1 **Breach detection**); **b** comparison between '*non-local*' (downstream rating curve and mass balance) and '*local*' estimates

Fig. 21 Mass detachment episode during the breaching evolution represented by the dashed red vertical line in Fig. 20: **a** crest view **b** Frontal view of the dam downstream slope

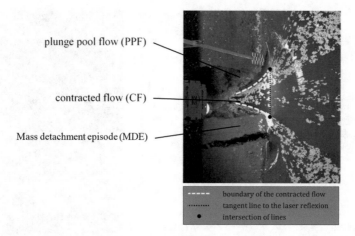

Fig. 22 Flow structure and morphology of the channel over the breach

being well measured, why do the discontinuous erosions not originate flow discontinuities? The analysis of the breach hydrodynamic and the corresponding morphologic evolution during the occurrence of mass detachments allows answering this question. This section presents the methodology adopted to perform this analysis and an illustrative result (Sect. 4.3, 3D reconstruction of the failed dam).

For this purpose, based on an idealized flow structure depicted in Fig. 22, the following variables were monitored:

1. *geometry of the contracted flow* (flow width—black dashed line in Fig. 22— measured above the trace of the laser sheet over the high quality images acquired with the CCD camera placed above the dam—Figs. 2a and b, 6 and Table 1);
2. *kinematic field next to mass detachments* (velocity vectors at specific locations— near the mass detachments—estimated through PTV as described in Sect. 4.2 and illustrated in Fig. 8, because PIV revealed to be insufficient for the accuracy required for this particular analysis and PTV-estimated velocities increased the amount of phenomenological information that could be extracted from a dam breach—[22]);
3. *mass detachments of soil—location, dimension, and cause of occurrence* → characterization of the erosion below the breach (based on the 3D reconstruction of the failed dam presented in Sect. 4.3).

Based on the results of the monitored variables, it was possible to evaluate the *impact of mass detachments on the breach flow*, i.e. the affectation of the kinematic field and of the contracted flow width. It was observed that the reference section where the breach effluent flow is controlled is not dependent on the balance between erosional and depositional processes, as usually adopted in parametric breaching models, but rather only on its purely erosive characteristics predominantly due to tractional erosion.

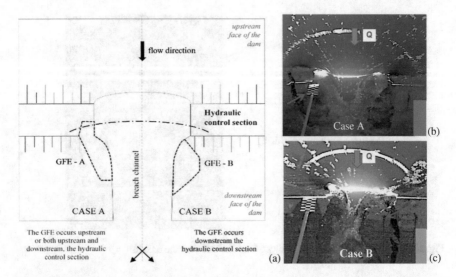

Fig. 23 **a** Conceptual model for the expected response of the breach flow hydrodynamics; **b, c**) mass detachments visualized from the top—cases A and B

It was also observed that the *breach effluent flow may, or may not, be affected by sudden mass detachments of soil* (marked with red arrows in Fig. 23) from the dam body, depending on whether these occur in a proximal (Case A—Fig. 23a, b) or distant zone (Case B—Fig. 23a, c) from the reference section (hydraulic control section), respectively, allowing to develop a conceptual model for the breach flow hydrodynamics.

In general, it was observed that mass detachment episodes occur alternately (from the right and left sides of the breach channel), as illustrated in the images of a dam breach test—Fig. 24. This sequence also reflects that the detached masses become larger as the breach failure progresses, i.e. as the breach effluent flow increases.

The hydrograph of Fig. 25 shows that at the first failure stages, when the discharge flow is approximately constant, the fall of the blocks occur at an approximately constant cadence, being only when BEF starts to increase, meaning that the failure is in full evolution stage, that the time interval between the falling of these blocks turns larger.

The study of breach morphology evidenced that the mass detachments are mostly originated by underscouring, which becomes more intense as discharge increases. This evidence was based on the 3D reconstructions of several instants of the failed dam, as exposed in Sect. 4.3, since it allowed to characterize and observe the phenomena occurring in depth in the dam body (in the underwater portion of the dam but also in the area underneath the overhangs created by the underscouring—Fig. 26c).

In general the qualitative step that was given in this work due to the use of novel instrumentation and methods allowed a deeper understanding of the embankments erosion evolution, showing the failure to occur predominantly due to headcut erosion

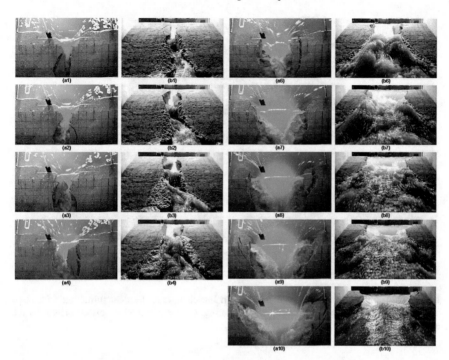

Fig. 24 Sequence of mass detachment episodes represented with red vertical lines in Fig. 26 illustrating its alternate behaviour. Crest view (**a**); downstream slope view (**b**)

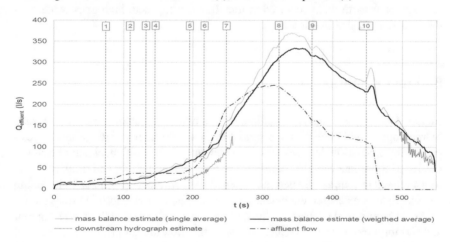

Fig. 25 Detail of the estimates of the breach effluent hydrograph of a dam breach experiment based on non-local measurements and reservoir inflow. Representation of the instants of occurrence of mass detachment events (dashed red vertical lines)

Fig. 26 Erosion processes observed in the dam breach tests: **a** headcut formation with steps and pools; **b** erosion cavity created by underscouring; **c** mass detachment episode caused by the underscouring

(Fig. 26a) with infra-excavation, with occasional appearance of lateral erosion cavities (Fig. 26b and c). This finding is in agreement with the breaching process of real homogeneous earth dams reported in literature (mostly with high fines content, as the earth dams tested in this study).

5 Final Remarks

In this Chapter, the instrumentation, and methods to characterize the breach morphology, flow kinematics and spatial surface velocimetry through image post-processing analysis techniques was fully expounded. In addition, signal filtering of level and inflow data was also exposed. The potential, limitations and results of those measurement techniques are herein examined and conveniently compared, including a description of measurement devices, reference points, and respective software.

Novel estimates of the breach effluent flow, the visualization of the breach contour and 3D morphology at a local scale, generic surface flow fields and more detailed local velocity vectors as well as 3D reconstructions of the failed dam were possible thanks to the experimental methods and instrumentation that composed the monitoring layout developed in this study.

In particular, the use of a high-power laser sheet combined with visualization through high-speed video cameras strategically placed and with a constant addition of seeding particles permitted to extract the breach area and the surface velocity field in the breach vicinity, allowing to develop novel estimates of the breach outflow discharge flow based on proximity measurements.

The use of Kinect sensor to reconstruct the 3D breach morphology during the failure revealed to be a further advance on the current abilities to characterize the process of embankment dams breaching allowing to observe the underscouring erosion mechanism. These image-based techniques allowed significant insights into the whole breach geometry evolution and the image analysis detection methods used revealed to be promising to extract important information from dam breach experiments difficult to obtain with the common methods.

The advanced instrumentation and monitoring techniques applied to dam breach allowed to model in experimental environment the main hydrological and geotechnical processes usually observed in the failure of homogeneous embankment dam's prototypes, which could not be highlighted by traditional measurement devices. Particularly, it was found that those advanced tools can capture previously hidden details of the mass flux trough the breach allowing to develop more accurate conceptual models for the breach flow hydrodynamics.

These tools were used and/or developed under the scope of this work envisaging to solve concrete problems in what concerns the measurement of important variables to characterize the dam failure process. Other studies that might lean over different problems from those studied here might require different combinations of the local variables here exposed but can still use the developed methods. As a general conclusion, the optical methods herein used proved to be an excellent choice for applications for non-intrusive and local dam breach measurements or reliable alternatives to more traditional approaches.

References

1. ICOLD: In: International Commission on Large Dams—Bulletin on Risk Assessment in Dam Safety Management (ICOLD) (2003)
2. ICOLD: https://www.icold-cigb.org/article/GB/world_register/general_synthesis/general-syn thesis (2019)
3. ICOLD: Dam failures statistical analysis, (Bulletin 53) (1995)
4. Powledge, G.R., Ralston, D.C., Miller, P., Chen, Y.H., Clopper, P.E., Temple, D.M.: Mechanics of overflow erosion on embankments. I—research activities. J. Hydraul. Eng. **115**(8), 1040–1055 (1989)
5. Bento, A.M., Amaral, S., Viseu, T., Cardoso, R., Ferreira, R.M.L.: Direct estimate of the breach hydrograph of an overtopped earth dam. J. Hydraul. Eng. **143**(6) (2017). https://doi.org/10.1061/(ASCE)HY.1943-7900.0001294
6. Amaral, S.: Experimental characterization of the failure by overtopping of embankment dams. Ph.D. Thesis, Instituto Superior Técnico, Universidade de Lisboa (2017)

7. Amaral, S., Caldeira, L., Viseu, T., Ferreira, R.M.L.: Designing experiments to study dam breach hydraulic phenomena. J. Hydraul. Eng. **146**(4), 04020014 (2020). https://doi.org/10.1061/(ASCE)HY.1943-7900.0001678
8. Amaral, S., Ferreira, R.M.L., Viseu, T.: Experimental methods for local-scale characterization of hydro-morphodynamic dam breach processes. Breach detection, 3D reconstruction, flow kinematics and spatial surface velocimetry. Flow Measure. Instrum. **70**, 101658 (2019). https://doi.org/10.1016/j.flowmeasinst.2019.101658
9. Amaral, S., Alvarez, T., Viseu, T., Ferreira, R.M.L.: Image analysis detection applied to dam breach experiments. In: 5th IAHR Europe Congress—New Challenges in Hydraulic Research and Engineering. Trento, Italy (2018)
10. Harpold, A.A., Mostaghimi, S., Vlachos, P.P., Brannan, K., Dillaha, T.: Stream discharge measurement using a large-scale particle image velocimetry (LSPIV) prototype. Trans. ASABE **49**(6), 1791–1805 (2006)
11. Kantoush, S.A., Schleiss, A.J., Sumi, T., Murasaki, M.: LSPIV implementation for environmental flow in various laboratory and field cases. J. Hydro-Environ. Res. **5**(4), 263–276 (2011)
12. Umeyama, M.: Coupled PIV and PTV measurements of particle velocities and trajectories for surface waves following a steady current. J. Waterw. Port Coast. Ocean Eng. **137**(2), 85–94 (2010)
13. Tropea, C., Yarin, A.L., Foss, J.F.: Springer Handbook of Experimental Fluid Mechanics, vol. 1. Springer Science & Business Media (2007)
14. Raffel, M., Willert, C.E., Kompenhans, J.: Particle Image Velocimetry: A Practical Guide. Springer Science & Business Media (2007)
15. Thielicke, W., Stamhuis, E.: PIVlab—towards user-friendly, affordable and accurate digital particle image velocimetry in MATLAB. J. Open Res. Softw. **2**(1) (2014)
16. Adrian, R.: Particle-imaging techniques for experimental fluid-mechanics. Annu. Rev. Fluid Mech. **23**(1), 261–304 (1991)
17. Amaral, S., Viseu, T., Ferreira, R.M.L.: Estimates of breach effluent flow from the failure by overtopping of homogeneous earthfill dams based on non-local measurements. In: 38th IAHR World Congress. 1–6 September 2019. Panama City, Panama (2019)
18. Coleman, S.E., Andrews, D.P., Webby, M.G.: Overtopping breaching of noncohesive homogeneous embankments. J. Hydraul. Eng. **128**(9), 829–838 (2002)
19. Feliciano Cestero, J.A., Imran, J., Chaudhry, M.H.: Experimental investigation of the effects of soil properties on levee breach by overtopping. J. Hydraul. Eng. **141**(4), 4014085 (2015)
20. Wasenmüller, O., Stricker, D.: Comparison of Kinect v1 and v2 depth images in terms of accuracy and precision. In: Asian Conference on Computer Vision. Springer, Cham (2016)
21. Zhao, G., Visser, P.J., Peeters, P.: Large Scale Embankment Breach Experiments in Flume. Report of Delft University of Technology, Flanders Hydraulics Research and Rijkswaterstaat (2013)
22. Orendorff, B., Rennie, C.D., Nistor, I.: Using PTV through an embankment breach channel. J. Hydro-Environ. Res. **5**(4), 277–287 (2011)

Bridging the Water Gap Between Neighboring Countries Through Hydrometeorological Data Monitoring and Sharing

Rui Raposo Rodrigues

Abstract Hydrology is a data driven science. Its birth as a science in the late seventeenth century is asserted by the measurements of rainfall and river flow in France establishing the first experimental relation between these two physical quantities. Today highly sophisticated models mimicking the atmosphere-soil-vegetation interactions are used to simulate the flows but its proper calibration depends on good hydrometeorological data. New measuring techniques, like satellite or radar rainfall estimation or Acoustic Doppler Current Profilers (ADCP) for river flow determination, have enhanced the spatial resolution of both estimates allowing, at the same time, a more real-time insight into the rainfall-runoff phenomena. Data storage and dissemination capabilities are also evolving, allowing on-line data transfer between people, institutions and countries. This latter capability combined with the real-time perception of the water status in wide areas coming from hydrometeorological data measurements, improves the capacity for better and faster water management decisions to be taken. Portugal and Spain have signed a treaty in 1998 on Shared Waters which is still being improved steered by new data findings and its sharing. The present paper tackles the issue of what the experience of using new data gathering and transmission brought to the water management coordination between Spain and Portugal (e.g. on flood and drought management), as a good example on finer water-related collaboration.

Keywords Hydrometeorology · Weather sensors · ADCP flow measurement · Transboundary water cooperation · Information sharing · Albufeira convention

R. R. Rodrigues (✉)
National Laboratory for Civil Engineering (LNEC), Lisbon, Portugal
e-mail: rjrodrigues@lnec.pt

© The Author(s), under exclusive license to Springer Nature Switzerland AG 2023
C. Chastre et al. (eds.), *Advances on Testing and Experimentation in Civil Engineering*, Springer Tracts in Civil Engineering,
https://doi.org/10.1007/978-3-031-05875-2_14

1 Historical Framework

1.1 On the Origin of Fountains

Hydrology is the science that encompasses the study of water on the Earth's surface and beneath the surface of the Earth, the occurrence and movement of water, the physical and chemical properties of water, and its relationship with the living and material components of the environment [1]. All these phases—from the occurrence and distribution to transport—take place in a water cycle, which is a continuous process by which water is lifted by evaporation and transported from the earth's surface (including the oceans) to the atmosphere and back to the land and oceans.

This simple and nowadays completely assimilated concept is however quite recent in terms of the Earth Sciences' history. Up until the end of the seventeenth century the theoretical concepts on the occurrence of water being taught at the Jesuits' colleges were still the same principles recovered during the Middle Ages where Plato and Aristotelian thoughts were mixed with Lucrecius and Seneca's own speculations as well as other post-Aristotelian thinkers.

In short, these medieval concepts, where speculation took precedence over observation, refused to link the occurrence of rain to the flowing waters in the surface water bodies, mainly because it seemed that there was much more water in the streams and for longer periods than in the rainfall events. Instead of exploring this link the view put forward by Athanasius Kircher in his geology textbook "Mundus Subterraneus" created an exaggerated reality that credited the sea as the main source capable of feeding the rivers by way of underground channels, where the seawater was lifted from underneath the ground until the surface, losing the salt on the way and acquiring water qualities in conformity with the mineral substances encountered enroute. Since Aristotle recognized that the caves helped to condensate moisture, the idea that the underground channels ended in caves was easily merged, as depicted in Fig. 1.

Additionally, various theoretical processes for getting water to flow uphill (from the sea level to mountain tops) by mechanical methods were explained by Kircher with elaborate diagrams [2].

From a speculative point of view one must recognize that the presence of water in the streams long after the occurrence of the last rains did not encourage any attempt to draw some relation between the two phenomena but a more attentive observation would have also shown that there is not a single spring issuing from any summit without an appreciable surrounding area above it.

But the second half of the seventeenth century was also the era of the 'Scientific Revolution' that witness a new worldview that followed Johann Kepler's principle '*to measure is to know*' based on commitment to observation and reason unconfined by the requirement to square with religious doctrine [3]. Curiously, just 10 years after Kircher's book drawn from scholastic Aristotelianism was printed, a revolutionary publication appeared dealing with the origin of fountains by Pierre Perrault [4] that sparked interest among the scientific community. On that seminal book it was proven that '*only about one-sixth part of the rain and snow water that falls is therefore needed*

Fig. 1 Conceptual model for keeping the rivers continuously flowing: *continua fluxus & refluxus* (in Athanasius Kircher's Mundus Subterraneus—Liber Quintus, 1665—books.google.com)

to cause this river [the Seine] to flow continually for one year'. This was done by measuring directly the rainfall over one year in the headwaters of the Seine River coupled with some judgement (by an upscaling of the flow measured in a smaller river near Versailles) about the amount of water flowing in the Seine from the headwater to the first tributary junction 13 km downstream.

The scientific method undertaken by Perrault can be translated in what may be labeled as the first known hydrological equation

$$Q = P/6 \tag{1}$$

where P is the annual rainfall and Q is the corresponding annual flow value. Perrault measured rainfall in French inches but translated the amounts in volume units by multiplying the area of the headwaters draining to the river reach by the accumulated inches of annual rainfall, thus introducing the concept of drainage basin as a storage reservoir. In this way the variables in Eq. (1) can either be set in volumetric flux units (in $L/m^2 = mm$) or in units of volume (in $dam^3 = mm \ km^2$).

A few years later Edmé Mariotte and Edmond Halley consolidated Perrault's achievements: the previous via his work "Treatise on the movement of water and other fluids" (1686) that brought into the calculations the field measurements of river flows in the Seine River at Paris—where floats were used to determine the velocities in the channel—whilst the latter through his paper presented to the Royal Society with measurements on evaporation, allowing him to ascertain that *'enough evaporation takes place from the oceans to more than replenish all springs and rivers'* [2], demonstrating the closure of the new concept of the water cycle.

The new supportive estimates coming from Edmé and Halley's research allowed to frame Eq. (1) in a more general water balance configuration as

$$Q = P - E \tag{2}$$

where P is the annual rainfall, Q is the corresponding annual flow and E is Evapotranspiration (a variable already considered by Perrault as a loss term affecting rainfall).

So, by the beginning of the eighteenth century three physical variables have been identified as important to be monitored when one wishes to describe the moisture flow in the hydrological cycle, overruling the Aristotelian principle on the fundamental moisture source.

Throughout the development of the 'theory of fountains' a new auxiliary concept was introduced to assist in the calculations: the hydrologic basin area delineation; although, at this inception stage, the watershed was not yet used within a 'contributing area' rationale for flow generation but just as a storage capacity concept for flow accumulation and rainfall losses. That was not a big problem at the end of the seventeenth century since the measurement accuracy was only expected to be precise enough to prove the volumetric disparity between rainfall totals and accumulated flow.

This undemanding characteristic allowed some acquiescence towards gross disparities when referring to areal determination of river basins' extension. That's exactly what happened when referring to the basin areas studied by Perrault and Mariotte. Biswas [2] reproduces an illustration with the limits of the region studied by Mariotte that encompasses an area of the Seine basin smaller than it should have been considered, once it does not comprise the Marne River as a tributary. As the flow measurements made by Mariotte were performed near Pont Royal [2], an extra area of almost 13,000 km^2 should have been considered. Since Mariotte described the basin area as 60 leagues by 50 leagues the only way these dimensions would match the area of the contour depicted in the illustration was assuming a very low value for the French lieue (1/6 less than the lieue de Paris). Probably the picture published by Biswas was just used as a sketch to point at the disparity between the area initially considered by Perrault (118 km^2) and the broader region considered by Mariotte (373 time bigger) without repercussions on the units conversion. Alternatively, it could be that the cartography at the end of the seventeenth century as well as the basins limits delineation was not accurate enough to allow for a better reliability as to areal quantification.

With these accuracy limitations Eq. (2) is still far from the more common equation for water balances from temperate climates on a time scale shorter than a year (generally for monthly time steps) which is

$$Q = P - E - \Delta M - \Delta GW \tag{3}$$

where the variation in both soil moisture content (ΔM) and groundwater (ΔGW), either by recharge or depletion, sets the relevance of the non-linear underground component contributing to the water balance.

The need to tackle with both time scales and spatial scales was beginning to steer the future of hydrological research and monitoring.

1.2 *Infiltration and Groundwater Components*

The other scholastic principle that was still pending for revision—after the contribution of Perrault, Mariotte and Haley to "Quantitative Hydrology" (as coined by James Dooge)—was thus the one related with the infiltration mechanism and groundwater recharge within the water cycle.

Seneca, who followed on the Greek mind setting over the origin of springs, assumed that water could not percolate deeper than 10 ft below the surface [2] and, Kircher did not conceived groundwater fluxes other than those ascending in conduits (Fig. 1).

Perrault did recognize that the rainfall accumulated in a river basin should serve not only to replenish the springs but also to supply losses from direct flow in the rivers (such as evaporation), but groundwater recharge was not one of such losses, since he did not believe in general infiltration of rain water [2]. Thus, the principle of massive solidity of the Earth underground was still very much alive at the end of the seventeenth century.

Dalton a century later (1802) made an important contribution to the validation of the percolation concept by building a lysimeter to measure the evaporation from the soil by controlling the water infiltrated [5] and Darcy, in his report from 1856 on the water supply for the city of Dijon, presented his formula of flow through a porous media with its linear relation between velocity and hydraulic gradient, based on careful field and laboratory observations on filtration [2].

Finally in the 1930's Horton made several contributions to the infiltration theory that culminated in his famous infiltration formula [6] of utmost importance to hydrological modeling once it allows to extract from the whole rainfall event the portion of rain solely responsible for the overland flow (called net rainfall or excess rainfall). This makes it a very powerful tool in flood studies as the fraction of rainfall that exceeds the infiltration threshold values is linearly correlated with the quick-flow hydrograph's values. This is especially important during flood events where the groundwater contribution to the peak flow is usually small.

When Horton's infiltration formula was coupled with another new concept developed in de 1930's—Snyder's Unit Hydrograph—Hydrological Modeling was finally born.

2 The Instrumentation Rise

2.1 Spatial Representativeness of Estimates

From the previous historical scattered notes it is perceivable how perfectly hydrology is depicted when coined as a data driven science. In line with this assertion emerge the issues of time and spatial scales.

As the time-step of analysis of the water cycle becomes shorter (progressively closer to the duration of the rainfall event itself) new perceptible volumes come into play in the water balance calculations, such as the amount of interception of rainfall by vegetation or the rainwater retention in depressions and ponds. Accordingly, this increases the number of variables that have to be considered in Eq. (3). The more generalized form of the water balance equation should thus be written as

$$dS/dt = P - Q - E \tag{4}$$

where dS/dt is the change in water storage (S) in the basin, which can be considered zero in the case of Eq. (2), or can translate the accumulation/depletion episodes of sub-surface and underground storage such as referred in Eq. (3), but it can also refer to other types of storage variations such as those occurring on lakes (natural or man-made) and in small surface depressions, or those impacting accumulated snowpack.

The same goes for the spatial scale: the smaller the area being modeled, the more physically meaningful the parameters of a flow model can be. As area increases more averaging takes place, and the more the parameters and the variables, such as infiltration or soil moisture, become averaged entities [7]. Thus, the conceptual structure of a model designed to simulate the flow is very dependent on the type and quantity of hydrological data available and of the way that data depicts the river basin moisture's flux exchange.

Usually when analyzing a water resources system within a data scarce region (and when short-term estimates from a hydrological study are to be made before a proper measuring network can be installed and enough data can be collected), the common procedure is to resort to models readily available, normally derived from data-rich river basins, without any chance to learn from your local hydrological system by *listening to the data*. Most of the time this environmental transfer procedure performs relatively well if regional values of precipitation are used in combination with simple adjustments made to the soil parameters of the model and to land cover specifics. But transferring models derived from data which are peculiar to a specific environment may not correctly reproduce the water fluxes existing in river basins from different geographical portions of the Earth, or the ground-water flow paths within geological specific environments. In some geographical zones snow fall plays an important role and snowmelt can be responsible for the highest percent of river flows. Karst or volcanic environments require also special conceptualizations and monitoring specifications [8–10]. This signals that in some particular cases models may not have been tested enough in their water flux conceptualization.

Dawdy [7] summarizes some of the most peculiar hydrological areas where research is still needed both in hydrological modeling and specific data gathering, such as: low-relief regions (where the infiltration drainage component overcomes the topographically sloping drainage); arid and semi-arid regions (where rainfall is sparser and less permanent in time than in humid regions, and river channel losses from infiltration are a critical factor in flow calculation); Artic regions (where thawing of the active layer in the permafrost region deepens continuously from spring to late summer due to the seasonality of its energy supply, causing a change in the storage capacity for soil moisture, and where the channels are choked with snow and ice when melt starts to run off); and, Snowmelt-runoff in temperate mountainous regions (which have distinct approaches from the Artic low relief regions).

2.2 The Need for Networking

As seen in the first section of this paper monitoring had an important role in disentangling science from speculation but, more importantly, it steered the conceptualization theory in Hydrology (and still does).

The systematic monitoring of meteorological and hydrometric variables of the hydrological cycle began with the creation of the first meteorological and hydraulic services in the second half of the nineteenth century in most European countries as well as in the American continent. The proliferation of measurements in the twentieth century contributed to the accumulation of data from different geographical settings that contributed in turn to a better knowledge of the hydrological processes.

Up to the present day, data series have been building up from an ever-increasing network of points enabling the records to encompass more extreme phenomena (floods and droughts) and in turn to perform better and more accurate statistical studies, allowing the level of uncertainty to decrease as the instrumented periods grow longer.

However, as the time-series expand the more likely is to arise some homogeneity problems on the data sets coming from each measuring point due to inevitable changes in the personnel attached to field information retrieval (usually long-time local residents of the area) and in the replacement of measuring instruments or changes in their exposure or obstruction.

There are also homogeneity issues related to the spatial representativeness of the hydrometeorological phenomena, once the majority of the measuring stations were historically placed on populated easy-access regions. Regions of difficult access, where it was not easy to find people capable of taking readings on a daily basis and changing the paper register from the devices, remained poorly monitored—in some remote areas it was sometimes possible to find a school teacher or a priest to make the readings. The bad quality of extrapolated estimates for mountain regions are a good example of this non-representativeness and truncated vision of reality, contributing to an underestimation of the magnitude of the hydrological phenomena in the headwaters of river basins, as these generally have higher rainfall amounts than low-lying areas.

The importance of the peculiar hydrological areas with specific data gathering needs (summarized by Dawdy [7]) was revealed using the meagre data retrieved from hydrological networks developed during WMO's Hydrological Decade Program (1965–1974). That program aimed among other objectives at developing technical regulations, the standardization of observing methods and instruments and the routine exchange of hydrological data, which contributed to develop what is known as Operational Hydrology, as well as to stimulate the increase on hydrological monitoring activities around the World and the share of hydrological information.

The Hydrological Decade Program was an invaluable project to exchange knowledge and data. Hydrological data, notably flow data, was not the object of current data exchange during the first half of the twentieth century as opposed to meteorological data where exchange had already been occurring since the second half of the nineteenth century. In fact, weather data networking was boosted in mid-nineteenth century by urgent necessity, as is usually the case, although international cooperation was already brewing in 1853 when the first International Meteorological Conference was held in Brussels with representatives from Belgium, Denmark, France, Great Britain, the Netherlands, Norway, Portugal, Russia, Sweden and the USA [11]. On this Conference the parties agreed on a standardized ship's meteorological logbook and instructions as well as to specify the parameters to be measured laying the foundation stone for the interchange of information. Curiously, several international projects have recently been involved with gathering and scanning handwritten ships logbooks' data (their location and detailed weather observations) into computer-accessible formats to study previous climate [11].

But what launched the development of an international network of synoptic weather stations was an unusually violent storm that crossed the Black Sea-Crimea area from the southwest toward the northeast, causing heavy losses and damage to the ships engaged in the siege of Sebastopol. The British lost 21 ships or vessels and additional ones were dismasted; the French lost 16, including the steamship Le Pluton and the battleship Henri IV the "pride of the French Navy" [12]. In the aftermath of the disaster the scientifically minded French Minister of War asked the newly inaugurated director of the Paris Observatory, Le Verrier, to determine if the path of the storm could have been foreshadowed and the Navy warned in time. At the end of November Le Verrier wrote to various European astronomers and meteorologists, requesting them to send any observations they might have made on atmospheric conditions between the 12th and the 16th of November 1854 and received 250 replies [11]. He then charged is chief of the Service Météorologique International at the Observatoire, Emmanuel Liais, to process the data who finishes by interpreting the Black Sea storm as the propagation throughout Europe of a 'wave of pressure' that could have been tracked and anticipated using the telegraph [13]. On 19 January 1855 Le Verrier submitted a preliminary analysis to the Minister for Public Instruction and at the end of the year, after another French warship was dashed onto the rocks between Corsica and Sardinia, the French observation network was established comprising 24 stations that were progressively augmented with other European countries stations using the telegraph as the communicating system [11].

Besides organizing a weather observation system that used the telegraph to centralize all relevant data gathering Le Verrier also took care of the important task of data broadcasting. In this regard he launched the Daily Bulletin of the Imperial Observatory of Paris providing an accurate, quantified summary of the weather in France and throughout Europe—from Lisbon to Moscow, and from Leghorn to Helsinki. By September 1863, the Observatory's Bulletin began to include morning maps of isobars and winds, and during the 1860s and 1870s most of the Paris daily press was receiving it [13].

Adding flow measurements to the list of shared data became also an important issue throughout the twentieth century, especially on river management of shared basins but only recently has it been taken into consideration by international agreements on shared waters covering large regional world areas. For the Convention on the Protection and Use of Transboundary Watercourses and International Lakes (known as the Helsinki Convention of 1992, initially impacting the ECE countries) data sharing is the subject of targeted legislation in articles 6 and 13 (the former for All Parties, whilst the latter to Riparian Parties), together with articles 4 and 11 dealing with joint programs for monitoring the conditions of transboundary waters, including floods and ice drifts, as well as transboundary impact. Similar provisions are contemplated in the Convention on the Law of the Non-navigational Uses of International Watercourses of 1997 (the global UN Watercourses Convention), as well as in the Albufeira Water Convention (see Sect. 3 ahead).

2.3 The Electronics Revolution

Today, owing to the autonomy provided by new communication systems and remote control, as well as new sensors and their standalone power source (combining solar panels and backup batteries), it is possible to cover rainfall fields inside hard-to-reach areas in a more representative way.

The same is true for evaporation studies on great lakes and the associated microclimate induced by the water bodies, once lakes can affect the regional heat energy balance [14–16]. Figure 2 presents one of the two floating rafts installed on the Alqueva dam reservoir, a man-made lake with a reservoir capacity of 4.15 km^3 (billions of cubic meters) and with a surface area covering 25,000 ha.

Each floating platform is equipped with meteorological instruments for measuring air temperature, relative humidity, wind speed and direction, wind vertical velocity, incoming solar radiation, pressure, precipitation and evaporation. The evaporation is measured by water-level readings on an evaporation pan embedded in the platform in a way that its bottom touches the water surface so as to reduce the effects of direct solar radiation on the walls of the pan and also to ensure greater proximity between the temperatures of the water in the pan—also measured—and the temperature of the water in the reservoir's top layer [14]. In order to get a good representation of the

Fig. 2 Assembling and setting up a floating meteorological station, with underwater sensors for temperature and water quality implemented in the man-made lake of Alqueva in southern Portugal

evolution of lake temperature and thus allow a better estimate of the thermal regime of the lake, temperature sensors were also placed at several depths beneath the raft (at five different levels: 1, 5, 10, 15 and 20 m). All measurements from the sensors are parameterized, pre-processed and temporarily stored in a data logger placed inside a weatherproof fiberglass enclosure located at the central mast. The data stored in the acquisition system is transmitted via GSM to the central database on a daily basis. The power source is made up by a solar panel and a battery. Although the platform is anchorage (by cables tied to three sunken blocks), some inevitable longitudinal displacement and rotation on itself occurs so a GPS is also installed in the central mast for positioning of estimates.

The opportunity and flexibility that electronics brought to hydrological measurements was also felt in the quantification of flow rates. The most common flow measuring method used in rivers throughout the twentieth century relied on current meters where discrete measurements proceeded from one riverbank to another at evenly spaced verticals [17, 18] (as depicted in the upper part of Fig. 3). Initially the meters used for measuring the flow applied the same principle as the wind current anemometers: they had a rotating end (with cups or propeller blades) and the velocity was taken from calibrated relations between the angular speed of the rotor and the velocity of the water. Later these mechanical current meters were replaced by electromagnetic probes where the movement of water passing the magnetic-inductive sensor causes an electric potential in the water that is detected by two electrodes on the probe to directly obtain the speed signal in meters per second.

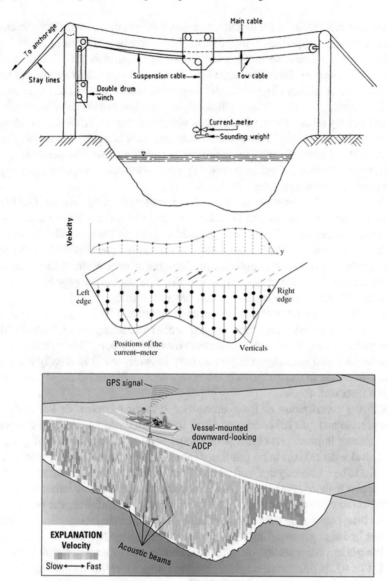

Fig. 3 Traditional flow measurement of rivers using current meter from a cableway system (on top), with unmanned instrument carriage (modified from [17]), to get discrete sampling of velocity (center) [18], as opposed to the continuous measurement of discharge with an acoustic Doppler current profiler from a moving boat (bottom—modified from [19])

From the last decade of the twentieth century onwards the acoustic Doppler current profilers (ADCPs), operated from a moving boat, have been increasingly used as the most common and reliable method for measuring streamflow [19]. These profilers measure the Doppler shift of acoustic signals that are reflected by suspended matter within the water column. The water and suspended matter are assumed to be moving at the same velocity [19]. By using the Doppler principle to measure water velocities this technology allows for the scanning of velocity along the cross-section area of a stream (lower end of Fig. 3) to produce flow estimates at a time-travel pace. In this way ADCP is a powerful measuring tool by reducing substantially the time spent in discharge measurements and by measuring water velocities at a spatial and temporal scale previously unattainable.

Also using the same technology, the acoustic velocity meters (AVM) are measuring devices that use the Doppler principle to measure water velocities in a two-dimensional plane. Usually, they are submerged and fixed to a pillar or abutment of a bridge to measure continuously the velocity in the area covered by the beam, enabling the continuous computation of discharge at sites where a stage-discharge relation can result in more than one possible discharge for a given gage height because of variable backwater conditions or drawdown conditions.

In order to avoid the measuring system to be harmed by sediments and floating refuse it is possible to use an alternative and similar measuring device to the AVM that employs Radar instead of acoustic beams to measure the flow. This system (coupling the radar with and optical probe (to measure the water level) is usually placed at a contact-free zone of the bridge (e.g. the bridge deck or higher) to avoid collisions by water transported debris.

With the contribution of these unmanned systems (acoustic or Radar devices) more precise and quicker continuous flow information can be retrieved as opposed to the former indirect procedure where water level readings (from staff gauges or water level sensors) had to be translated into runoff by means of flow rating curves that need to be constantly updated by new time-consuming measurements. Although the Doppler technology does not avoid altogether the need for updating the effect of proportionality of the cross section's area on the flow calculation, the process is by far less time consuming than the number of flow measurements required to account for river bed changes resulting from erosion or deposition trends.

Changes in a cross section due to sedimentation will progressively alter the rating curve (Fig. 4): as time progresses the river bed surges and the net area available in a cross section for the water to flow beneath a reference water mark is reduced, thus diminishing the flow associated to that water level; this translates in a "shifting" of the rating curve to the left on a Cartesian coordinate system. If the analytical benchmark is not a water level freeze in time but an unchanged flood magnitude throughout time (a design flood), this requires adopting an incremental "lifting" of the rating curve on the coordinate system to compensate for the average height loss associated with the thickness of the deposits. The reverse of these considerations (using Q_2–H_0 as the starting point) is valid for describing the erosion process of the cross section when the river bed is being lowered by scouring, as well as the rating curves with it (Fig. 4).

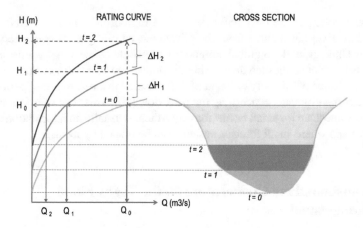

Fig. 4 Q–H changes in the rating curve of a cross section

Thanks to the capacity for real-time data transmission, allowing on-line feeding of data bases and forecasting models, it is possible to make more reliable studies and issue warnings to downstream population with greater anticipation lag-time regarding the impact situation.

In terms of precipitation estimates the use of Radar has made it possible to do not only better detailing of the rainfall fields but also to track the convective cells movement to and within the basin, making it a powerful tool for flood forecasting [20]. Nevertheless what the Radar measures de facto is reflectivity of the precipitation targets in the volume of atmosphere being observed and not precipitation itself. To that end reflectivity values have to be translated into rainfall values through a converting power function that needs ground rainfall values (captured in traditional rain gauges) to adjust the estimates.

From the late 1980's onwards a new trend focusing on climate induced changes has emerged that have further broadened the type of data requirements needed for water management: from local or regional data sets on a world-wide basis to data sets aimed at solving global scale problems, where oceans, plants and soils interact. This added further parameter regionalization problems to the hydrologic models (usually land surface schemes that simulate the energy balance at soil, atmosphere and vegetation interfaces) applied from continental to global scales [21]. It also created the need for larger scale meteorological data to the models that only satellite imagery is able to provide.

Social network data. The steady evolution of electronics that lead us to the satellites also pushed to the limit the concept of networking. What is now known as *crowdsourced data assimilation* is beginning to be considered as a good supplementary inflow data to forecast models [22]. Although some of the low-cost sensors can display low reliability and varying accuracy in time and space, adding more uncertainty to forecast estimates, their data value can be tremendous when there is no other information available. During the flood of November 2015 in Albufeira

(Southern Portugal) the only real-time on-line rainfall information available near the 30 km^2 river basin came from the private owned rain gauges connected to the Weather Underground—a global community of people connecting hyperlocal data. The possibility of using such data during the flood allowed a quick evaluation of the abnormal degree of intensity of the rainfall (measured in terms of return period) and to make some estimates concerning the peak flow [20]. This type of information is especially adequate for small basins for which there is usually no information readily available and where small forecast systems can be funded by locals.

3   Transboundary Cooperation on the Tagus Flood Management

3.1   Preliminary Context

Portugal and Spain share five river basins accounting as a whole for 45% of the total area of the Iberian Peninsula. While the portion of these basins under Portuguese management contributes only 22% to the joint drainage area, it encapsulates roughly $^2/_3$ of the Portuguese continental territory.

The two countries have been establishing water sharing agreements since the nineteenth century focusing on equitable water uses (such as hydropower) at specific boundary river reaches or basins. These partial agreements were crowned by the Albufeira Water Convention (hereafter called Convention) signed between Portugal and Spain in 1998 reflecting a more holistic approach on water resources much in pace with the Water Framework Directive that was being negotiated at the time between the European Union Member States and the European Commission. This meant that a new concept of ecological status preservation and restoration in shared waters was being considered as a framework to the spirit of the Convention, which was translated in its text. This was done by assigning for the first time to the downstream country mandatory minimum flows intended to be maintained at the river mouth in order to secure freshwater inflows to estuary ecosystems in addition to the minimum flows assigned to the upstream country at the border reaches.

Except for one small basin (the Lima River basin) all the remaining four major river basins shared between Portugal and Spain have mandatory minimum flow volumes to be deliver to the water bodies each year and every three-month period at the control points (borders and estuaries). In two of the river basins (the Douro and Tagus basins) where hydropower generation is an economic relevant activity on both sides of the border the mandatory flow volumes are further specified for weekly time periods. All these minimum assured flows (annual, seasonal and weekly) are not to be fulfilled if a very dry year occurs when ad hoc values have to be established. To this end exemption procedures were built for each river basin in order to detect in advance during the year if it might be classified as a dry one, avoiding the maintenance of higher values in the rivers than those naturally occurring during a dry year.

In general terms the "exemption verification" is an accounting procedure built on the sum of all the rainfall falling in the wet months of each hydrological year (nine-to eight-month periods in Northern basins; six- to five-month periods in Southern basins) to be subsequently confronted with the minimum dry-threshold values defined as percentages of the mean values in those same nine to five months.

The reason for varying the duration period for rainfall accumulation from nine to five months is related to the southwards intensification of the climate's dryness and the torrential nature of the flow regimes in the Southern basins—making them more unlikely to recover from wet semester's rainfall deficits during the dry semester. In order to compensate for this effect a specific timing for issuing a decision on whether an exemption is due or not each year is anticipated towards the Southern basins. The accumulation period for rainfall is set at six months for the Tagus and at five months to the Guadiana, and the process of verification becomes more complex than the northern basins' standard, by adding an extra check into the dryness of the previous year for the Tagus basin (multiyear dry spell) or the verification of a possible mitigation of the severity provided by the amount of water stored in the six upstream reservoirs in the case of the Guadiana basin [23].

A detailed description of the threshold values and exemption's verification procedures for each river basin can be found in the text of the Revision of the Convention adopted by the parties in 2008. The text of the Convention also identifies the 13 rain gauges, 10 flow gauges and six reservoirs (with a joint capacity of 7.25 km^3, e.g. billions of cubic meters), that assure a steady interchange of information for the "exemption verification" and water management control. All the information exchange is carried out by web server protocols.

The Convention has a special article dealing with floods (article 18) where the type of data to be exchange and the coordinating procedures to be undertaken are described. When managing extreme hydrological conditions such as floods the amount of information exchanged is substantially enlarged, due not only to hourly interchange time spans but also to the increasing number of rain gauges, flow gauges and reservoir data considered (much larger than the number used for regular exchange).

The transboundary cooperation between the Portuguese water authority and its Spanish counterpart on flood management coordinated activities developed firstly for the Tagus river basin prior to the other shared basins due to the specifics of the hydromorphology (as they enable the deployment and exploitation of a significant water storage potential close to the border) and benefitting from the digital capabilities and telemetering expertise on remote data gathering that were precociously developed in the Tagus basin on both sides of the border. This is why the Tagus basin is being use as "case-study" in this text.

The zoomed area of interest that brings into context the transboundary flood dynamics within the Tagus basin is depicted in Fig. 5 with the relevant river action points flagged-in (consisting of flow gauge stations and dams—where the incoming and outgoing flow is calculated).

The information in these points is crucial both for the detection and tracing of the routing flood waves (see example in Fig. 6) and for the delineation of the areal extent

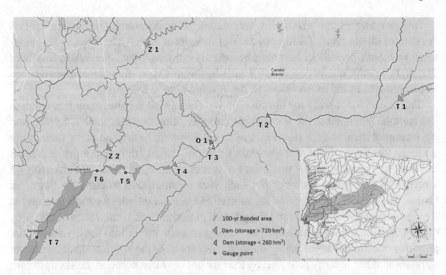

Fig. 5 Relevant hydrometric points for flood management control in the transboundary zone of the Tagus basin: T-points are placed in the main Tagus channel and are numbered from upstream Spanish dams (T1 and T2) to the downstream flood plain; Z-points in the Zêzere River represent the two dams implanted on the Tagus' major tributary inside the Portuguese territory, both in area and flow, with its confluence immediately upstream the flood plain; O-point controls the Ocreza River, smaller tributary from the right-hand margin of the Tagus River; other contributing areas are controlled through hydrological modeling

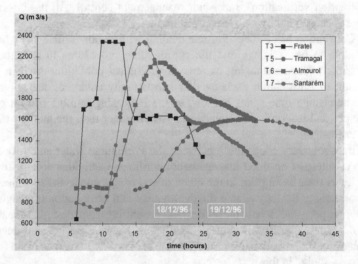

Fig. 6 Propagation of a flood wave generated by a dam discharge hydrograph (T3) along the Tagus flood plain

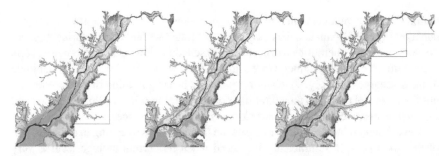

Fig. 7 GIS representation of the propagation of a flood wave along the Tagus flood plain given by an hydraulic model (from left to right)

of the flooded area downstream (see Fig. 7). Three of these action points are also actionable points for managing the flood: point T1—Alcantara dam, with a storage capacity of 3200 hm^3; point Z2—Castelo do Bode dam, with a storage capacity of 1000 hm^3; and point Z1—Cabril dam, with a storage capacity of 720 hm^3 at the top of the reservoir cascade. All the remaining dams in the Tagus main channel are run-of-the-river dams, meaning they have limited capacity to significantly attenuate flood flows, also a handicap for Pracana dam (point O1).

In managing jointly flood events throughout decades in the Tagus, the water authorities of both countries were able to develop a plan for data exchange in real time that was then used as standard for the implementation of article 18 of the Convention. The scheme considers packages of relevant data for each country's flood management (especially incoming flow data for the downstream country and storm movement and rainfall data for the upstream country, since storm paths move inland from the ocean to Portugal and then to Spanish territory) and is supported by new technological developments encompassing sensors and data loggers, hydrological modeling with GIS capabilities, dam operation, data bases and data communication protocols.

But the most important achievement consolidated in the data exchange plan is the adoption of the consultation between the parties about reservoir release options during the flood event as a standard procedure. The relevance of such measure is paramount enabling to sense the potential consequences that certain discharge options considered upstream can cause downstream due to the flood situation that is taking place there and how it is expected to evolve—paving the way for an optimization of the decision making. This approach was enshrined in article 18 of the Convention.

3.2 Coordinated Flood Management

Owing to the channel of cooperation on flood management developed prior to the signing of the Convention but consolidated under its aegis, it is possible to manage the lag and route of flood discharge hydrographs along the Portuguese Tagus stretch in such a way that peak flows can be significantly attenuated by avoiding to overlap, within the Tagus main channel, two major flood components (the upstream component coming from Spain and the component generated within the tributaries' basins). To carry out these operations it is necessary to schedule flow releases from reservoirs in a preemptive or delayed mode (see Fig. 8) jointly with reductions of the magnitude of the discharges (see Fig. 9) whenever possible, bearing in mind that reservoirs are not flood control dams, as they have their own operating rules for floods events, and that their security cannot be put at risk under any circumstances.

The capacity of a reservoir to mitigate the flood impacts by acting in the reduction of the flood peak is significantly increased when its current storage level is very

Fig. 8 Flood flow gauged at point T6 showing a two-peaked hydrograph resulting from: the outflow from Z2 dam (estimated from the spillway discharge plus the hydropower production); coupled with the delayed flood released from Spain and routed downstream. If this delay of 24 h had not been possible the overlapping of both flood waves around the first peak would have risen the peak to a value above 5000 m³/s

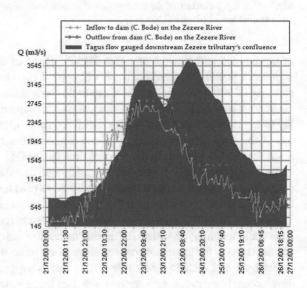

Fig. 9 Flood volume withheld in point T1 by reducing substantially the outflow spill downstream during the rising limb and peak of the flood hydrograph (for almost 30 h) and then resuming the outflow spill at a lower discharge rate

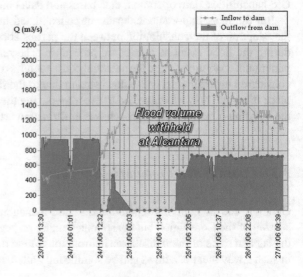

low, which usually happens at the end of the summer season and during prolonged droughts. In these low-level scenarios not only the peak can be attenuated but also an important amount of flood volume can be subtracted from the flood hydrograph using the occasional storage void to keep that flood portion blocked until the end of the flood event instead of just lagging it to the recession limb of the flood hydrograph.

One example of how a gigantic water holding potential can attenuate both flood peak and flood volume occurred in November 1997 when a violent storm hit the border area of the Tagus basin. Due to a prolonged summer the first rainfalls of the hydrological year occurred only in November which contributed to a further lowering of the storage level during early fall (although not abnormally low as during drought events). That turned out to be beneficial in relation to what happened over the next two days enabling an amount of 700 hm^3 of flood water to be withheld at Alcantara dam (point T1 from Fig. 5) whose stored level went from 60 to 80% of its capacity in that short amount of time. This water retention exercise avoided the routing of a peak flow with a magnitude of 15,000 m^3/s along the river stretch that defines the border (until point T2) that would have in turn captured the additional contributing flow from the intermediate sub-catchment area along the way, producing a flood hydrograph with a peak flow around 18,000 m^3/s (see Fig. 10).

This '*flood that never was*' is a record breaker within the long time-series of peak flow annual maxima. Together with an historical flood occurred in December 1876 and a similar hypothetical flood that was attenuated in January 1996, these three maxima helped to better define the tail of the flood probability density function to help detect the return period of each flood event (see further ahead in Sect. 3.3).

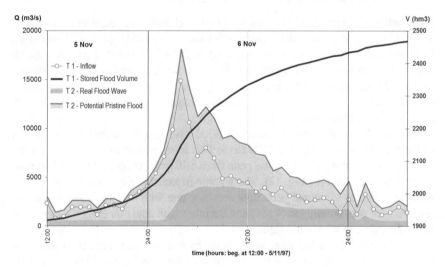

Fig. 10 Record breaking peak flood at point T2 reconstructed from the inflow hydrograph captured at point T1 overlaid with the flood hydrograph from the intermediate drainage area between T1 and T2, registered at point T2, and with the evolution of stored volume in hm^3

During the last decades several flood events occurred in the Tagus basin whose management was handled according to the basic principles exemplified above which are: combination of lag and attenuation of hydrographs with routing modeling on one side, and test of storage retention or discharge anticipation on the other [23–25].

The common procedure is to use the available meteorological forecasts on a regular basis to simulate preemptive flow conditions, and as these estimates evolve into threatening situations the data exchange between the main players intensifies. The list of players involved comprises, other than the water agencies' personnel, each country's flood committees, the hydropower companies' technicians responsible for dam management on both sides of the border, civil protection agents from each country and meteorological technicians. It is important that all the information concerning each party is gathered and centralized in the respective water authority to be conveyed to the other side only through one exchange channel. By using this procedure, the noise resulting from the proliferation of cross channels of information with different data is avoided.

3.3   Challenges Coming from Collected Data

The data collected during the last three decades is essential to the follow-up of the strategies envisaged. In this sub-section a quick view over three examples illustrates the type of issues raised when confronting new data sets to earlier ones that demand to be further investigated and hopefully reexamined.

False Sense of Security. The last relevant flood (relevant meaning with a recurrence $T > 2.33$ years) impacting the Tagus border zone until today, year 2021, occurred in an unusual time of the year (early spring) of 2013. The synoptic map revealed a situation where a surface front after crossing the Portuguese territory had moved into Spain and intensified over the border and beyond. The result of the flood regulation performed at that time in order to keep the flow magnitude not higher than 4000 m³/s at point T2 is presented in Fig. 11.

The transformation operated over the flood hydrograph resulted in a reduction of the peak flow from 7350 to 4100 m³/s. This is equivalent to a reduction on the severity (or recurrence) of the flood peak from a 10-year return period to a less than 4-year return period (see Fig. 12). The graphic depicted in Fig. 12 was constructed with two samples of annual peak flow maxima: one with data recorded before the beginning of dam building in Spain; and the other with data reflecting the impact of the impoundments on natural floods.

Also perceptible on the same graph is the motive why downstream populations feel more secure today than used to feel in the 1940's and mid-1950's when there were no dams on the basin. In fact, during the first half of the century a 6000 m³/s peak flood was a common flood (with recurrence $T = 2.5$ years); today, after the boom in dam construction on the Spanish Tagus basin during the third quarter of the twentieth century, it has become a 7-year return period flood on average (see Fig. 12).

Q (m3/s)

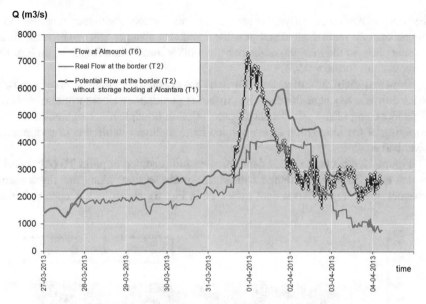

Fig. 11 Reconstitution of the natural flood at the border (point T2—Cedillo dam) that might have happened had the incoming flood wave to point T1 (Alcantara dam) not been attenuated and lagged

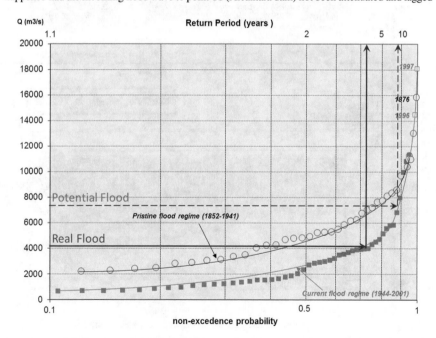

Fig. 12 Reduction of the severity (e.g. recurrence) of the flood on March/April 2013 at the border point T2 due to upstream dam storage capacity engagement

But this attenuation capacity is not permanent since for floods with recurrence periods higher than 20 years there are no possible significant changes that can be made on a regular basis, as the merging of the two probability distribution functions on Fig. 12 demonstrates (prob. > 0.95).

Dam sediment retention and downstream erosion. The false sense of security related to less frequent floods brings riverside populations closer to the river bed. This loosening of the safe distance's perception in floodplain is heightening as the lowering of the inundation levels became more distinguishable due to continuous river bed erosion.

Figure 13 depicts the scour depth progression recorded at point T6 (Almourol) which is a characteristic shared with several cross sections along the Tagus main channel downstream Belver dam (point T4 in Fig. 5), signaling a deepening trend

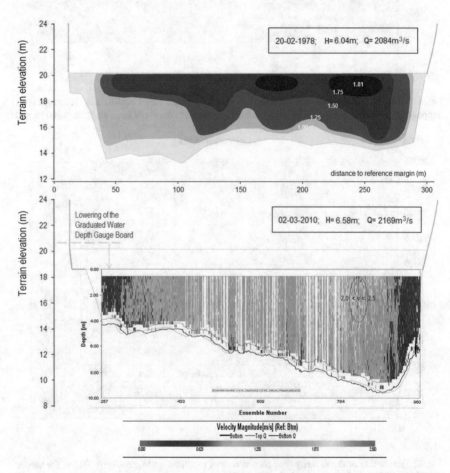

Fig. 13 Profile of Almourol cross section (point T6 in Fig. 5) on two occasions (1978 and 2010) with similar flood flows. The contour lines of equal velocity on top correspond to the current meter gauging procedure and the digital velocities on bottom were measured with an ADCP

that is currently propagating further downstream. This trend was first detected in gauge section T5 (Tramagal) following the start of operation of Alcantara reservoir in the 1970's, making it necessary to lower by two meters the graduated water depth's gauge boards of both point T5 and point T6 during the 1990's (see Fig. 13). The maximum depth in point T6 cross section for the 2100 m³/s flood has gone from 6 to 9 m, with repercussions on the increase of the bottom-current velocities that contribute to further destabilizing the cross section.

The cause-effect association between the downstream erosion and upstream sediments' retention due to damming must be further investigated. To this end more regular bathymetric sounding and sediment transport measurements (not considered in this article) needs to be done to keep track of changes.

The new flow regime. As stated at the beginning of Sect. 3, the Convention tried to come out with a solution for the systematic low flows' display in the shared watercourses by establishing threshold minima on four basins, only to be exempted during very dry years (that could also be drought years). These minimum flow regimes were not supposed to last long since a "Committee on the Application and Development of the Convention" was created inside the Convention to develop flow regimes in accordance with good water status, among other pressing issues.

Failing to produce such flow regimes soon the perseverance of the minimum threshold regime in the Tagus has created a situation where a range of median annual flow values was wiped out from the river regime (see Fig. 14). This has to do with the type of management in place, which is driven by a growing concern among the

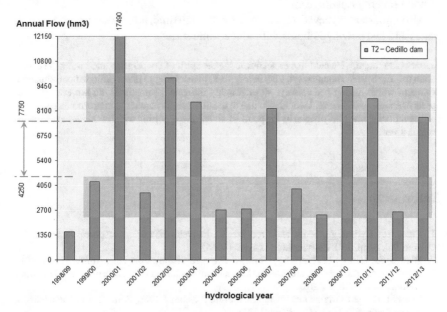

Fig. 14 Annual flow totals at Cedillo dam (point T2 of Fig. 5) after the convention enter into force, depicting the flow-gap between 4250 and 7750 millions of cubit meters

water authorities at the beginning of each hydrological year, on whether the possible dryness of the first few months will develop into an exception situation or whether it will remain slightly above the threshold. In the hydrological year of 2018/2019 this course of action led to a situation where the border dam of Cedilho had to be completely drained in September to supplement the minimum flow amount still due downstream.

4   Conclusion

Hydrology is a data driven science as shown by its history and development and, as such, the accurate and tailored collection of hydrological data for each particular case-study is determinant in the validation or rejection of hypotheses on the functioning of natural or human-modified systems focusing on water resources management.

Spatial and time scales are important issues in hydrology and require appropriate data and modeling tools for specific analyses, especially when dealing with extreme events in a transboundary context.

The recent advances in telecommunications and sensors have led to an unimaginable insight capacity and real time management control that was here depicted in a transboundary context using the international cooperation on flood management as an example. The several flood events used as case-studies demonstrate how the coordination of water management action is of paramount importance on the weakening of flood severity downstream.

Also important, as new data accumulates through time, is to reexamine space–time trends and test new hypotheses with longer time series.

Acknowledgements I would like to acknowledge the spirit of cooperation and invaluable contribution that my Spanish counterpart, Luiz Perez, made in resolving so many difficult flood situations upstream, which having been always so thoroughly mitigated, sometimes did not even produce upsets or shock downstream. I would also like to recall his frank manner of approaching each difficult scenario that invariably began by the point-of-situation's reporting with his famous exclamation "vamos a ver …"

References

1. Bales, R.: Hydrology, floods and droughts. In: Encyclopedia of Atmospheric Sciences, 2nd edn. Vol. 3, pp. 180. Elsevier Ltd., London (2015)
2. Biswas, A.: History of Hydrology. North-Holland Publishing Company, Amsterdam (1970)
3. Grayling, A.: The Age of Genius—The Seventeenth Century and the Birth of the Modern Mind. Bloomsbury, London (2016)
4. Perrault, P.: De l'Origine des Fontaines, édition fac-similé 1966, Série: Textes fondateurs de l'hydrologie n° 2, UNESCO, Paris (1674)
5. Dooge, J.: Concepts of the hydrological cycle, ancient and modern. In: International Symposium OH2 'Origins and History of Hydrology', pp. 1–10. Université de Bourgogne, Dijon (2001)

6. Horton, R.: Analysis of runoff plat experiments with varying infiltration capacity. Trans. Am. Geophys. Union **20**, 693–711 (1939)
7. Dawdy, D.: Problems of runoff modeling which are particular to the area or climate being modeled. In: Recent Advances in the Modeling of Hydrologic Systems—Chapter 24 NATO ASI Series 345, pp. 541–547. Springer Science+Business Media, Oordrecht (1991)
8. Rodrigues, R.: Flow modeling in volcanic massifs. In: Advances in Water Resources Technology and Management, pp. 335–341. Tsakiris & Santos (eds.), A.A. Balkema, Rotterdam (1994)
9. Rodrigues, R.: Precipitation mapping in volcanic islands. In: Advances in Water Resources Technology and Management, pp. 411–419. Tsakiris & Santos (eds.), A.A. Balkema, Rotterdam (1994)
10. Rodrigues, R.: Isotopic hydrology as a relevant mechanism for the assessment of water resources in volcanic massifs. In: Sustainable Development of Water Resources, n°1, vol. 2, pp. 87–95 ABRH Publications (in Portuguese) (1995)
11. Hontarrede, M.: Meteorology and the maritime world: 150 years of constructive cooperation. WMO Bulletin **47**(1), 15–26 (1998)
12. Locher, F.: Atmosphere of globalisation. depressions, the astronomer and the telegraph (1850–1914). Revue d'histoire Moderne et Contemporaine **56**(4), 77–103 (2009)
13. Lindgrén, S., Neumann, J.: Great historical events that were significantly affected by the weather, part 5—some meteorological events of the crimean war and their consequences. Bull. Am. Meteorol. Soc. **61**(12), 1570–1583 (1980)
14. Rodrigues, C.: Calculation of evaporation from highly regulating-capacity reservoirs in southern Portugal, PhD Thesis, Évora University, Évora, Portugal. Available at: http://hdl.handle.net/10174/11108. Last Access 5 April 2021 (in Portuguese) (2009)
15. Rodrigues, C., Rodrigues, R., Salgado, R.: Reservoir evaporation estimates in the south of Portugal by the FLAKE Model. In: Proceedings of the 6ª Portuguese–Spanish Assembly of Geodesy and Geophysics. Tomar, Portugal (2008) (in Portuguese)
16. Rodrigues, C., Moreira, M., Guimarães, R., Potes, M.: Reservoir evaporation in a Mediterranean climate: comparing direct methods in Alqueva reservoir, Portugal. Hydrol. Earth Syst. Sci. **24**, 5973–5984 (2020)
17. Holland, P.: Current metering. In: Encyclopedia of Hydrology and Lakes, 1st edn. p. 146. Kluwer Academic Publishers, Dordrecht the Netherlands (1998)
18. Le Coz, J., Camenen, B., Peyrard, X., Dramais, G.: Uncertainty in open-channel discharges measured with the velocity–area method. Flow Meas. Instrum. **26**, 18–29 (2012)
19. USGS: Measuring discharge with acoustic Doppler current profilers from a moving boat. Techniques and Methods 3–A22, Version 2.0. U.S. Department of the Interior/U.S. Geological Survey (2013)
20. Rodrigues, R.: The effect of rainstorm movement dynamics on the shape of the flood hydrograph and its peak occurring in Albufeira on the 1st of November 2015, and its relevance for the sustainability of drainage solutions. In: 13th Water Congress (pub. # 194), APRH, Lisbon, 12p (2016) (in Portuguese)
21. Beck, H., van Dijk, A., de Roo, A., Miralles, D., McVicar, T., Schellekens, J., Bruijnzeel, L.: Global-scale regionalization of hydrologic model parameters. Water Resour. Res. **52**(5), 3599–3622 (2016)
22. Mazzoleni, M.: Improving flood prediction assimilating uncertain crowdsourced data into hydrologic and hydraulic models, (IHE Delft PhD Thesis Series) (2016)
23. Rodrigues, R.: Portuguese–Spanish transboundary water management cooperation. In "Future Perfect", Rio+20 United Nations Conference on Sustainable Development, Chapter on Environment, pp. 87–88 (2012)
24. Rodrigues, R., Brandão, C., Costa, J., Saramago, M., Almeida, M.: Brief characterization of the autumn floods of 2006. In: 2nd Journey of the Presidency's Roadmap for Science, INAG press (2007) (in Portuguese)
25. Rodrigues, R.: Flood surveillance and early warning system for Portugal. In: Precautionary Flood Protection in Europe International Workshop. Bonn (2003)

Quality of Measurement in Urban Water Services: A Metrological Perspective

Maria do Céu Almeida⬤, **Alvaro Silva Ribeiro**⬤, **and Rita Salgado Brito**⬤

Abstract Urban water services make use of extensive networked infrastructures—water supply and drainage systems. Efficient management of these services relies on the quality of measurements of hydraulic variables in pipes and sewers to support robust analysis of systems' performance, needed to many decision-making processes. The quality of measurements is key for the sustainability of urban water services, having a growing impact on decisions and on trade relations. Often, quantities of interest (e.g., flow rate, volume, level, and velocity), affecting net balances (e.g., inflow and outflow net balances, water losses, undue inflows and trade volumes) and used to support decision making, are obtained indirectly from a wide diversity of instrumentation, using different principles and methods. To reach acceptable accuracy, required for the measurement itself, and because of implications when used for management, three elements are critical: good measurement practices, traceability, and measurement uncertainty. Measurement uncertainty plays a relevant role today, to capture the random nature of measurement, and is considered the most reliable parameter able to provide information about the accuracy of measurement within a certain degree of confidence, allowing comparisons between instruments, practices, and methods. Advances in synergies of knowledge from Hydrology, Hydraulics, Metrology and Data Science are reported, as well as perspectives of future developments.

Keywords Accuracy · Measurement · Quality · Uncertainty · Urban water systems

M. do Céu Almeida (✉) · A. S. Ribeiro · R. S. Brito
Laboratório Nacional de Engenharia Civil, Lisboa, Portugal
e-mail: mcalmeida@lnec.pt

A. S. Ribeiro
e-mail: asribeiro@lnec.pt

R. S. Brito
e-mail: rsbrito@lnec.pt

© The Author(s), under exclusive license to Springer Nature Switzerland AG 2023
C. Chastre et al. (eds.), *Advances on Testing and Experimentation in
Civil Engineering*, Springer Tracts in Civil Engineering,
https://doi.org/10.1007/978-3-031-05875-2_15

1   Measurement in Urban Water Systems

Monitoring of urban water systems is essential for management of water supply and drainage services. The regular collection of information on hydraulics, hydrology and water quality allows understanding systems' functioning, assessing performance and supporting the setting of management targets, responding to regulatory requirements, and enables the identification of inefficiencies and opportunities for improvement.

Efficient management of these services relies on the quality of measurements in pipes and sewers, to support robust analysis of systems' performance, needed to many decision-making processes. Furthermore, sustainability of these essential and universal services depends on fair competition and confidence, especially with the transition from state owned organizations to models introducing competition in services that are natural monopolies. In the later, issues regarding fair trade and consumer protection have a close link to measurement quality.

However, what should be measured in water supply and drainage systems, considering the associated extensive networked infrastructures, mostly underground?

In all these water systems, quantification of inflows and outflows is paramount for supporting sound management and commercial relations with costumers and between neighbour systems.

Water utilities may need to monitor systems due to technical needs, contractual obligations, and financial, regulatory, or legal requirements. Technical needs usually aim at better understanding the functioning of the system (namely for performance assessment, to support decision-making or to implement mathematical modelling). For contractual or financial obligations, data is a need as input to apply tariffs or to quantify transactions between different utilities. Regulatory or legal requirements include audits by regulators, compliance of self-control monitoring or to the control of withdrawals or discharges from or into natural water bodies. Regardless of the specific purpose, the quality of the data is of uttermost importance and a common requirement for any measurement. Only with evidence of measurement quality, confidence is ensured. Confidence on data, along with data adequacy to the objectives, are the main drivers for data reliability.

For water supply systems, the widely used water balance scheme [1] allows identification of the typical measurements location (Fig. 1).

Water supply is mostly provided by pressure flow networks, although in some situations surface water flows can be used for raw water, usually before treatment. Variables of interest include water flow rate, water level (for instance, in abstraction works and storage tanks), pressure, and water quality parameters.

The measurement locations vary from costumer connections, in smaller diameter pipes, to network sectors or water mains, typically in conduits with larger cross sections and flows. Technology used may vary significantly but given the good quality of the water, use of mechanic instruments is practicable.

For wastewater and stormwater systems, including combined systems, the typical measurement locations are represented in Fig. 2. In wastewater and stormwater

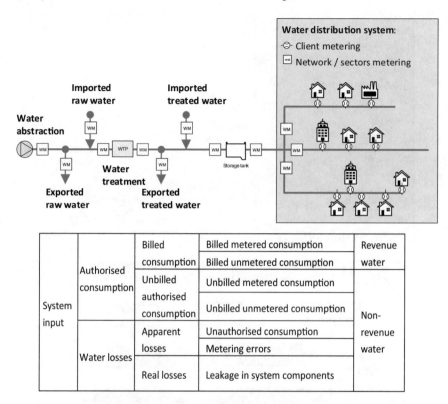

Fig. 1 Schematic representation of typical measurement needs for water supply systems and water balance components. Adapted from Hirner and Lambert [1]

systems, flows are mostly under free surface; pressure flows usually occur downstream of pumping stations, in pressurised systems (not commonly used) or when inflows exceed free surface capacity of sewers, for instance because of a high intensity rainfall. Variables of interest include flow rate, water level (for instance, in weirs), flow velocity, water pressure and water quality parameters. Water quality characteristics are measured for compliance with environmental legislation and process control, namely for treatment processes. In parallel, it is recommended to measure rainfall.

For urban areas, requirements for rainfall measurements are higher than for meteorological purposes, both in time resolution of records and spatial density of measurements. Traditionally, precipitation is measured by means of rain gauges, giving the time series associated with location. Spatial integration can be carried out if enough gauges are available. Combination of rain gauges with radar allows improving knowledge on spatial rainfall distribution, but at scales of interest for urban studies, X-band radar are often considered more adequate to cover smaller areas with higher spatial resolution and more cost effective than the alternative S-band and C-band radars [3, 4]. Relevant issues associated with rainfall measurement include the number

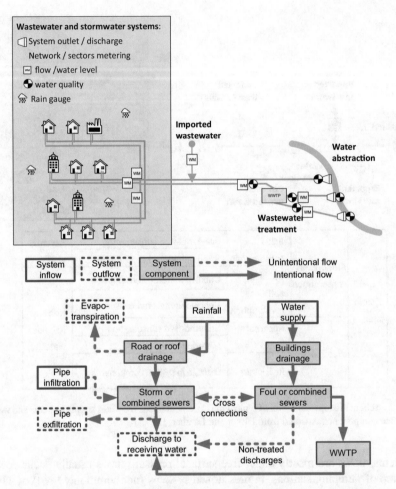

Fig. 2 Schematic representation of typical measurement needs for wastewater, stormwater, and combined systems (left) and water cycle components (right). Adapted from Butler and Davies [2]

and density of measurement locations and usually these are insufficient for ensuring lower uncertainties and representativeness of spatial variation of rainfall.

Unlike water supply systems, measurement locations in drainage pipes are not feasible at regular costumer connections and hardly practicable in small diameter sewers. Costs for free surface measurement equipment are significantly higher than for pressure flows. Therefore, typical measurement locations are downstream locations such as at entrance of pumping stations or wastewater treatment plants, in main sectors of the network and at boundaries between water utilities. Despite the costs, it is highly recommended to monitor untreated discharges from the systems, such as emergency bypasses and combined sewer overflows.

Technology used may vary significantly. Characteristics of flows, carrying a large number of solids, often biochemically aggressive to materials, limits the use of some technologies. Therefore, local measurement conditions in free surface flow in urban drainage are far from optimal [5–7]. Moreover, sensors are usually installed in an aggressive environment, exposed to flow turbulence, to accumulation of grease or deposits, or to obstruction of sensors. The occurrence of unstable flow is rather frequent due to changes in pipes characteristics (e.g., slope, diameter, curves, drops) the existence of manholes regularly and connection of pipes. The relative size of the sensors is not negligible, interfering with the flow conditions, as pipe diameters between 200 and 400 mm are quite common in sewer drainage [8].

Measurement of low flows also presents constraints. Besides the relative size of the sensors, in lower flows, the water height or velocity are frequently below the minimum limits of the equipment's measurement range. Turbulence may be aggravated due to deposits in the pipe's invert or to the biological layer in the walls.

Additionally, flow in urban drainage may present high variability. Dry weather flows are characterised by a typical daily and weekly pattern, relatively predictable, depending on water consumption. However, in combined, stormwater or sanitary systems with undue rain induced inflows, hydrographs vary significantly, leading to rapid variations in flow velocity and water height. Flow can abruptly alternate between free surface and pressure flow. These factors can all interfere with the integrity of the equipment and with data quality.

In the following sections, applications are illustrated using quantity measurements, but similar principles and procedures apply to water quality measurements.

2 Good Measurement Practices

In the last decades, many water utilities have made considerable investments in monitoring but ensuring measurement data reliability is still a challenge. Designing, installing, and operating measurement systems overlooking good practices, essential to overall data quality, has a direct consequence on data reliability. Reliance on practices not duly implemented, not allowing verification of conformity with recognised good practices, frequently results in a biased perception of confidence by decision makers and data users. The implications include inadequate decisions and uses of the data, which can be significant.

Confidence on measurement data to achieve the monitoring objectives must be at the core of monitoring activities (as on any economic sector other than the water industry) and have been supported by the development and improvement of international standards. The ISO 17000 standards series establishes the framework for conformity assessment, being the ISO/IEC 17025:2017 [9] specifically intended to demonstrate the competence of organizations carrying out measurement activities, provided by testing and calibration laboratories, usually called accreditation. This recently revised standard introduces a new structure with a technical part, related with the resource and process requirements, and a management and system requirements part, in line with ISO 9001:2015 [10], promoting the aim of ISO of having a common management system approach for the different standards series.

The ISO/IEC 17025 [9] provides a ground for the development of competence in measurement activities, with a set of technical requirements to evaluate the proper implementation of the measurement process (including the human resources, technical equipment and software, methods, and its validation, traceability, and measurement uncertainty). Good measurement practices go beyond these requirements, which usually consider the basic procedures of the measurement process as ensured. There are known references on good measurement practices, namely related with laboratory activity [11], describing quality systems regarding organizational processes and the conditions for quality on its development (planning, performance, monitoring, recording, archive, and report).

Confidence in a monitoring process can be expressed by the uncertainty associated with measurement data, which is sometimes misinterpreted as the uncertainty of the measuring equipment. The latter is only one of the contributions to the overall measurement uncertainty. To improve measurement data quality, identifying all possible sources of measurement uncertainty raises awareness to the problem and motivates monitoring teams to tackle each of the identified sources and to minimize its effect.

In specific measurement conditions, as in sewer systems, the sensors and measurement chains withstand a very aggressive environment, being critical to make a robust analysis of the influence of quantities affecting the quality of measurement signals and processing, given by the sources of uncertainty. Several sources of uncertainty are reported, and can be grouped as follows [5, 8, 9, 12, 13]:

- equipment uncertainties (generally known, provided by the manufacturer);
- calibration uncertainty;
- context factors of the measurement environment, such as temperature, corrosion potential or humidity;
- inadequate sensor installation or site characteristics, uncompliant with the manufacturer's geometric instructions or hydraulic conditions;
- absence of a pre-campaign;
- inadequate representation of the phenomena by the sampling options;
- absence or non-compliance with equipment maintenance;
- inadequate technical skills in hydraulics, electronics, computing and metrology;
- uncertainties associated with the method for calculating derived variables (e.g., how to calculate flow based on water height and peak velocity); absent or incomplete knowledge of the measurement model;
- data processing; and
- absence of document management.

In practice, estimation of uncertainty is not yet widespread in urban water systems monitoring. The implementation of good measurement practices is inherent to the reduction of measurement uncertainty.

Measurement practices may be site specific (technical aspects related to the selected equipment for each site and its suitability, its installation context and site adequacy) or common to several sites (those that may affect the monitoring system, usually applied to several sites or to the monitoring system).

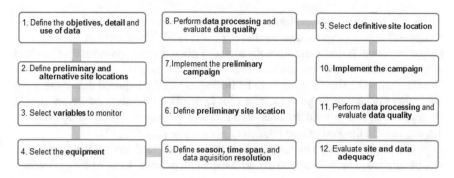

Fig. 3 Monitoring program steps

A preliminary definition of a monitoring program should be the foundation of good measurement in urban water systems. An example of steps to follow in designing a monitoring program is presented in Fig. 3.

In step 1, clear stating the objective of the monitoring program and the intended use of data is essential for many subsequent decisions. Further, several management issues regarding the monitoring system and site characteristics are recognized, mainly addressing the already mentioned sources of uncertainty. Steps 2, 3 and 4 can be applied at the same time.

Selecting the variables to monitor depends on several features, such as the monitoring objectives, the site location and hydraulic conditions, the available equipment, its installation and operation costs, and the required reliability. Aiming at hydraulic variables, in pressurised pipes both flow or volume and pressure are commonly chosen; in free surface flow, mostly water height and velocity are monitored. Precipitation usually complements hydraulic monitoring in sewers. To detect and characterize variations in the water quality matrix, specific water quality variables can be selected. For the characterization of the surrounding water environment, data on tidal height, water quality in local water bodies and groundwater level in piezometers can also the necessary.

To avoid undesirable risks related with measurements, selecting the equipment for hydraulic monitoring should take into consideration the following aspects:

- variables to be measured and expected measuring ranges;
- metrological requirements;
- whether the measurements are temporary or permanent;
- physical characteristics of the installation site;
- robustness and resistance requirements, given local conditions;
- expected time interval between local maintenance actions (which could also condition data storage capacity);
- required local data storage, the remote communication conditions and the availability of battery power supply to the equipment;
- available budget;
- technical capacity of internal human resources.

For the preliminary selection of the location for the hydraulic monitoring sites, equipment technical specifications must be considered. The correct positioning of a sensor in a sewer is important since the accuracy often depends on the proper position of the device [14]. However, sometimes even when fulfilling the equipment technical specifications, the site may prove to be inadequate to the required data quality. To overcome this, accepted good practices for hydraulic monitoring in sewers should be applied; typical criteria are presented in Table 1, some of which based on [5, 8, 12, 15, 16] and the Portuguese regulatory decree-law n. 23/95 [17].

While implementing a monitoring program, several good measurement practices can be applied. These focus mostly on whether steps from 1 till 7 of Fig. 3 where

Table 1 Typical criteria for selection of locations in sewers for hydraulic monitoring

1. Conditions of access to the manhole
Site accessibility (possibility of access by car)
External security (prevention of vandalism, security of the installation against third parties)
Accessible manhole cover with no need for traffic diversion and easy handling
2. Conditions inside the manhole
Viable access to the flow inside the manhole (allowing the equipment installation and maintenance)
Staff safety ensured to hazards, such as sudden increase of flow, manhole height, falling objects, pathogens and disease vectors, explosive gases, insufficient oxygen, toxic gases
Adequate equipment for local conditions (occurrence of pressure flow, explosive atmosphere)
3. Adequacy of hydraulic and environmental conditions
Regular pipe cross-section
Conditions in accordance with equipment specifications, usually ensuring near uniform flow regime, e.g., minimum length of pipe upstream without singularities, bends, obstacles, pipe confluences
Expected water height and flow velocity values within the equipment specifications
No significant turbulence, solids deposition, accumulation of sediment or grease on sensors No interference of the sensors with the normal flow in the pipe (variable water surface over submerged sensors, presence of water splashes, air bubbles, side waves in the flow)
Absence of other environmental constraints, i.e., presence of external agents preventing site suitability for the purpose, such as electromagnetic fields or mechanical devices
4. Operation and maintenance records
Absence of frequent obstruction records
Verification of flow altering between free surface and pressurized flow and of variations in flow direction (e.g. due to downstream tidal changes or other downstream intermittent influence)
Lack of structural damage in nearby upstream pipe
5. Additional information
Need for construction works to adapt the site
Possibility of in site simultaneous measurement of hydraulic and water quality variables
Ease of implementation of ancillary connections (power supply, communication)

adequately carried out and if the identified sources of uncertainty were minimized during this process. Therefore, in alignment with ISO/IEC 17025:2017 [9], specific requirements are identified for the following criteria: *objectives and use of data*; *document management and traceability*; *human resources*; *influencing conditions*; *methods*; *equipment*; *sampling*. Most criteria focus on each monitoring site, but some focus on the monitoring system, and are transversal to various monitoring sites. This is the case of *document management* and *human resources*. Regarding the *methods*, some aspects are site specific; others are common to several sites. Figure 4 details the requirements for good management practices to be verified for each criterion. Document management and traceability are addressed in Sect. 3.

About human resources, it should be noted that some activities require specific skills, such as equipment parameterization, installation and programming, maintenance, data processing or securely accessing the installation sites.

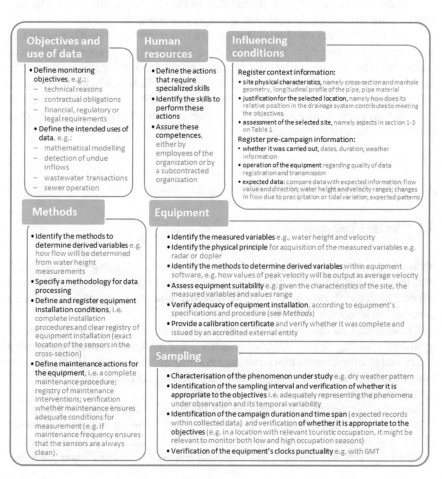

Fig. 4 Requirements for good management practices

Sampling intervals for hydraulic variables in a sewer are typically of few minutes to capture sudden variations in flow; a common value is 5 min. For water supply, higher values are often used, e.g., 15 min. Often, analysis requires the use of several variables, e.g., water consumption and wastewater or the influence of precipitation on flow, so time synchronisation of records is essential, as well as keeping to one time reference, not changing with season as is common in many countries. As for precipitation, for tipping buckets, record of each tip occurrence ought to be registered with a resolution below 1 s.

Human resources play a very important role both within technical and management strands. Even if skilled, with many of the competences required, water utilities' teams frequently have low confidence in data from monitoring and struggle with how to improve data reliability. They can benefit from a structured procedure towards good measurement practices. When these monitoring services are outsourced, these procedures help to structure the overall data quality approach and are an adequate way to guide the setting of requirements for service providers.

The ISO17025:2017 standard [9] has a robust structure dealing with both technical and management aspects, to ensure the overall data quality and common sources of uncertainty. Despite the primary aim of this standard being the quality assurance in laboratories, there are clear benefits in extending the application to other areas within water utilities. Knowledge sharing and collaboration between departments such as systems management and laboratories can bring clear advantages for the organisation in terms of data quality management.

3 Measurement Traceability

3.1 Basic Concepts

Measurement of quantities can be carried out in many ways, using instrumentation based on different principles and methods, under distinct procedures, conditions and affected by external quantities. With this diversity, how can measurement results be compared in an objective way?

Consider that an observation of a quantity x is made, using two measurement approaches leading to two different results x_a, and x_b, having similar uncertainty (expressed by normal probability distribution functions), as presented in Fig. 5a. The interpretation of this figure allows to compare both results regarding its "precision" but not to say which is the most "accurate" result.

Measurement precision is defined in the International Vocabulary of Metrology [13] as: *closeness of agreement between indications or measured quantity values obtained by replicate measurements on the same or similar objects under specified conditions.* In this case, result x_a has larger dispersion of values and lower precision.

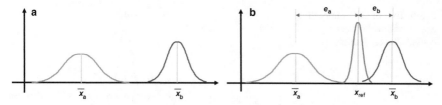

Fig. 5 **a** (left) Two experimental results of a quantity (estimates and its uncertainty); **b** (right) comparison with a reference value and its uncertainty

Measurement accuracy is also defined in the International Vocabulary of Metrology [JCGM 200, 2012] as: *closeness of agreement between a measured quantity value and a true quantity value of a measurand.* In this case, is a conceptual definition, being appropriate to consider instead of "true value" the "conventional true value" taken as a reference value, "x_{ref}", of the quantity to be measured, being characterized by an estimate near the "true value" and lower dispersion of values. In Fig. 5b, this reference value is added, to obtain, for each measurement, an estimate of the (systematic) errors (e_a and e_b) and an evaluation of the combined dispersion (further called *combined uncertainty*, explained in the next section).

3.2 Calibration Key Role to Assure Measurement Quality

Measurement usually requires the use of instrumentation able to make the transduction of the observation of objects or phenomena, to measurable quantities, being the measurement results affected by intrinsic and external effects, systematic and random, which can happen during the measurement process and over time. These changes have impact in the quality of the results, being the main reason to perform the calibration.

The definition of *calibration* is also found in the International Vocabulary of Metrology [13]: *operation that, under specified conditions, in a first step, establishes a relation between the quantity values with measurement uncertainties provided by measurement standards and corresponding indications with associated measurement uncertainties and, in a second step, uses this information to establish a relation for obtaining a measurement result from an indication.*

Having this definition in mind, calibration allows to correct the measurement results obtained with a measuring instrument and to evaluate its measurement uncertainty. Periodic calibration is recommended [18] for most types of equipment as it is acknowledged that instrumentation develops bias with time (see Fig. 6).

Accepting the principle that measurement instruments need to be calibrated, then the reference standards need also to be more accurate. This stepwise process ends with the most accurate realization of the units, according to the international BIPM standards for the quantities.

Fig. 6 Bias of uncertainty in
a scale of time between
consecutive calibrations

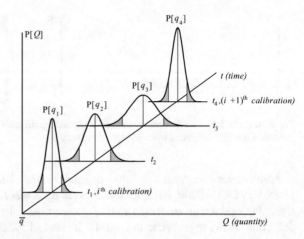

3.3 Metrological Traceability

Metrological traceability is the path that relates the metrological quality of a measurement results with the highest international standard of a certain quantity through a chain of calibrations [19]. The International Vocabulary of Metrology [13] provides the definitions of:

- *Metrological traceability* as "*property of a measurement result whereby the result can be related to a reference through a documented unbroken chain of calibrations, each contributing to the measurement uncertainty*"; and
- *Metrological traceability chain* as the "*sequence of measurement standards and calibrations that is used to relate a measurement result to a reference*".

Figure 7 shows a traceability chain that provides traceability to a measurement result, being each step from measurement equipment to the SI related with calibration operations. In this process, downwards, every step will incorporate the uncertainty of the previous calibration, meaning that the uncertainty of the measurement results includes all the uncertainty contributions of previous calibrations in the chain.

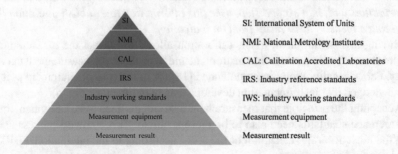

Fig. 7 Example of a traceability chain

Calibration processes usually originate certificates, containing the general information needed to establish the traceability of measurement. It should provide information required for the metrological management of measuring instruments and standards, especially those found in sections 7.8.2.1 and 7.8.4.1 in ISO/IEC 17025:2017 [9], namely regarding *Common requirements for reports (test, calibration, or sampling)* and *Specific requirements for calibration certificates.* The lists presented in the referred sections should be used to evaluate if calibration certificates issued by calibration laboratories provide the complete information needed to the metrological management of measuring equipment and standards used in measurement activities. Metrological traceability is a major requirement in the accreditation found in ISO/IEC 17025:2017 [9], sections 6.5.1 regarding documented unbroken chain of calibrations, and linkage to linking them to an appropriate reference, and 6.5.2 for traceability to the International System of Units (SI).

In practice, how to express the metrological traceability of a measurand (measurable quantity)? Consider the measurement of flow rate in an industrial infrastructure using an electromagnetic flowmeter, as in Fig. 8a, calibrated in a laboratory as in Fig. 8b. The calibration uses the gravimetric method, which calculates volumetric flow rate, Q_V, based on the measurement of the weight of a volume, m, of water running through the flowmeter during a time interval, Δt, and considering the fluid density, ρ, using the expression (1):

$$Q_V = \frac{1}{\rho} \cdot \frac{m}{\Delta t} \tag{1}$$

The weight is measured using a weighing platform traceable to NMI reference standard weights and time is measured using a time universal counter traceable to the NMI reference clock. In this case, it should be stated that the *flow rate measurement is traceable to the SI units of mass and time.*

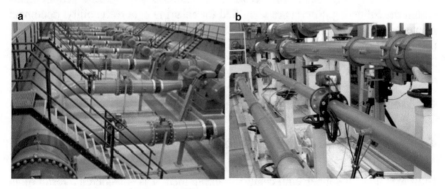

Fig. 8 Electromagnetic flowmeters installed in **a** (left) an industrial infrastructure and **b** (right) in a reference gravimetric calibration system

Metrological traceability is a modern concept for Metrology, providing international recognition based on consistency, comparability, and confidence of measurement results. Together with standardization and accreditation, it is a key part of the conformity assessment having an important role for the testing, inspection and certification industry, stakeholders, governance, and society.

3.4 Data Traceability

Data processing refers specifically to the actions that are triggered after data collection and, to some extent, are oblivious to this earlier process. It is sought to detect the occurrence of anomalous situations, and possibly exclude data originated in these conditions, aiming to reduce uncertainty and increase confidence in the resulting information [20]. Exclusion of data should be based on clear and well-structured rejection and acceptance criteria [12]. Briefly, for data processing underlying principles:

- It should be framed by context information on the monitoring program and sites [21];
- A data processing protocol must be set [16];
- At any stage, detection of anomalies and its diagnosis must always be included [20];
- It must be carried out internally, with respect to the sensor or measurement site itself, and externally, compared with other sources of information [20].

Data processing procedures and results must be stored to allow traceability, process repeatability and transparency. Preferably, this process should be automated [16].

For such, in alignment with ISO/IEC 17025:2017 [9], specific requirements within the following criteria are identified for ensuring data traceability: document management and traceability, data processing and presentation of results adequacy (Fig. 9).

4 Measurement Uncertainty

4.1 Concept of Measurement Uncertainty

Until the end of the nineteenth century, measurement was considered a result of an experimental task or set of tasks, ruled by a deterministic perspective, and the result was expressed by an estimate and its error. In a deterministic approach, the correction of the error would provide an estimate of the real value of a measurand.

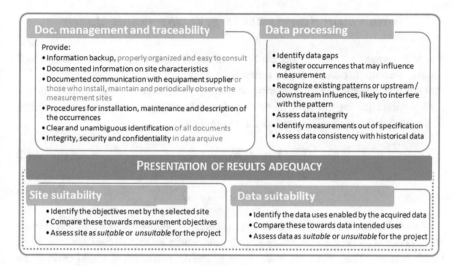

Doc. management and traceability

Provide:
- Information backup, properly organized and easy to consult
- Documented information on site characteristics
- Documented communication with equipament supplier or those who install, maintain and periodically observe the measurement sites
- Procedures for installation, maintenance and description of the occurrences
- Clear and unambiguous identification of all documents
- Integrity, security and confidentiality in data arquive

Data processing

- Identify data gaps
- Register occurrences that may influence measurement
- Recognize existing patterns or upstream / downstream influences, likely to interfere with the pattern
- Assess data integrity
- Identify measurements out of specification
- Assess data consistency with historical data

PRESENTATION OF RESULTS ADEQUACY

Site suitability

- Identify the objectives met by the selected site
- Compare these towards measurement objectives
- Assess site as *suitable* or *unsuitable* for the project

Data suitability

- Identify the data uses enabled by the acquired data
- Compare these towards data intended uses
- Assess data as *suitable* or *unsuitable* for the project

Fig. 9 Requirements for measurement traceability

Since the beginning of the twentieth century, scientific advances lead to the introduction of a stochastic perspective associated with measurement. Within these advances, it is worth mentioning the conceptual study of measurement scales [22], the development of statistical concepts applied to micro analysis of phenomena in comparison with the conventional macro view (e.g., in Thermodynamics), the development of quantum mechanics, information theory and measurement applied to social sciences.

In 1979, Leaning and Finkelstein [22] proposed a new concept for measurement, creating a formal definition of measurement introducing the new idea of "uncertainty of measurement" understood as the expression of the stochastic nature of measurement. This idea was rapidly recognized as a major transformation in the measurement conceptual nature and the major international entities (ISO, IEC, among others) promoted a cooperation action that resulted in the publication of the GUM (Guide to the Expression of Uncertainty in Measurement) in 1993 with small corrections introduce in 1995, being the latest version published by BIPM [23]. The GUM allowed the wide dissemination of the new concept, later adopted by the scientific community and by the industry in many ways, being probably the most relevant the impact given in traceability where measurement uncertainty is the parameter relevant in the calibration transition at each level and in accreditation, being the reference for the measurement comparability and, therefore, a basis for the international recognition.

Today, different definitions of measurement uncertainty can be found, but in this document, the one presented in the GUM [23] is adopted: *parameter, associated with the result of a measurement, that characterizes the dispersion of the values that could reasonably be attributed to the measurand.*

4.2 Measurement Representation by Mathematical Models

Every measurement can be generally expressed by a mathematical model, which relates the output quantity (to be measured), Y, with a set of n input quantities, X_i, that are measurands (2). Considering that measurement values can be obtained for each input quantities, allowing to estimate for each one, x_i, the experimental estimate of the output quantity takes the form of expression (3).

$$Y = f(X_1, \ldots, X_n) \tag{2}$$

$$y = f(x_1, \ldots, x_n) \tag{3}$$

This is the simpler approach that considers only one output quantity, which in some cases is not truly representative, requiring more complex approaches not developed in this context. This, however, is the basis of the development of GUM and will be further used as the practical mathematical representation of a measurement.

An example of application considers that there is a need to obtain the electrical voltage, V, from the measurement of the electrical resistance, R, and of the electrical current, I, using Ohm's Law (4). In practice, a sample of observations should be obtained to provide an average value of the electrical current, $\bar{i}$, and an average value of the electrical resistance, $\bar{r}$, in order to estimate the electrical voltage using expression (5).

$$V = f(R, I) = R \cdot I \tag{4}$$

$$v = \bar{r} \cdot \bar{i} \tag{5}$$

4.3 Measurement Uncertainty Related to the Measurement Model

In a stochastic approach every estimate of a measurement has uncertainty, usually represented by a normal (probability) distribution function (PDF). Considering the mathematical model (1), each estimate, x_i, will have its own uncertainty, $u(x_i)$, and the problem to solve is to obtain the uncertainty of the output quantity, $u(y)$. To facilitate understanding the problem, a simple approach is given from a mathematical model like Eq. (6).

$$Y = f(X_1, X_2) = X_1 + X_2 \tag{6}$$

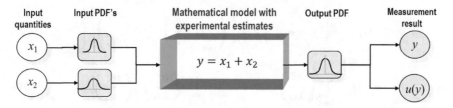

Fig. 10 Probabilistic measurement model for a given relational function

If the measurand (quantity) x_1, has an uncertainty with normal PDF, $u(x_1)$, and the measurand (quantity) x_2, has an uncertainty with normal PDF, $u(x_2)$, then the output quantity, y, will have an uncertainty, $u(y)$, and also a normal PDF obtained from the input quantities PDF's being applied the mathematical model, as represented in Fig. 10.

Using mathematical terminology, the problem consists in making the convolution of two (probability) functions to obtain the output (probability) function, which have an analytical solution known from Mathematics [24]. However, if several input quantities and more complex mathematical models are considered, the analytical approach to convolution becomes much more difficult to apply and will require knowledge not so common. This reason leaded to the development of a more friendly approach, which is given by GUM and is presented in the next section.

4.4 The GUM Law of Propagation of Uncertainty

The development of the LPU—Law of Propagation of Uncertainty [23] was made, starting from expression (2) and taking four steps.

The first step is to develop Eq. (2) using the estimates of the quantities, as a 1st order Taylor series expansion (being μ_i the statistical estimate of the average value of the input quantity x_i), as presented in (7). The second step is, to transfer the first term of the second member of the equation to the first member, presented in (8). The third step is to introduce the statistical variance parameter in both terms of the equation and neglecting the 2nd and higher terms of the Taylor series expansion, $r_2(x_i)$ (9), and the final step, is to make the sum of diagonal terms and non-diagonal terms to obtain the expression of the LPU, given in (10).

$$y = f(\mu_1, \mu_2, \ldots, \mu_N) + \sum_{i=1}^{N} \left(\frac{\partial f}{\partial x_i}\right) \cdot (x_i - \mu_i) + r_2(x_i) \tag{7}$$

$$(y - \mu_y) = \sum_{i=1}^{N} \left(\frac{\partial f}{\partial x_i}\right) \cdot (x_i - \mu_i) + r_2(x_i) \tag{8}$$

$$\sigma_y^2 = E\left[(y - \mu_y)^2\right] = E\left[\left(\sum_{i=1}^{N}\left(\frac{\partial f}{\partial x_i}\right) \cdot (x_i - \mu_i)\right)^2\right] \tag{9}$$

$$s^2(y) = \sum_{i=1}^{N}\left(\frac{\partial f}{\partial x_i}\right)^2 \cdot s_i^2 + 2\sum_{i=1}^{N-1}\sum_{j=i+1}^{N}\left(\frac{\partial f}{\partial x_i}\right) \cdot \left(\frac{\partial f}{\partial x_j}\right) \cdot s_{ij} \tag{10}$$

Using the GUM nomenclature, the previous expression can be written as (11) and in the cases in which correlation between input quantities is null (does not exist) or can be neglected, the LPU can be simplified to (12). The LPU shows the need to have a mathematical model (that must be differentiable) and can be applied to any model of single output quantity (as presented in (2) above).

$$u_c^2(y) = \sum_{i=1}^{N} c_i^2 \cdot u^2(x_i) + 2\sum_{i=1}^{N-1}\sum_{j=i+1}^{N} c_i \cdot c_j \cdot u(x_i) \cdot u(x_j) \cdot r(x_i, x_j) \tag{11}$$

$$u_c^2(y) = \sum_{i=1}^{N}\left(\frac{\partial f}{\partial x_i}\right)^2 \cdot u^2(x_i) = \sum_{i=1}^{N} c_i^2 \cdot u^2(x_i) \tag{12}$$

Using the previous example of the Ohm's Law, the LPU allows to calculate the combined measurement uncertainty, $u_c(v)$, applying expression (12) to (5).

$$u_c^2(v) = \left(\frac{\partial v}{\partial r}\right)^2 \cdot u^2(r) + \left(\frac{\partial v}{\partial i}\right)^2 \cdot u^2(i) = (i)^2 \cdot u^2(r) + (r)^2 \cdot u^2(i) \tag{13}$$

Using (13), this evaluation requires to previously obtain the measurement uncertainties of the input quantities, usually related with measurement instruments, being the topic for the next section.

4.5 Evaluation of Measurement Uncertainty Related with Input Quantities

The measurement of input quantities as measurands is usually made using measuring instruments, traceable to SI through calibration processes, being its performance influenced by random and systematic effects that, combined, produce the errors of measurement. The combination of random effects gives the measurement uncertainty of the estimates. Taking this into account, the evaluation of the measurement uncertainty due to the use of measurement instruments needs to be found, being defined by Eq. (14):

$$x_i = x_{eq} + \varepsilon_{cal} + \sum_{i=1}^{k} \delta x_i \qquad (14)$$

being, x_i, the estimate of the measurement; x_{eq}, the value of the reading using the measurement instrument (equipment); ε_{cal}, the calibration error correction (by definition, the error of calibration is given by the difference between the readings and reference values); δx_i, the errors that influence the measurement process. The LPU applied to this expression, considering that there is no correlation between quantities, gives Eq. (15).

$$u^2(x_i) = u^2(x_{eq}) + u^2(\varepsilon_{cal}) + \sum_{i=1}^{k} u^2(\delta x_i) \qquad (15)$$

Regarding the uncertainty contribution for each input quantity, the calibration error is found in the calibration certificate, being the others evaluated according to the approach generally described in the GUM and adapted in the following procedure:

a. identify the sources of error;
b. select the PDF for each source of error and quantify the contributions for the uncertainty budget; and
c. apply (15) using the variables of (14) to obtain the measurement uncertainty of the specific input quantity.

The following sections provide a detailed explanation of steps a. and b. according with the prescribed in the GUM, being c. only a mathematical implementation.

4.6 Identify the Sources of Uncertainty

Measurement can be affected by a diversity of conditions being difficult to provide a universal approach for this purpose. However, GUM provides some information that helps the process of identification of the sources of uncertainty. Based on this approach, LNEC uses a 2-step procedure that defines first five macro classes:

- Materialization of the quantity
- Measurement method
- Instrumentation (metrological characteristics of instrumentation)
- Personnel
- Data processing.

The second step is to select different sources of uncertainty related with each macro class. Often, lists can be provided to support this decision, like the ones provided in Fig. 11 (not exhaustive).

As an example of *materialization of the quantity*, in the case of a block gauge as a reference dimension, there are imperfect conditions of flatness and parallelism

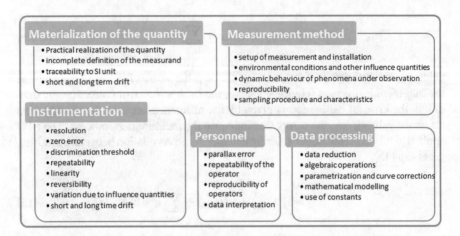

Fig. 11 Examples of sources of uncertainty related to measurement conditions

between the measurement faces of the reference. As an example of *measurement method*, in water sampling procedure, the process, size and conditions of storage can affect the measurement results.

Instrumentation metrological performance is key in the identification of sources of error. In this case, the International Vocabulary of Metrology provides a description of common sources of error found in measuring instruments and measuring systems.

In the case of *personnel*, the parallax error is probably the most common, given the different shapes of meniscus of two liquid-in-glass thermometers of alcohol and mercury.

4.7 Select the PDF for Each Source of Error and Quantify Its Contribution

Having identified the sources of error, two decisions must be taken: to consider then as applicable to type A or type B (GUM) approach; and to select the PDF and find the parameters to its quantification.

For the clarification of this procedure, should be said that type A means that there is an experimental sample allowing to use statistical estimates, being a typical example the repeatability evaluated using the standard-deviation formula. In this case, the selected PDF is the normal distribution function, and the contribution is given by the calculation of the sampling standard-deviation of the average as follows by Eq. (16):

$$u^2(x_i) = \frac{\sum_{i=1}^{n}(x_i - \overline{x})^2}{n(n-1)} \tag{16}$$

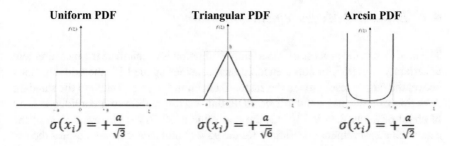

Uniform PDF

$$\sigma(x_i) = +\frac{a}{\sqrt{3}}$$

Triangular PDF

$$\sigma(x_i) = +\frac{a}{\sqrt{6}}$$

Arcsin PDF

$$\sigma(x_i) = +\frac{a}{\sqrt{2}}$$

Fig. 12 Common PDF's used in uncertainty calculation and related estimates of the standard deviation

In the case of type B approach, is considered that the information to select the PDF and its parameters is known from sources different from the experimental data (from literature, certificates, equipment manual or other). In this case, the "a priori" (sometimes also called Bayesian) information should be used to select the PDF and the parameters that allows the calculation of the contribution using the estimate of standard deviation related to the PDF. For this purpose, common PDF's are considered beyond normal PDF, together with the estimate of standard deviation (Fig. 12).

A common example of application: consider a flowmeter that can measure flow with a resolution of 0.1 l/s. Its uncertainty can be estimated adopting a uniform PDF varying between -0.05 l/s (parameter corresponding to $-a$ in Fig. 13) and $+0.05$ l/s (parameter corresponding to $+a$ in Fig. 13). 0.05 l/s is the halfwidth of the flowmeter resolution and the adoption of the uniform PDF means that all possible reading outcomes in the interval $[-0.05$ l/s, $+0.05$ l/s$]$ have equal probability. The estimated contribution to the uncertainty budget will be given by Eq. (17). A similar approach should be taken for the other two PDF's presented in Fig. 13.

$$u(Q_{\text{res}}) = \frac{0.05}{\sqrt{3}}\,1/s \approx 0.029\,1/s \tag{17}$$

Fig. 13 Normal standard PDF with confidence intervals of 68 and 95%

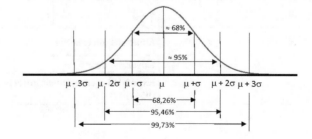

4.8   Expanded Measurement Uncertainty

The final step in the process of uncertainty evaluation is to calculate the measurement uncertainty with 95% of confidence. In fact, considering the LPU method described above, the output result, given the standard uncertainty, is equivalent to the standard deviation of a normal standardized PDF, thus corresponding to a confidence level of about 68% (as seen in Fig. 13, the real value is 68.26%). The evaluation of the expanded measurement uncertainty for the 95% confidence interval requires the use of a coverage factor, which is usually considered as 2 (in fact, for normal standardized PDF, the value corresponding to 2σ is 95.46%, being the factor of 2 considered appropriate as a good approach).

In the previous example, to obtain the expanded measurement uncertainty, the standard uncertainty should be multiplied by the coverage factor of 2 (Eq. 18).

$$U_{95}(Q_{res}) = 2 \cdot u(Q_{res}) = 2 \cdot 0.029\,l/s = 0.058\,l/s \qquad (18)$$

The evaluation of measurement uncertainty has a useful tool presented in the GUM, which is the uncertainty budget table, giving a support to the data for the evaluation and being widely used internationally for this purpose, also allowing a quick perception of the data used for the calculation. Table 2 presents an application of the structure of this table assuming a normal PDF.

Table 2 Example of an uncertainty budget table (assuming a normal PDF)

Input quantities X_i	Estimates x_i	Standard uncertainty (PDF and type A/B) $u(x_i)$	Sensitivity coefficients $c_i = (\partial f/\partial x_i)$	Contributions for the standard uncertainty of the output quantity $u_i(y)$
X_1	x_1	$u(x_1)$	c_1	$u_1(y) = \sqrt{\left[c_1^2 u^2(x_1)\right]}$
X_2	x_2	$u(x_2)$	c_2	$u_2(y) = \sqrt{\left[c_2^2 u^2(x_2)\right]}$
...	...	...	...	
X_n	x_n	$u(x_n)$	c_n	$u_n(y) = \sqrt{\left[c_n^2 u^2(x_n)\right]}$
Y	y	**Standard uncertainty**		$u_c(y) = \sqrt{\left[\sum u_i(y)^2\right]}$
		Coverage factor		2.00
		Expanded measurement uncertainty (95%)		$u_{95}(y) = 2.00\,u_c(y)$

4.9 Approaches for Added Accuracy and Complex Problems

Some additional remarks and suggestions for further reading are required to complete the approach presented. The first remark is related with sampling being often made with small sets of observations, typically between three to five, somehow in conflict with the use of a normal PDF to represent the output quantity. GUM presents an alternative, more robust approach, to the evaluation of measurement uncertainty using t-Student PDF and including the degrees of freedom for each uncertainty contribution of input quantities. This approach will evaluate the effective decrees of freedom of the output quantity and uses this result to adopt a coverage factor for the expanded measurement uncertainty of the output quantity. This way, sampling related with the sensitivity of the input quantities contributions are considered in the result. The GUM presents an uncertainty budget table, alternative to the one presented in Table 2, which is practical for this implementation.

About the nature of the mathematical models, it should be noticed that the approach given by the LPU in the GUM [23] provides an exact solution for linear models because it is developed from a first order Taylor series. Often, mathematical models are nonlinear, requiring more complex approaches with higher order terms (as presented in [23]) but, in some cases, numerical simulation is used namely using Monte Carlo methods. The supplement [25] provides the framework for the application of numerical simulation as a tool to the evaluation of measurement uncertainty.

Other supplements in GUM provide procedures to apply in specific conditions related with the mathematical models, namely explanation of the approach to deal with mathematical models with any number of output quantities [26], related with modelling [27], explaining the concepts and basic principles [28], introducing the role of measurement uncertainty in conformity assessment [29] and applications of the least-squares method [30]. For additional information refer to www.bipm.org.

5 Perspectives of Future Developments

Measurement quality is closely associated with instrumentation and increasingly sophisticated measurement systems, which require specific technical knowledge by users, applicable to good measurement practices and skills, that allow the control of conditions that, direct and indirectly, affect equipment performance.

In the current context, the digital transition is strongly related with new technologies, materials, and methods that, in turn, respond to new needs for security and quality of life. Indeed, the reality presented today induces different solutions, involving 5G communication, Internet-of-Things, Artificial Intelligence, Big data, cloud computing, among others, with measurement approaches supported on models involving nanomaterials, characteristics and structures that require the application of Metrology in a different context from the current one.

Investigation and industry must work together in this path. The work with utilities allowed to recognise that the selection of the appropriate site for measuring and of the adequate equipment, for the site and monitoring objectives, are some of the most challenging issues in the field. The need to improve data traceability in utilities is also evident.

The creation of large-scale sensor networks, which must be autonomous, with intrinsic traceability and providing permanent information on measurable quantities, is an example of this new nature. Other examples are the growing use of optical measurement and non-invasive digital processing for activities associated with dimensional and geometric measurement, monitoring, inspection, and the development of system components (e.g., pipes) with embedded sensors.

The implementation of good measurement practices and technological innovation is under consolidation, and synergies between data science and applied engineering are sought.

Evolution of measurement requires a strong intervention in the context of what is called data science, and where there will be a growing need for interdisciplinary knowledge, associating the classic disciplines of Engineering and Physics, Applied Mathematics and Computing, among others, as key elements to achieve the required levels of knowledge and rigor essential to the measurement activity.

References

1. Hirner, W., Lambert, A.: Losses from water supply systems: standard terminology and recommended performance measures. IWA Blue Pages (2000)
2. Butler, D., Davies, J.: Urban Drainage. E&FN Spon Press, London (2000). ISBN: 0419223401. https://doi.org/10.4324/9780203244883
3. Cifelli, R., Chandrasekar, V., Chen, H., Johnson, L.E.: High resolution radar quantitative precipitation estimation in the San Francisco bay area: rainfall monitoring for the urban environment. J. Meteorol. Soc. Jpn. **96A**, 141–155 (2018). https://doi.org/10.2151/jmsj.2018-016
4. Shucksmith, P.E., Sutherland-Stacey, L., Austin, G.L.: The spatial and temporal sampling errors inherent in low resolution radar estimates of rainfall. Meteorol. Appl. **18**(3), 354–360 (2011)
5. Bonakdari, H., Larrarte, F., Joannis, C., Levacher, D.: Une méthodologie d'aide à l'implantation de débitmètres en réseaux d'assainissement. In: 25e RENCONTRES DE L'AUGC, pp. 1–8. AUGC, France (2007)
6. Baures, E., Helias, E., Junqua, G., Thomas, O.J.: Fast characterization of non domestic load in urban wastewater networks by UV spectrophotometry. J. Environ. Monit. **9**, 959–965 (2007)
7. Campisano, A., Cabot Ple, J., Muschalla, D., Pleau, M., Vanrolleghem, P.A.: Potential and limitations of modern equipment for real time control of urban wastewater systems. Urban Water J. **10**(5), 300–311 (2013). https://doi.org/10.1080/1573062X.2013.763996
8. WaPug: Code of practice for the hydraulic modelling of urban drainage systems. Version 01. In: Wastewater Planning Users Group. The Chartered Institution of Water and Environmental Management (CIWEM), United Kingdom (2017)
9. ISO/IEC 17025:2017: General Requirements for the Competence of Testing and Calibration Laboratories. International Organization for Standardization. Genève, Switzerland (2017)
10. ISO 9001:2015: Quality Management Systems—Requirements. International Organization for Standardization. Genève, Switzerland (2015)

11. OECD: Series on Principles of Good Laboratory Practice and Compliance Monitoring. Number 1 OECD Principles on Good Laboratory Practice (as revised in 1997). ENV/MC/CHEM (98)17. Organisation for Economic Cooperation and Development, Paris (1998)
12. Bertrand-Krajewski, J.L., Laplace, D., Joannis, C., Chebbo, G.: Mesures en hydrologie urbaine et assainissement. Editions Technique & Documentation, Lavoisier, France (2000). ISBN: 9782743003807
13. JCGM: GUM 200.2012. Evaluation of Measured Data—Guide to the Expression of Uncertainty in Measurement. Joint Committee for Guidelines in Metrology. Sèvres, France (2012)
14. Mignot, E., Bonakdari, H., Knothe, P., Lipeme Kouyi, G., Bessette, A., Rivière, N., Bertrand-Krajewski, J.L.: Experiments and 3D simulations of flow structures in junctions and their influence on location of flowmeters. Water Sci. Technol. **66**(6), 1325–1332 (2012)
15. Henriques, J., Pires da Palma, J., Silva Ribeiro, A.: Medição de caudal em sistemas de abastecimento de água e de saneamento de águas residuais urbanas Guia técnico do IRAR. IRAR-LNEC, Lisboa (2007). ISBN 978-989-95392-1-1
16. Environmental Protection Agency: Smart Data Infrastructure for Wet Weather Control and Decision Support. EPA830-B-17-004. U.S. EPA Office of Wastewater Management, USA (2018)
17. Portuguese Decree Law DR 23/95, of 23rd August 1995. Regulation for Public and Building Water Distribution and Wastewater Drainage Systems (in Portuguese) (1995)
18. LAC G24, OIML D10: 2007—Guidelines for the Determination of Calibration Intervals of Measuring Instruments. ILAC, Silverwater, Australia (2007)
19. ILAC P10:07/2020—ILAC Policy on Metrological Traceability of Measurement Results. ILAC, Silverwater, Australia (2020)
20. OTHU: Rapport d'activité scientifique 2006/2008. Féderation d'equipes de recherche OTHU Observatoire de terrain en hydrologie urbaine. INSA, BRGM, CEMAGREF, ECL, ENTPE, Lyon, France (2009)
21. Deletic, A., Fletcher, T.: Selecting variables to monitor, Chap. 4. In: Data Requirements for Integrated Urban Water Management. Taylor & Francis—Urban Water Series. UNESCO—IHP, London (2008)
22. Sydenham, P., Thorn, R.: Handbook of Measurement Science, Vol. 2: Practical Fundamentals. In: P.H. Sydenham, R. Thorn (eds.). Wiley-Interscience Publication (1991)
23. JCGM: 100:2008: Evaluation of Measurement Data—Guide to the Expression of Uncertainty in Measurement. Joint Committee for Guidelines in Metrology. Sèvres, France (2008)
24. Dietrich, C.F.: Uncertainty, Calibration and Probability: The Statistics of Scientific and Industrial Measurement, 2nd edn. Adam Hilger, Bristol, UK (1991)
25. JCGM 101: 2008: Evaluation of Measurement Data—Supplement 1 to the "Guide to the Expression of Uncertainty in Measurement"—Propagation of Distributions Using a Monte Carlo Method. Joint Committee for Guidelines in Metrology. Sèvres, France (2008)
26. JCGM 102: 2011: Evaluation of Measurement Data—Supplement 2 to the "Guide to the Expression of Uncertainty in Measurement"—Extension to Any Number of Output Quantities. Joint Committee for Guidelines in Metrology. Sèvres, France (2011)
27. JCGM 103: Guide to the Expression of Uncertainty in Measurement—Developing and Using Measurement Models. JCGM WG1 (under development)
28. JCGM 104: 2009: Evaluation of Measurement Data—An Introduction to the "Guide to the Expression of Uncertainty in Measurement" and Related Documents. Joint Committee for Guidelines in Metrology. Sèvres, France (2009)
29. JCGM 106: 2012: Evaluation of Measurement Data—The Role of Measurement Uncertainty in Conformity Assessment. Joint Committee for Guidelines in Metrology. Sèvres, France (2012)
30. JCGM 107: Evaluation of Measurement Data—Applications of the Least-Squares Method. JCGM WG1 (under development)

Printed in the United States
by Baker & Taylor Publisher Services